Use R!

Series Editors:
Robert Gentleman   Kurt Hornik   Giovanni Parmigiani

For other titles published in this series, go to
http://www.springer.com/series/6991

Use R!

Series Editors:
Robert Gentleman · Kurt Hornik · Giovanni Parmigiani

For other titles published in this series, go to
http://www.springer.com/series/6991

Giovanni Petris · Sonia Petrone · Patrizia Campagnoli

Dynamic Linear Models with R

Giovanni Petris
Department of Mathematical Sciences
University of Arkansas
Fayetteville, AR 72701
USA
gpetris@uark.edu

Sonia Petrone
Department of Decision Sciences
Bocconi University
Via Roentgen, 1
20136 Milano
Italy
sonia.petrone@unibocconi.it

Patrizia Campagnoli
Department of Decision Sciences
Bocconi University
Via Roentgen, 1
20136 Milano
Italy
patrizia.campagnoli@unibocconi.it

ISBN 978-0-387-77237-0 e-ISBN 978-0-387-77238-7
DOI 10.1007/b135794
Springer Dordrecht Heidelberg London New York

Library of Congress Control Number: 2009926480

Printed on acid-free paper

Springer is part of Springer Science+Business Media (www.springer.com)

A Mary e Giulio
G.P.

A Francesca, ai miei nipoti
S.P.

A Andrea e Michele
P.C.

Preface

This book aims at introducing the reader to statistical time series analysis by dynamic linear models. We have tried to be precise and rigorous in discussing the main concepts and tools, yet keeping a simple and friendly style of presentation. The main methods and models are widely illustrated with examples based on real data, implemented in R. Together with the book, we developed an R package for inference and forecasting with dynamic linear models; the dlm package is available as a contributed package in the Comprehensive R Archive Network at `http://www.r-project.org/`.

In the recent years, there has been an enormous growth of interest for statistical applications of dynamic linear models and, more generally, state-space models, in a wide range of applied fields, such as biology, economics, finance, marketing, quality control, engineering, demography, climatology, to mention only a few. State space models provide a very flexible yet fairly simple tool for analyzing dynamic phenomena and evolving systems, and have significantly contributed to extend the classical domains of application of statistical time series analysis to non-stationary, irregular processes, to systems evolving in continuous-time, to multivariate, continuous and discrete data. An extremely wide range of applied problems can be treated inside the framework of dynamic linear models or, more generally, state-space models.

The book covers the basic notions of dynamic linear models and state space models, the celebrated Kalman filter for estimation and forecasting in a dynamic linear model with known parameters, and maximum likelihood estimation. It also presents a wide array of specific dynamic linear models particularly suited for time series analysis, both for univariate and multivariate data. But these topics are of course also covered in other very good books in the rich literature on dynamic linear models, and several statistical softwares include packages for time series analysis through maximum likelihood and Kalman filtering. What we felt was somehow missing was an up to date, rigorous yet friendly reference—and software—for applied Bayesian time series analysis through dynamic linear models and state space models. This seemed to be missing despite the fact that the Bayesian approach has become more and more popular in applications, due to the availability of modern and efficient computational tools. So, while also covering maximum likelihood methods, our focus in the book is on *Bayesian* time series analysis based on dynamic linear models.

We do not expect the reader to be an expert in Bayesian inference, so we begin with a short introduction to the Bayesian approach in Chapter 1. Also for a Bayesian reader, this is useful to set the notation and to underline some basic concepts that are used in the following chapters: for example, in presenting the simplest notions, such as Bayesian conjugate inference for a Gaussian model, we underline the recursive structure of the estimates, that will be one of the basic aspects of inference for dynamic linear models. Chapter 2 introduces the general setting of state space models and dynamic linear models, including the fundamental algorithms to sequentially update estimates and forecasts and the Kalman filter. Chapters 3 and 4 are in a sense the core of the book. In Chapter 3 the reader will find a discussion of a broad spectrum of specific models suited for the analysis of many kinds of data showing different features. Thus, Chapter 3 should be considered as a toolbox, illustrating a set of models from which the user can select the most appropriate for the application at hand. Chapter 4 covers maximum likelihood and Bayesian inference for dynamic linear models containing unknown parameters—which is always the case in practice. Many of the models introduced in Chapter 3 are discussed again there in this perspective. For most of the covered models we provide detailed examples of their use, corredated with the relevant R code. When possible, Bayesian estimates are evaluated using closed form algorithms. But in more elaborate models, analytical computations become intractable and simulation techniques are used to approximate the Bayesian solutions. We describe Markov chain Monte Carlo methods for Bayesian inference in dynamic linear models. The R package dlm provides functions for one of the basic steps in Bayesian computations in dynamic linear models, the so-called *forward filtering-backward sampling* algorithm, and other computational tools, with many examples, are provided. In Chapter 5, we present modern sequential Monte Carlo and particle filter algorithms for on-line estimation and forecasting.

Of course we cannot cover all of the extremely rich variety of models, applications, and problems in Bayesian inference with dynamic linear models, and many things will be missing. However, we hope to give a solid background on the main concepts and notions, leading the reader to acquire the skills for specific, personal elaborations, for which the flexibility of R and the dlm package will provide convenient, helpful tools. On the web site of the book, `definetti.uark.edu/~gpetris/dlm`, the reader will find data sets not included in the package and the code to run all the examples in the book. In addition, we plan to post there an updated list of errata.

The motivation for this book came from the authors' teaching experience in courses on time series analysis. We wanted to teach a course including—besides the classical ARMA models, descriptive techniques, exponential smoothing, and so on—more modern approaches, in particular Bayesian inference for time series through dynamic linear models. Again, we felt that a textbook, and a friendly but flexible software, were missing. So we started working on this project. We hope students, researchers, and practitioners will find the book and the software that resulted from our effort of some help.

We would like to thank Springer-Verlag's referees for their encouragement and valuable suggestions. Our thanks go also to our editor, John Kimmel, for his patience and support.

The dlm package would not exist without R, for which we thank R-core. Several people on `r-help`, the general R mailing list, have contributed their suggestions and feedback during the development of the package: we thank all of them. In particular, we thank Spencer Graves and Michael Lavine for their comments and suggestions on earlier versions of the package. Michael Lavine taught a course at the University of Massachusets using R and dlm from an early draft of the book, and we thank him for the valuable feedback he gave us. One of the authors (GP) taught some short courses based on preliminary versions of the book at Bocconi University and the University of Roma 3 and would like to thank Pietro Muliere, Carlo Favero, Julia Mortera, and his coauthor, Sonia Petrone, for the kind invitations and the hospitality. SP used draft versions of the book in her graduate courses on time series analysis at Bocconi University: students' feedback has been precious. We thank all our students at the University of Arkansas, Bocconi University, and the University of Roma 3 who, with their comments, questions, suggestions, interest and enthusiasm, have contributed to the development of this book. Among them, a special thanks goes to Paolo Bonomolo and Guido Morandini.

Needless to say, the responsibility for any remaining mistakes, obscurities, or omissions—in the book and in the package—lies solely with us.

Fayetteville, Arkansas
and
Milano, Italy
December 15, 2008

Giovanni Petris
Sonia Petrone
Patrizia Campagnoli

Contents

1

Introduction: basic notions about Bayesian inference

Dynamic linear models were developed in engineering in the early 1960's, to monitor and control dynamic systems, although pioneer results can be found in the statistical literature and go back to Thiele (1880). Early famous applications have been in the Apollo and Polaris aerospace programs (see, e.g., Hutchinson; 1984), but in the last decades dynamic linear models, and more generally state space models, have received an enormous impulse, with applications in an extremely vast range of fields, from biology to economics, from engineering and quality control to environmental studies, from geophysical science to genetics. This impressive growth of applications is largely due to the possibility of solving computational difficulties using modern Monte Carlo methods in a Bayesian framework. This book is an introduction to Bayesian modeling and forecasting of time series using dynamic linear models, presenting the basic concepts and techniques, and illustrating an R package for their practical implementation.

Statistical time series analysis using dynamic linear models was largely developed in the 1970-80's, and state space models are nowadays a focus of interest. In fact, the reader used to descriptive time series analysis or to ARMA models and Box–Jenkins model specification, may find the state space approach a bit difficult at first. But the powerful framework offered by dynamic linear models and state space models reveals to be a winning asset. ARMA models can be usefully regarded in terms of dynamic linear models. But dynamic linear models offer much more flexibility in treating nonstationary time series or modeling structural changes, and are often more easily interpretable; and the more general class of state space models extends the analysis to non-Gaussian and non-linear dynamic systems. There are, of course, different approaches to estimate dynamic linear models, via generalized least squares or maximum likelihood for example, but we believe that a Bayesian approach has several advantages, both methodological and computational. Kalman (1960) already underlines some basic concepts of dynamic linear models that we would say are proper to the Bayesian approach. A first step is moving from a deterministic to a stochastic system; the uncertainty,

G. Petris et al., *Dynamic Linear Models with R*, Use R, DOI: 10.1007/b135794_1,

which is always present due to forgotten variables, measurement errors, or imperfections, is described through probability. Consequently, the estimation of the quantities of interest (in particular, the state of the system at time t) is solved by computing their conditional distribution, given the available information. This is a general, basic concept in Bayesian inference. Dynamic linear models are based on the idea of describing the output of a dynamic system, for example a time series, as a function of a nonobservable state process (which has a simple, Markovian dynamics) affected by random errors. This way of modeling the temporal dependence in the data, by conditioning on latent variables, is simple and extremely powerful, and again it is quite natural in a Bayesian approach. Another crucial advantage of dynamic linear models is that computations can be done recursively: the conditional distributions of interest can be updated, incorporating the new data, without requiring the storage of all the past history. This is extremely advantageous when data arrive sequentially in time and on-line inference is required, and the reduction of the storage capacity needed becomes even more crucial for large data sets. The recursive nature of computations is a consequence of the Bayes formula in the framework of dynamic linear models.

However, analytical computations are often not manageable, but Markov chain Monte Carlo algorithms can be applied to state space models to overcome computational difficulties, and modern, sequential Monte Carlo methods, which have been enormously improved in the last years, are successfully used for on-line analysis.

We do not expect that the reader is already an expert in Bayesian statistics; therefore, before getting started, this chapter briefly reviews some basic notions, with a look to the concepts that are important in the study of dynamic linear models. Reference books on Bayesian statistics are Bernardo and Smith (1994), DeGroot (1970), Berger (1985), O'Hagan (1994), Robert (2001), Cifarelli and Muliere (1989), or Zellner (1971), Poirier (1995) and Geweke (2005) for a more econometric viewpoint.

1.1 Basic notions

In the analysis of real data, in economics, sociology, biology, engineering and in any field, we rarely have perfect information on the phenomenon of interest. Even when an accurate deterministic model describing the system under study is available, there is always something that is not under our control, such as effects of forgotten variables, measurement errors, or imperfections. We always have to deal with some uncertainty. A basic point in Bayesian statistics is that all the uncertainty that we might have on a phenomenon should be described by means of *probability*. In this perspective, probability has a *subjective* interpretation, being a way of formalizing the incomplete information that the researcher has about the events of interest. Probability theory prescribes how to assign probabilities coherently, avoiding contradictions and undesirable consequences.

The Bayesian approach to the problem of "learning from experience" about a phenomenon moves from this crucial role played by probability. The learning process consists of the application of probability rules: one simply has to compute the *conditional probability* of the event of interest, given the experimental information. Bayes' theorem is the basic rule to be applied to this aim. Given two events A and B, probability rules say that the joint probability of A and B occurring is given by $\mathrm{P}(A \cap B) = \mathrm{P}(A|B)\mathrm{P}(B) = \mathrm{P}(B|A)\mathrm{P}(A)$, where $\mathrm{P}(A|B)$ is the conditional probability of A given B and $\mathrm{P}(B)$ is the (marginal) probability of B. Bayes' theorem, or the theorem of inverse probability, is a simple consequence of the above equalities and says that

$$\mathrm{P}(A|B) = \frac{\mathrm{P}(B|A)\mathrm{P}(A)}{\mathrm{P}(B)}.$$

This is an elementary result that goes back to Thomas Bayes (who died in 1761). The importance of this theorem in Bayesian statistics is in the interpretation and scope of the inputs of the two sides of the equation, and in the role that, consequently, Bayes' theorem assumes for formalizing the inductive learning process. In Bayesian statistics, A represents the event of interest for the researcher and B an experimental result which she believes can provide information about A. Given $\mathrm{P}(A)$ and consequently $\mathrm{P}(\bar{A}) = 1 - \mathrm{P}(A)$, and having assigned the conditional probabilities $\mathrm{P}(B|A)$ and $\mathrm{P}(B|\bar{A})$ of the experimental fact B conditionally on A or $\bar{A}$, the problem of learning about A from the "experimental evidence" B is solved by computing the conditional probability $\mathrm{P}(A|B)$.

The event of interest and the experimental result depend on the problem. In statistical inference, the experimental fact is usually the result of a sampling procedure, and it is described by a random vector Y; it is common to use a parametric model to assign the probability law of Y, and the quantity of interest is the vector θ of the parameters of the model. Bayesian inference on θ consists of computing its conditional distribution given the sampling results. More specifically, suppose that, based on her knowledge of the problem, the researcher can assign a conditional distribution $\pi(y|\theta)$ for Y given θ, the *likelihood*, and a *prior distribution* $\pi(\theta)$ expressing her uncertainty on the parameter θ. Upon observing $Y = y$, we can use a generalization of the elementary Bayes' theorem, known as Bayes' formula, to compute the conditional density of θ given y:

$$\pi(\theta|y) = \frac{\pi(y|\theta)\pi(\theta)}{\pi(y)},$$

where $\pi(y)$ is the marginal distribution of Y,

$$\pi(y) = \int \pi(y|\theta)\pi(\theta)\, d\theta.$$

Thus, Bayesian statistics answers an inference problem by computing the relevant conditional distributions, and the Bayes formula is a basic tool to achieve this aim. It has an elegant, appealing coherence and simplicity. Differently from Bayesian procedures, frequentist statistical inference does not use a probability distribution for the unknown parameters, and inference on θ is based on the determination of estimators with good properties, confidence intervals, and hypothesis testing. The reason is that, since the value of the parameter θ does not "vary," θ is not interpretable as a random "variable" in a frequentist sense, neither can the probability that θ takes values in a certain interval have a frequentist interpretation. Adopting subjective probability instead, θ is a random quantity simply because its value is uncertain to the researcher, who should formalize the information she has about it by means of probability. This seems, indeed, quite natural. We refer the reader to the fundamental works by de Finetti (1970a,b) and Savage (1954) for a much deeper discussion.

In many applications, the main objective of a statistical analysis is *forecasting*; thus, the event of interest is the value of a future observation Y^*. Again, prediction of a future value Y^* given the data y is solved in the Bayesian approach simply by computing the conditional distribution of Y^* given $Y = y$, which is called *predictive distribution.* In parametric models it can be computed as

$$\pi(y^*|y) = \int \pi(y^*, \theta|y)\, \mathrm{d}\theta = \int \pi(y^*|y, \theta)\pi(\theta|y)\, \mathrm{d}\theta.$$

The last expression involves again the posterior distribution of θ. As a matter of fact, apart from controversies about frequentist or subjective probability, a difficulty with (prior or posterior) probability distributions on model parameters is that, in some problems, they do not have a clear physical interpretation, so that assigning to them a probability law is debatable, even from a subjective viewpoint. According to de Finetti, one can assign a probability only to "observable facts"; indeed, the ultimate goal of a statistical analysis is often forecasting the future observations rather than learning on unobservable parameters. Taking a *predictive approach*, the parametric model is to be regarded just as a tool to facilitate the task of specifying the probability law of the observable quantities and, eventually, of the predictive distribution. The choice of the prior should be suggested, in this approach, by predictive considerations, that is, by taking into account its implications on the probability law of Y. We discuss this point further in the next section.

Before moving on to the next, more technical, sections, let us introduce some notation and conventions that will be used throughout. Observable random variables or random vectors will be denoted by capital letters – most of the times by Y, possibly with a subscript. A possible value of the random variable or vector will be denoted by the corresponding lower-case letter. Note that we are not making any notational distinction between vectors and scalars, or between random variables and random vectors. This is true also when writing integrals. For example, $\int f(x)\, \mathrm{d}x$ denotes a univariate integral if

f is a function of one variable, but a multivariate integral if f is a function of a vector argument. The correct interpretation should be clear from the context. A univariate or multivariate *time series* is a sequence of random variables or vectors and will be denoted by $(Y_t : t = 1, 2, \ldots)$, $(Y_t)_{t \geq 1}$, or just (Y_t) for short. When considering a finite sequence of consecutive observations, we will use the notation $Y_{r:s}$ for the observations between the rth and sth, both inclusive. Similarly, $y_{r:s}$ will denote a sequence of possible values for those observations. Probabiity densities will be generically denoted by $\pi(\cdot)$. We will adopt the sloppy but widespread convention of using the same symbol π for the distribution of different random variables: the argument will make clear what distribution we are referring to. For example, $\pi(\theta)$ may denote a prior distribution for the unknown parameter θ and $\pi(y)$ the marginal density of the data point Y. Appendix A contains the definitions of some common families of distributions. We are going to use the same symbol for a distribution and its density, in this case adding an extra argument. For example, $\mathcal{G}(a, b)$ denotes the gamma distribution with shape parameter a and rate parameter b, but $\mathcal{G}(y; a, b)$ denotes the density of that distribution at the point y. The k-dimensional normal distribution is $\mathcal{N}_k(m, C)$, but we will omit the subscript k whenever the dimension is clear from the context.

1.2 Simple dependence structures

Forecasting is one of the main tasks in time series analysis. A univariate or multivariate time series is described probabilistically by a sequence of random variables or vectors $(Y_t : t = 1, 2, \ldots)$, where the index t denotes time. For simplicity, we will think of equally spaced time points (daily data, monthly data, and so on); for example, (Y_t) might describe the daily prices of m bonds, or monthly observations on the sales of a good. One basic problem is to make forecasts about the value of the next observation, Y_{n+1} say, having observed data up to time n, $Y_1 = y_1, \ldots, Y_n = y_n$ or $Y_{1:n} = y_{1:n}$ for short. Clearly, the first step to this aim is to formulate reasonable assumptions about the dependence structure of the time series. If we are able to specify the probability law of the time series (Y_t), we know the joint densities $\pi(y_1, \ldots, y_n)$ for any $n \geq 1$, and Bayesian forecasting would be solved by computing the predictive density

$$\pi(y_{n+1} | y_{1:n}) = \frac{\pi(y_{1:n+1})}{\pi(y_{1:n})}.$$

In practice, specifying the densities $\pi(y_1, \ldots, y_n)$ directly is not easy, and one finds it convenient to make use of parametric models; that is, one often finds it simpler to express the probability law of $(Y_1, \ldots, Y_n)$ conditionally on some characteristic θ of the data generating process. The relevant characteristic θ can be finite- or infinite-dimensional, that is, θ can be a random vector or, as is the case for state space models, a stochastic process itself. The researcher often finds it simpler to specify the conditional density $\pi(y_{1:n}|\theta)$ of $Y_{1:n}$ given θ, and a density $\pi(\theta)$ on θ, then obtain $\pi(y_{1:n})$ as $\pi(y_{1:n}) = \int \pi(y_{1:n}|\theta)\pi(\theta)\, d\theta$. We will proceed in this fashion when introducing dynamic linear models for time series analysis. But let's first study simpler dependence structures.

Conditional independence

The simplest dependence structure is conditional independence. In particular, in many applications it is reasonable to assume that $Y_1, \ldots, Y_n$ are conditionally independent and identically distributed (i.i.d.) given θ: $\pi(y_{1:n}|\theta) = \prod_{i=1}^n \pi(y_i|\theta)$. For example, if the Y_i's are repeated measurements affected by a random error, we are used to think of a model of the kind $Y_i = \theta + \epsilon_i$, where the ϵ_i's are independent Gaussian random errors, with mean zero and variance σ^2 depending on the precision of the measurement device. This means that, conditionally on θ, the Y_i's are i.i.d., with $Y_i|\theta \sim \mathcal{N}(\theta, \sigma^2)$.

Note that $Y_1, Y_2, \ldots$ are only conditionally independent: the observations $y_1, \ldots, y_n$ provide us information about the unknown value of θ and, through θ, on the value of the next observation Y_{n+1}. Thus, Y_{n+1} depends, in a probabilistic sense, on the past observations $Y_1, \ldots, Y_n$. The predictive density in this case can be computed as

$$\begin{aligned}
\pi(y_{n+1}|y_{1:n}) &= \int \pi(y_{n+1}, \theta|y_{1:n})\, d\theta \\
&= \int \pi(y_{n+1}|\theta, y_{1:n})\pi(\theta|y_{1:n})\, d\theta \\
&= \int \pi(y_{n+1}|\theta)\pi(\theta|y_{1:n})\, d\theta,
\end{aligned}$$

the last equality following from the assumption of conditional independence, where $\pi(\theta|y_{1:n})$ is the posterior density of θ, conditionally on the data $(y_1, \ldots, y_n)$. As we have seen, the posterior density can be computed by the Bayes formula:

$$\pi(\theta|y_{1:n}) = \frac{\pi(y_{1:n}|\theta)\pi(\theta)}{\pi(y_{1:n})} \propto \prod_{t=1}^{n} \pi(y_t|\theta)\, \pi(\theta)\,. \tag{1.1}$$

Note that the marginal density $\pi(y_{1:n})$ does not depend on θ, having the role of normalizing constant, so that the posterior is proportional to the product of the likelihood and the prior[1].

It is interesting to note that, with the assumption of conditional independence, the posterior distribution can be computed *recursively*. This means that one does not need all the previous data to be kept in storage and reprocessed every time a new measurement is taken. In fact, at time $(n-1)$, the information available about θ is described by the conditional density

[1] The symbol $\propto$ means "proportional to".

$$\pi(\theta|y_{1:n-1}) \propto \prod_{t=1}^{n-1} \pi(y_t|\theta)\pi(\theta),$$

so that this density plays the role of prior at time n. Once the new observation y_n becomes available, we have just to compute the likelihood, which is $\pi(y_n|\theta, y_{1:n-1}) = \pi(y_n|\theta)$ by the assumption of conditional independence, and update the "prior" $\pi(\theta|y_{1:n-1})$ by the Bayes rule, obtaining

$$\pi(\theta|y_{1:n-1}, y_n) \propto \pi(\theta|y_{1:n-1})\pi(y_n|\theta) \propto \prod_{t=1}^{n-1} \pi(y_t|\theta)\pi(\theta)\pi(y_n|\theta),$$

which is (1.1). The recursive structure of the posterior will play a crucial role when we study dynamic linear models and the Kalman filter in the next chapters.

To illustrate the idea, let us use a simple example. Suppose that, after a wreck in the ocean, you landed on a small island, and let θ denote your position, the distance from the coast, say. When studying dynamic linear models, we will consider the case when θ is subject to change over time (you are on a life boat in the ocean and not on an island, so that you slowly move with the stream and the waves, being at distance θ_t from the coast at time t). However, for the moment let's consider θ as fixed. Luckily, you can see the coast at times; you have some initial idea of your position θ, but you are clearly interested in learning more about θ based on the measurements y_t that you can take. Let us formalize the learning process in the Bayesian approach.

The measurements Y_t can be modeled as

$$Y_t = \theta + \epsilon_t, \quad \epsilon_t \overset{\text{iid}}{\sim} \mathcal{N}(0, \sigma^2),$$

where the ϵ_t's and θ are independent and, for simplicity, σ^2 is a known constant. In other words:

$$Y_1, Y_2, \ldots |\theta \overset{\text{iid}}{\sim} \mathcal{N}(\theta, \sigma^2).$$

Suppose you agree to express your prior idea about θ as

$$\theta \sim \mathcal{N}(m_0, C_0),$$

where the prior variance C_0 might be quite large if you are very uncertain about your guess m_0. Given the measurements $y_{1:n}$, you update your opinion about θ computing its posterior density, using the Bayes formula. We have

$$\begin{aligned}\pi(\theta|y_{1:n}) &\propto \text{likelihood} \times \text{prior} \\ &= \prod_{t=1}^{n} \frac{1}{\sqrt{2\pi}\sigma} \exp\left\{-\frac{1}{2\sigma^2}(y_t - \theta)^2\right\} \frac{1}{\sqrt{2\pi C_0}} \exp\left\{-\frac{1}{2C_0}(\theta - m_0)^2\right\} \\ &\propto \exp\left\{-\frac{1}{2\sigma^2}\left(\sum_{t=1}^{n} y_t^2 - 2\theta\sum_{t=1}^{n} y_t + n\theta^2\right) - \frac{1}{2C_0}(\theta^2 - 2\theta m_0 + m_0^2)\right\} \\ &\propto \exp\left\{-\frac{1}{2\sigma^2 C_0}\left((nC_0 + \sigma^2)\theta^2 - 2(nC_0\bar{y} + \sigma^2 m_0)\theta\right)\right\}.\end{aligned}$$

The above expression might appear complicated, but in fact it is the kernel of a Normal density. Note that, if $\theta \sim \mathcal{N}(m, C)$, then $\pi(\theta) \propto \exp\{-(1/2C)(\theta^2 - 2m\theta)\}$; so, writing the above expression as

$$\exp\left\{-\frac{1}{2\sigma^2 C_0/(nC_0 + \sigma^2)}\left(\theta^2 - 2\frac{nC_0\bar{y} + \sigma^2 m_0}{(nC_0 + \sigma^2)}\theta\right)\right\},$$

we recognize that

$$\theta|y_{1:n} \sim \mathcal{N}(m_n, C_n),$$

where

$$m_n = \mathrm{E}(\theta|y_{1:n}) = \frac{C_0}{C_0 + \sigma^2/n}\bar{y} + \frac{\sigma^2/n}{C_0 + \sigma^2/n} m_0 \tag{1.2a}$$

and

$$C_n = \mathrm{Var}(\theta|y_{1:n}) = \left(\frac{n}{\sigma^2} + \frac{1}{C_0}\right)^{-1} = \frac{\sigma^2 C_0}{\sigma^2 + nC_0}. \tag{1.2b}$$

The posterior *precision* is $1/C_n = n/\sigma^2 + 1/C_0$, and it is the sum of the precision n/σ^2 of the sample mean and the initial precision $1/C_0$. The posterior precision is always larger than the initial precision: even data of poor quality provide some information. The posterior expectation $m_n = \mathrm{E}(\theta|y_{1:n})$ is a weighted average between the sample mean $\bar{y} = \sum_{i=1}^{n} y_i/n$ and the prior guess $m_0 = \mathrm{E}(\theta)$, with weights depending on C_0 and σ^2. If the prior uncertainty, represented by C_0, is small compared to σ^2, the prior guess receives more weight. If C_0 is very large, then $m_n \simeq \bar{y}$ and $C_n \simeq \sigma^2/n$.

As we have seen, the posterior distribution can be computed recursively. At time n, the conditional density $\mathcal{N}(m_{n-1}, C_{n-1})$ of θ given the previous data $y_{1:n-1}$ plays the role of prior, and the likelihood for the current observation is

$$\pi(y_n|\theta, y_{1:n-1}) = \pi(y_n|\theta) = \mathcal{N}(y_n; \theta, \sigma^2).$$

We can update the prior $\mathcal{N}(m_{n-1}, C_{n-1})$ on the basis of the observation y_n using (1.2), with m_{n-1} and C_{n-1} in place of m_0 and C_0. We see that the resulting posterior density is Gaussian, with parameters

$$\begin{aligned}m_n &= \frac{C_{n-1}}{C_{n-1} + \sigma^2} y_n + \left(1 - \frac{C_{n-1}}{C_{n-1} + \sigma^2}\right) m_{n-1} \\ &= m_{n-1} + \frac{C_{n-1}}{C_{n-1} + \sigma^2}(y_n - m_{n-1})\end{aligned} \tag{1.3a}$$

and variance

$$C_n = \left(\frac{1}{\sigma^2} + \frac{1}{C_{n-1}}\right)^{-1} = \frac{\sigma^2 C_{n-1}}{\sigma^2 + C_{n-1}}. \tag{1.3b}$$

Since $Y_{n+1} = \theta + \epsilon_{n+1}$, the *predictive distribution* of $Y_{n+1}|y_{1:n}$ is Normal, with mean m_n and variance $C_n + \sigma^2$; thus, m_n is the posterior expected value of θ and also the one-step-ahead "point prediction" $\mathrm{E}(Y_{n+1}|y_{1:n})$. Expression (1.3a) shows that m_n is obtained by correcting the previous estimate m_{n-1} by a term that takes into account the forecast error $e_n = y_n - m_{n-1}$, weighted by

$$\frac{C_{n-1}}{C_{n-1} + \sigma^2} = \frac{C_0}{\sigma^2 + nC_0}. \tag{1.4}$$

As we shall see in Chapter 2, this "prediction-error correction" structure is typical, more generally, of the formulae of the Kalman filter for dynamic linear models.

Exchangeability

Exchangeability is the basic dependence structure in Bayesian analysis. Consider again an infinite sequence $(Y_t : t = 1, 2, \ldots)$ of random vectors. Suppose that the order in the sequence is not relevant, in the sense that, for any $n \geq 1$, the vector $(Y_1, \ldots, Y_n)$ and any permutation of its components, $(Y_{i_1}, \ldots, Y_{i_n})$, have the same distribution. In this case, we say that the sequence $(Y_t : t = 1, 2, \ldots)$ is *exchangeable*. This is a reasonable assumption when the Y_t's represent the results of experiments repeated under similar conditions. In the example of the previous paragraph, it is quite natural to consider that the order in which the measurements Y_t of the distance from the coast are taken is not relevant. There is an important result, known as de Finetti's representation theorem, that shows that the assumption of exchangeability is equivalent to the assumption of conditional independence and identical distribution that we have discussed in the previous paragraph. There is, however, an important difference. As you can see, here we move from a quite natural assumption on the dependence structure of the observables, that is, exchangeability; we have not introduced, up to now, parametric models or prior distributions on parameters. In fact, the hypothetical model, that is the pair likelihood and prior, arises from the assumption of exchangeability, as shown by the representation theorem.

Theorem 1.1. ***(de Finetti representation theorem).*** *Let $(Y_t : t = 1, 2, \ldots)$ be an infinite sequence of exchangeable random vectors. Then*

1. *With probability one, the sequence of empirical distribution functions*

$$F_n(y) = F_n(y; Y_1, \ldots, Y_n) = \frac{1}{n}\sum_{i=1}^{n} I_{(-\infty, y]}(Y_i)$$

converges weakly to a random distribution function F, as $n \to \infty$;

2. for any $n \geq 1$, the distribution function of $(Y_1, \ldots, Y_n)$ can be represented as

$$\mathrm{P}(Y_1 \leq y_1, \ldots, Y_n \leq y_n) = \int \prod_{i=1}^{n} \pi(y_i) \, \mathrm{d}\pi(F)$$

where π is the probability law of the weak limit F of the sequence of the empirical distribution functions.

The fascinating aspect of the representation theorem is that the hypothetical model results from the assumptions on the dependence structure of the observable variables (Y_t). If we assume that the sequence (Y_t) is exchangeable, then we can think of them as i.i.d. conditionally on the distribution function (d.f.) F, with common d.f. F. The random d.f. F is the weak limit of the empirical d.f.'s. The prior distribution π (also called, in this context, de Finetti measure) is a probability law on the space $\mathcal{F}$ of all the d.f.s on the sample space $\mathcal{Y}$ and expresses our beliefs on the limit of the empirical d.f.s. In many problems we can restrict the support of the prior to a parametric class $\mathcal{P}_\Theta = \{\pi(\cdot|\theta), \theta \in \Theta\} \subset \mathcal{F}$, where $\Theta \subseteq \mathbb{R}^p$; in this case the prior is said *parametric.* We see that, in the case of a parametric prior, the representation theorem implies that $Y_1, Y_2, \ldots$ are conditionally i.i.d., given θ, with common d.f. $\pi(\cdot|\theta)$, and θ has a prior distribution $\pi(\theta)$. This is the conditional i.i.d. dependence structure that we have discussed in the previous subsection.

Heterogeneous data

Exchangeability is the simplest dependence structure, which allows us to enlighten the basic aspects of Bayesian inference. It is appropriate when we believe that the data are homogeneous. However, in many problems the dependence structure is more complex. Often, it is appropriate to allow some heterogeneity among the data, assuming that

$$Y_1, \ldots, Y_n | \theta_1, \ldots, \theta_n \sim \prod_{t=1}^{n} f_t(y_t|\theta_t),$$

that is, $Y_1, \ldots, Y_n$ are conditionally independent given a vector $\theta = (\theta_1, \ldots, \theta_n)$, with Y_t depending only on the corresponding θ_t. For example, Y_t could be the expense of customer t for some service, and we might assume that each customer has a different average expense θ_t, introducing heterogeneity, or "random effects," among customers. In other applications, t might denote time; for example, each Y_t could represent the average sales in a sample of stores, at time t; and we might assume that $Y_t|\theta_t \sim \mathcal{N}(\theta_t, \sigma^2)$, with θ_t representing the expected sales at time t.

In these cases, the model specification is completed by assigning the probability law of the vector $(\theta_1, \ldots, \theta_n)$. For modeling random effects, a common assumption is that $\theta_1, \ldots, \theta_n$ are i.i.d. according to a distribution G. If there is uncertainty about G, we can model $\theta_1, \ldots, \theta_n$ as conditionally i.i.d. given G, with common distribution function G, and assign a prior on G.

If $(Y_t : t = 1, 2, \ldots)$ is a sequence of observations over time, then the assumption that the θ_t's are i.i.d., or conditionally i.i.d., is generally not appropriate, since we want to introduce a temporal dependence among them. As we shall see in Chapter 2, in state space models we assume a Markovian dependence structure among the θ_t's.

We will return to this problem in the next section.

1.3 Synthesis of conditional distributions

We have seen that Bayesian inference is simply solved, in principle, by computing the conditional probability distributions of the quantities of interest: the posterior distribution of the parameters of the model, or the predictive distribution. However, especially when the quantity of interest is multivariate, one might want to present a summary of the posterior or predictive distribution. Consider the case of inference on a multivariate parameter $\theta = (\theta_1, \ldots, \theta_p)$. After computing the joint posterior distribution of θ, if some elements of θ are regarded as nuisance parameters, one can integrate them out to obtain the (marginal) posterior of the parameters of interest. For example, if $p = 2$, we can marginalize the joint posterior $\pi(\theta_1, \theta_2|y)$ and compute the marginal posterior density of θ_1:

$$\pi(\theta_1|y) = \int \pi(\theta_1, \theta_2|y) \, d\theta_2.$$

We can provide a graphical representation of the marginal posterior distributions, or some summary values, such as the posterior expectations $E(\theta_i|y)$ or the posterior variances $\mathrm{Var}(\theta_i|y)$, and so on. We can also naturally show intervals (usually centered on $E(\theta_i|y)$) or bands with high posterior probability.

The choice of a summary of the posterior distribution (or of the predictive distribution) can be more formally regarded as a decision problem. In a statistical decision problem we want to choose an *action* in a set $\mathcal{A}$, called the action space, on the basis of the sample y. The consequences of action a are expressed through a *loss function* $L(\theta, a)$. Given the data y, a *Bayesian decision rule* selects an action in $\mathcal{A}$ that minimizes the conditional expected loss, $E(L(\theta, a)|y) = \int L(\theta, a)\pi(\theta|y) \, d\theta$. Bayesian point estimation can be seen as a decision problem in which the action space coincides with the parameter space. The choice of the loss function depends on the problem at hand, and, of course, different loss functions give rise to different Bayes estimates of θ. Some commonly used loss functions are briefly discussed below.

Quadratic loss. Let θ be a scalar. A common choice is a quadratic loss function $L(\theta, a) = (\theta - a)^2$. Then the posterior expected loss is $E((\theta - a)^2|y)$, which is minimized at $a = E(\theta|y)$. So, the Bayes estimate of θ with quadratic loss is the posterior expected value of θ. If θ is p-dimensional, a quadratic loss function is expressed as $L(\theta, a) = (\theta - a)'H(\theta - a)$, for a symmetric positive definite matrix H. Then the Bayes estimate of θ is the vector of posterior expectations $\mathrm{E}(\theta|y)$.

Linear loss. If θ is scalar and

$$L(\theta, a) = \begin{cases} c_1 \mid a - \theta \mid & \text{if } a \leq \theta, \\ c_2 \mid a - \theta \mid & \text{if } a > \theta, \end{cases}$$

where c_1 and c_2 are positive constants, then the Bayes estimate is the $c_1/(c_1 + c_2)$ quantile of the posterior distribution. As a special case, if $c_1 = c_2$, the Bayes estimate is a posterior median.

Zero-one loss. If θ is a discrete random variable and

$$L(\theta, a) = \begin{cases} c & \text{if } a \neq \theta, \\ 0 & \text{if } a = \theta, \end{cases}$$

then the Bayes estimate is a mode of the posterior distribution.

For example, if $Y_1, \ldots, Y_n|\theta$ are i.i.d. with $Y_t|\theta \sim \mathcal{N}(\theta, \sigma^2)$ and $\theta \sim \mathcal{N}(m_0, C_0)$, the posterior density is $\mathcal{N}(m_n, C_n)$, where m_n and C_n are given by (1.2). The Bayes estimate of θ, adopting a quadratic loss, is $\mathrm{E}(\theta|y_{1:n}) = m_n$, a weighted average between the prior guess m_0 and the sample mean $\bar{y}$. Note that, if the sample size is large, then the weight of the prior guess decreases to zero, and the posterior density concentrates around $\bar{y}$, which is the maximum likelihood estimate (MLE) of θ.

This asymptotic behavior of the posterior density holds more generally. Let $(Y_t : t = 1, 2, \ldots)$ be a sequence of conditionally i.i.d. random vectors, given θ, with $Y_t|\theta \sim \pi(y|\theta)$ and $\theta \in \mathbb{R}^p$ having prior distribution $\pi(\theta)$. Under general assumptions, it can be proved that the posterior distribution $\pi(\theta|y_1, \ldots, y_n)$, for n large, can be approximated by a Normal density centered at the MLE $\hat{\theta}_n$. This implies that, in these cases, Bayesian and frequentist estimates tend to agree for a sufficiently large sample size. For a more rigorous discussion of asymptotic normality of the posterior distribution, see Bernardo and Smith (1994, Section 5.3), or Schervish (1995, Section 7.4).

As a second example, linking Bayes estimators and classical decision theory, consider the problem of estimating the mean of a multivariate Normal distribution. In its simplest formulation, the problem is as follows. Suppose that $Y_1, \ldots, Y_n$ are independent r.v.s, with $Y_t \sim \mathcal{N}(\theta_t, \sigma^2), t = 1, \ldots, n$, where σ^2 is a known constant. This is the case of heterogeneous data, discussed in Section 1.2. For instance, the Y_t's could be sample means, in n independent experiments; however, note that here $\theta = (\theta_1, \ldots, \theta_n)$ is regarded as a vector of unknown constants. Thus we have

$$Y = (Y_1, \ldots, Y_n) \sim \mathcal{N}_n(\theta, \sigma^2 I_n),$$

where I_n denotes the n-dimensional identity matrix, and the problem is estimating the mean vector θ. The MLE of θ, which is also the uniform minimum variance unbiased estimator, is given by the vector of sample means:

$\hat{\theta} = \hat{\theta}(Y) = Y$. However, an important result, which had a great impact when Stein proved it in 1956, shows that the MLE is not optimal with respect to the quadratic loss function $L(\theta, a) = (\theta - a)'(\theta - a)$ if $n \geq 3$. The overall expected loss, or mean square error, of $\hat{\theta}$ is

$$\mathrm{E}\left((\theta - \hat{\theta}(Y))'(\theta - \hat{\theta}(Y))\right) = \mathrm{E}\left(\sum_{t=1}^{n}(\theta_t - \hat{\theta}_t(Y))^2\right)$$

where the expectation is with respect to the density $\pi_\theta(y)$, i.e., the $\mathcal{N}_n(\theta, \sigma^2 I_n)$ distribution of the data. Stein (1956) proved that, if $n \geq 3$, there exists another estimator $\theta^* = \theta^*(Y)$, which is more efficient than the MLE $\hat{\theta}$ in the sense that

$$\mathrm{E}((\theta - \theta^*(Y))'(\theta - \theta^*(Y))) < \mathrm{E}((\theta - \hat{\theta}(Y))'(\theta - \hat{\theta}(Y)))$$

for every θ. For $\sigma^2 = 1$, the Stein estimator is given by $\theta^*(Y) = (1 - (n - 2)/Y'Y)Y$; it shrinks the sample means $Y = (Y_1, \dots, Y_n)$ towards zero. More generally, *shrinkage estimators* shrink the sample means towards the overall mean $\bar{y}$, or towards different values. Note that the MLE of θ_t, that is $\hat{\theta}_t = Y_t$, does not make use of the data Y_j, for $j \neq t$, which come from the other independent experiments. Thus, Stein's result seems quite surprising, showing that a more efficient estimator of θ_t can be obtained using the information from "independent" experiments. Borrowing strength from different experiments is in fact quite natural in a Bayesian approach. The vector θ is regarded as a random vector, and the Y_t's are *conditionally* independent given $\theta = (\theta_1, \dots, \theta_n)$, with $Y_t|\theta_t \sim \mathcal{N}(\theta_t, \sigma^2)$, that is

$$Y|\theta \sim \mathcal{N}_n(\theta, \sigma^2 I_n).$$

Assuming a $\mathcal{N}_n(m_0, C_0)$ prior density for θ, the posterior density is $\mathcal{N}_n(m_n, C_n)$ where

$$m_n = (C_0^{-1} + \sigma^{-2} I_n)^{-1}(C_0^{-1} m_0 + \sigma^{-2} I_n y)$$

and $C_n = (C_0^{-1} + \sigma^{-2} I_n)^{-1}$. Thus the posterior expectation m_n provides a shrinkage estimate, shrinking the sample means towards the value m_0. Clearly, the shrinkage depends on the choice of the prior; see Lindley and Smith (1972).

Similarly to a Bayes point estimate, a Bayes point forecast of Y_{n+1} given $y_{1:n}$ is a synthesis of the predictive density with respect to a loss function, which expresses consequences of the forecast error of predicting Y_{n+1} with a value $\hat{y}$, say. With the quadratic loss function, $L(y_{n+1}, \hat{y}) = (y_{n+1} - \hat{y})^2$, the Bayes forecast is the expected value $\mathrm{E}(Y_{n+1}|y_{1:n})$.

Again, point estimation or forecasting is coherently treated in the Bayesian approach on the basis of statistical decision theory. However, in practice the computation of Bayes estimates or forecasts can be difficult. If θ is multivariate and the model structure complex, posterior expectations or, more generally, integrals of the kind $\int g(\theta)\pi(\theta|y)d\theta$, can be analytically untractable. In fact, despite its attractive theoretical and conceptual coherence, the diffusion of Bayesian statistics in applied fields has been hindered, in the past, by computational difficulties, which had restricted the availability of Bayesian solutions to rather simple problems. As we shall see in Section 1.6, these difficulties can be overcome by the use of modern simulation techniques.

1.4 Choice of the prior distribution

The explicit use of prior information, besides the information from the data, is a basic aspect of Bayesian inference. Indeed, some prior knowledge of the phenomenon under study is always needed: data never speak entirely by themselves. The Bayesian approach allows us to explicitly introduce all the information we have (from experts' opinions, from previous studies, from the theory, and from the data) in the inferential process. However, the choice of the prior can be a delicate point in practical applications. Here we briefly summarize some basic notions, but first let us underline a fundamental point, which is clearly enlightened in the case of exchangeable data: the choice of a prior is in fact the choice of the *pair* $\pi(y|\theta)$ and $\pi(\theta)$. Often, the choice of $\pi(y|\theta)$ is called *model specification*, but in fact it is part, with the specification of $\pi(\theta)$, of the subjective choices that we have to do in order to study a phenomenon, based of our prior knowledge. At any rate, given $\pi(y|\theta)$, the prior $\pi(\theta)$ should be an honest expression of our beliefs about θ, with no mathematical restrictions on its form.

That said, there are some practical aspects that deserve some consideration. For computational convenience, it is common practice to use *conjugate priors*. A family of densities on θ is said to be conjugate to the model $\pi(y|\theta)$ if, whenever the prior belongs to that family, so does the posterior. In the example in Section 1.2, we used a Gaussian prior density $\mathcal{N}(m_0, C_0)$ on θ, and the posterior resulted still Gaussian, with updated parameters, $\mathcal{N}(m_n, C_n)$; thus, the Gaussian family is conjugate to the model $\pi(y|\theta) = \mathcal{N}(y; \theta, \sigma^2)$ (with σ^2 known). In general, a prior will be conjugate when it has the same analytic form of the likelihood, regarded as a function of θ. Clearly this definition does not determine uniquely the conjugate prior for a model $\pi(y|\theta)$. For the exponential family, we have a more precise notion of *natural conjugate prior*, which is defined from the density of the sufficient statistics; see for example Bernardo and Smith (1994, Section 5.2). Natural conjugate priors for the exponential family can be quite rigid in the multivariate case, and *enriched* conjugate priors have been proposed (Brown et al.; 1994; Consonni and Veronese; 2001). Furthermore, it can be proved that any prior for an exponential family parameter can be approximated by a mixture of conjugate priors (Dalal and Hall; 1983; Diaconis and Ylvisaker; 1985). We provide some examples below and in the next section. Anyway, computational ease has become less stringent in recent years, due to the availability of simulation-based approximation techniques.

In practice, people quite often use *default priors* or *non-informative priors*, for expressing a situation of "prior ignorance" or vague prior information. The problem of appropriately defining the idea of "prior ignorance," or of a prior with "minimal effect" relative to the data on the inferential results, has a long history and is quite delicate; see Bernardo and Smith (1994, Section 5.6.2) for a detailed treatment; or also O'Hagan (1994) or Robert (2001). If the parameter θ takes values in a finite set, $\{\theta_1^*, \ldots, \theta_k^*\}$ say, then the classical notion of a non-informative prior, since Bayes (1763) and Laplace (1814), is that of a uniform distribution, $\pi(\theta_j^*) = 1/k$. However, even in this simple case it can be shown that care is needed in defining the quantity of interest (see Bernardo and Smith; 1994). Anyway, extending the notion of a uniform prior when the parameter space is infinite clearly leads to *improper* distributions, which cannot be regarded as probability distributions. For example, if $\theta \in (-\infty, +\infty)$, a uniform prior would be a constant, and its integral on the real line would be infinite. Furthermore, a uniform distribution for θ implies a nonuniform distribution for any nonlinear monotone transformation of θ, and thus the Bayes–Laplace postulate is inconsistent in the sense that, intuitively, "ignorance about θ" should also imply "ignorance" about one-to-one transformations of it. Priors based on invariance considerations are Jeffreys priors (Jeffreys; 1998). Widely used are also *reference priors*, suggested by Bernardo (1979a,b) on an information-decisional theoretical base (see for example Bernardo and Smith; 1994, Section 5.4). The use of improper priors is debatable, but often the posterior density from an improper prior turns out to be proper, and improper priors are anyway widely used, also for reconstructing frequentist results in a Bayesian framework. For example, if $Y_t|\theta$ are i.i.d. $\mathcal{N}(\theta, \sigma^2)$, using an improper uniform prior $\pi(\theta) = c$ and formally applying Bayes' formula gives

$$\pi(\theta|y_{1:n}) \propto \exp\left\{-\frac{1}{2\sigma^2}\sum_{t=1}^{n}(y_t - \theta)^2\right\} \propto \exp\left\{-\frac{n}{2\sigma^2}(\theta^2 - 2\theta\bar{y})^2\right\},$$

that is, the posterior is $\mathcal{N}(\bar{y}, \sigma^2/n)$. In this case, the Bayes point estimate under quadratic loss is $\bar{y}$, which is also the MLE of θ. As we noted before, starting with a proper Gaussian prior would give a posterior density centered around the sample mean only if the prior variance C_0 is very large compared to σ^2, or if the sample size n is large.

Another common practice is to have a hierarchical specification of the prior density. This means assuming that θ has density $\pi(\theta|\lambda)$ conditionally on some hyperparameter λ, and then a prior $\pi(\lambda)$ is assigned to λ. This is often a way for expressing a kind of uncertainty in the choice of the prior density. Clearly, this is equivalent to the prior $\pi(\theta) = \int \pi(\theta|\lambda)\pi(\lambda)\,\mathrm{d}\lambda$.

In order to avoid theoretical and computational difficulties related to the use of improper priors, in this book we will use only proper priors. It is important, however, to be aware of the effect of the prior on the analysis. This can be assessed using sensitivity analysis, which, in one of its basic forms, may simply consist in comparing the inferences resulting from different prior hyperparameters.

We conlude this section with an important example of conjugate prior. In Section 1.2 we considered conjugate Bayesian analysis for the mean of a Gaussian population, with known variance. Let now $Y_1, \ldots, Y_n|\theta, \sigma^2$ be i.i.d. $\mathcal{N}(\theta, \sigma^2)$, where both θ and σ^2 are unknown. It is convenient to work with the *precision* $\phi = 1/\sigma^2$ rather than with the variance σ^2. A conjugate prior for (θ, ϕ) can be obtained noting that the likelihood can be written as

$$\pi(y_{1:n}|\theta, \phi) \propto \phi^{(n-1)/2} \exp\left\{-\frac{1}{2}\phi\, ns^2\right\} \phi^{1/2} \exp\left\{-\frac{n}{2}\phi(\mu - \bar{y})^2\right\}$$

where $\bar{y}$ is the sample mean and $s^2 = \sum_{t=1}^n (y_i - \bar{y})^2/n$ is the sample variance (add and subtract $\bar{y}$ in the squared term and note that the cross product is zero). We see that, as a function of (θ, ϕ), the likelihood is proportional to the kernel of a Gamma density in ϕ, with parameters $(n/2 + 1, ns^2/2)$ times the kernel of a Normal density in θ, with parameters $(\bar{y}, (n\phi)^{-1})$. Therefore, a conjugate prior for (θ, σ^2) is such that ϕ has a Gamma density with parameters (a, b) and, conditionally on ϕ, θ has a Normal density with parameters $(m_0, (n_0\phi)^{-1})$. The joint prior density is

$$\begin{aligned}\pi(\theta, \phi) &= \pi(\phi)\, \pi(\theta|\phi) = \mathcal{G}(\phi; a, b)\, \mathcal{N}(\theta; m_0, (n_0\phi)^{-1}) \\ &\propto \phi^{a-1} \exp\{-b\phi\}\; \phi^{1/2} \exp\left\{-\frac{n_0}{2}\phi(\theta - m_0)^2\right\},\end{aligned}$$

which is a Normal-Gamma, with parameters $(m_0, (n_0)^{-1}, a, b)$ (see Appendix A). In particular, $\mathrm{E}(\theta|\phi) = m_0$ and $\mathrm{Var}(\theta|\phi) = (n_0\phi)^{-1} = \sigma^2/n_0$, that is, the variance of θ, given σ^2, is expressed as a proportion $1/n_0$ of σ^2. Marginally, the variance $\sigma^2 = \phi^{-1}$ has an Inverse Gamma density, with $\mathrm{E}(\sigma^2) = b/(a-1)$, and it can be shown that

$$\theta \sim \mathcal{T}(m_0, (n_0\, a/b)^{-1}, 2a),$$

a Student-t with parameters $m_0, (n_0\, a/b)^{-1}$ and $2a$ degrees of freedom, with $\mathrm{E}(\theta) = \mathrm{E}(\mathrm{E}(\theta|\phi)) = m_0$ and $\mathrm{Var}(\theta) = \mathrm{E}(\sigma^2)/n_0 = (b/(a-1))/n_0$.

With a conjugate Normal-Gamma prior, the posterior of (θ, ϕ) is still Normal-Gamma, with updated parameters. In order to show this, we have to do some calculations. Start with

$$\begin{aligned}&\pi(\theta, \phi|y_{1:n}) \propto \\ &\quad \phi^{\frac{n}{2}+a-1} \exp\left\{-\frac{1}{2}\phi(ns^2 + 2b)\right\} \phi^{\frac{1}{2}} \exp\left\{-\frac{1}{2}\phi n\left((\theta - \bar{y})^2 + n_0(\theta_0)^2\right)\right\}.\end{aligned}$$

After some algebra and completing the square that appears in it, the last exponential term can be written as

$$\exp\left\{-\frac{1}{2}\phi\left(nn_0\frac{(m_0-\bar{y})^2}{n_0+n}+(n_0+n)\left(\theta-\frac{n\bar{y}+n_0m_0}{n_0+n}\right)^2\right)\right\},$$

so that

$$\pi(\theta,\phi|y_{1:n}) \propto$$
$$\phi^{\frac{n}{2}+a-1}\exp\left\{-\frac{1}{2}\phi\left(ns^2+2b+nn_0\frac{(m_0-\bar{y})^2}{n_0+n}\right)\right\}$$
$$\cdot\,\phi^{\frac{1}{2}}\exp\left\{-\frac{1}{2}\phi(n_0+n)(\theta-m_n)^2\right\}.$$

From the previous expression, we see that the parameters of the posterior Normal-Gamma distribution are

$$\begin{aligned} m_n &= \frac{n\bar{y}+n_0m_0}{n_0+n} \\ n_n &= n_0+n \\ a_n &= a+\frac{n}{2} \\ b_n &= b+\frac{1}{2}ns^2+\frac{1}{2}\frac{nn_0}{n_0+n}(\bar{y}-m_0)^2. \end{aligned} \tag{1.5}$$

This means that

$$\begin{aligned} \phi|y_{1:n} &\sim \mathcal{G}(a_n,b_n); \\ \theta|\phi,y_{1:n} &\sim \mathcal{N}(m_n,(n_n\phi)^{-1}). \end{aligned}$$

Clearly, conditionally on ϕ, we are back to the case of inference on the mean of a $\mathcal{N}(\theta,\sigma^2)$ with known variance; you can check that the expressions of $E(\theta|\phi,y_1,\ldots,y_n)=m_n$ and $V(\theta|\phi,y_1,\ldots,y_n)=((n_0+n)\phi)^{-1}=\sigma^2/(n_0+n)$ given above correspond to (1.2), when $C_0=\sigma^2/n_0$. Here, n_0 has a role of "prior sample size." The marginal density of $\theta|y_1,\ldots,y_n$ is obtained by marginalizing the joint posterior of (θ,ϕ) and results to be Student-t, with parameters m_n, $(n_n\,a_n/b_n)^{-1}$ and $2a_n$ degrees of freedom.

The predictive density is also Student-t:

$$Y_{n+1}|y_1,\ldots,y_n \sim \mathcal{T}\left(m_n,\frac{b_n}{a_nn_n}(1+n_n),2a_n\right).$$

The recursive formulae to update the distribution of (θ,ϕ) when a new observation y_n becomes available are

$$\begin{aligned} m_n &= m_{n-1}+\frac{1}{n_{n-1}+1}(y_n-m_{n-1}), \\ n_n &= n_{n-1}+1, \\ a_n &= a_{n-1}+\frac{1}{2}, \\ b_n &= b_{n-1}+\frac{1}{2}\frac{n_{n-1}}{n_{n-1}+1}(y_n-m_{n-1})^2. \end{aligned}$$

1.5 Bayesian inference in the linear regression model

Dynamic linear models can be regarded as a generalization of the standard linear regression model, when the regression coefficients are allowed to change over time. Therefore, for the reader's convenience we remind briefly here the basic elements of Bayesian analysis for the linear regression model.

The linear regression model is the most popular tool for relating the variable Y to explanatory variables x. It is defined as

$$Y_t = x_t'\beta + \epsilon_t, \quad t = 1, \ldots, n, \quad \epsilon_t \overset{\text{iid}}{\sim} \mathcal{N}(0, \sigma^2) \tag{1.6}$$

where Y_t is a random variable and x_t and β are p-dimensional vectors. In its basic formulation, the variables x are considered as deterministic or exogenous; while in stochastic regression x are random variables. In the latter case we have in fact, for each t, a random $(p+1)$-dimensional vector (Y_t, X_t), and we have to specify its joint distribution and derive the linear regression model from it. A way for doing this (but more general approaches are possible) is to assume that the joint distribution is Gaussian

$$\begin{bmatrix} Y_t \\ X_t \end{bmatrix} \Bigg| \mu, \Sigma \sim \mathcal{N}(\mu, \Sigma), \qquad \mu = \begin{bmatrix} \mu_y \\ \mu_x \end{bmatrix}, \quad \Sigma = \begin{bmatrix} \Sigma_{yy} & \Sigma_{yx} \\ \Sigma_{xy} & \Sigma_{xx} \end{bmatrix}.$$

From the properties of the multivariate Gaussian distribution (see Appendix A), we can decompose the joint distribution into a marginal model for X_t and a conditional model for Y_t given $X_t = x_t$ as follows:

$$\begin{aligned} X_t | \mu, \Sigma &\sim \mathcal{N}(\mu_x, \Sigma_{xx}), \\ Y_t | x_t, \mu, \Sigma &\sim \mathcal{N}(\beta_1 + x_t'\beta_2, \sigma^2), \end{aligned}$$

where

$$\begin{aligned} \beta_2 &= \Sigma_{xx}^{-1} \Sigma_{xy}, \\ \beta_1 &= \mu_y - \mu_x' \beta_2, \\ \sigma^2 &= \Sigma_{yy} - \Sigma_{yx} \Sigma_{xx}^{-1} \Sigma_{xy}. \end{aligned}$$

If the prior distribution on (μ, Σ) is such that the parameters of the marginal model and those of the conditional model are independent, then we have a *cut* in the distribution of $(Y_t, X_t, \beta, \Sigma)$; in other words, if our interest is mainly on the variable Y, we can restrict our attention to the conditional model. In this case the regression model describes the conditional distribution of Y_t given (β, Σ) and x_t.

Model (1.6) can be rewritten as

$$Y | X, \beta, V \sim \mathcal{N}_n(X\beta, V), \tag{1.7}$$

where $Y = (Y_1, \ldots, Y_n)$ and X is the $n \times p$ matrix with tth row x_t'. Equation (1.6) implies a diagonal covariance matrix, $V = \sigma^2 I_n$; i.e., the Y_t's are conditionally independent, with the same variance σ^2. More generally, V can be a symmetric positive-definite matrix.

We describe Bayesian inference with conjugate priors for the regression model, for three cases: inference on the regression coefficients β, assuming that V is known; inference on the covariance matrix V when β is known; and inference on β and V.

Inference on the regression coefficients

Here we suppose that V is known and we are interested in inference about the regression coefficients β given the data y. As briefly discussed in the previous section, a conjugate prior for β can be obtained by looking at the likelihood as a function of β. The likelihood for the regression model (1.7) is

$$\begin{aligned} \pi(y|\beta, V, X) &= (2\pi)^{-n/2}|V|^{-1/2} \exp\left\{-\frac{1}{2}(y - X\beta)'V^{-1}(y - X\beta)\right\} \\ &\propto |V|^{-1/2} \exp\left\{-\frac{1}{2}(y'V^{-1}y - 2\beta'X'V^{-1}y + \beta'X'V^{-1}X\beta)\right\} \end{aligned} \tag{1.8}$$

where $|V|$ denotes the determinant of V. Now, note that, if $\beta \sim \mathcal{N}_p(m, C)$ then

$$\pi(\beta) \propto \exp\left\{-\frac{1}{2}(\beta - m)'C^{-1}(\beta - m)\right\} \propto \exp\left\{-\frac{1}{2}(\beta'C^{-1}\beta - 2\beta'C^{-1}m)\right\}.$$

Therefore, we see that the likelihood, as a function of β, is proportional to a multivariate Gaussian density, with mean $(X'V^{-1}X)^{-1}X'V^{-1}y$ and variance $(X'V^{-1}X)^{-1}$. Thus, a conjugate prior for β is the Gaussian density, $\mathcal{N}_p(m_0, C_0)$, say. As usual, m_0 represents a prior guess about β; the elements on the diagonal of C_0 express prior uncertainty on the prior guess m_0, and the off-diagonal elements of C_0 express the prior opinion about the dependence among the regression coefficients, β_t's.

With a conjugate Gaussian prior, the posterior will be Gaussian as well, with updated parameters. In order to derive the expression of the posterior parameters, we can compute the posterior density using Bayes' formula:

$$\begin{aligned} \pi(\beta|Y, X, V) &\propto \exp\left\{-\frac{1}{2}(\beta'X'V^{-1}X\beta - 2\beta'X'V^{-1}y\right\} \\ &\qquad \cdot \exp\left\{-\frac{1}{2}(\beta - m_0)'C_0^{-1}(\beta - m_0)\right\} \\ &\propto \exp\left\{-\frac{1}{2}\left(\beta'(X'V^{-1}X + C_0^{-1})\beta - 2\beta'(X'V^{-1}y + C_0^{-1}m_0)\right)\right\}. \end{aligned}$$

We recognize the kernel of a p-variate Gaussian density with parameters

$$\begin{aligned} m_n &= C_n(X'V^{-1}y + C_0^{-1}\mu_0) \\ C_n &= (C_0^{-1} + X'V^{-1}X)^{-1}. \end{aligned}$$

The Bayes point estimate of β, with respect to a quadratic loss function, is the posterior expected value $\mathrm{E}(\beta|X,y) = m_n$. Note that we do not require the assumption that $(X'V^{-1}X)^{-1}$ exists, which is instead necessary for computing the classical generalized least square estimate of β, that is $\hat{\beta} = (X'V^{-1}X)^{-1}X'V^{-1}y$. However, when $(X'V^{-1}X)$ is non-singular, the Bayes estimate m_n can be written as

$$m_n = (C_0^{-1} + X'V^{-1}X)^{-1}(X'V^{-1}X\hat{\beta} + C_0^{-1}m_0),$$

that is, as a matrix-weighted linear combination of the prior guess m_0, with weight proportional to the prior precision matrix C_0^{-1}, and of the generalized least square estimate $\hat{\beta}$, whose weight is proportional to the precision matrix $X'V^{-1}X$ of $\hat{\beta}$. Clearly, m_n is a shrinkage estimator of the regression coefficients; see Lindley and Smith (1972).

The posterior precision matrix is the sum of the prior precision C_0^{-1} and the precision of $\hat{\beta}$, $X'V^{-1}X$. Of course, one can integrate the joint posterior density of β to obtain the marginal posterior density of one or more coefficients β_j.

For the analysis that we will do in the next chapter, when studying dynamic linear models, it is useful to provide an alternative "recursive" expression of the posterior parameters. It can be proved that the posterior variance can be rewritten as

$$C_n = (X'V^{-1}X + C_0^{-1})^{-1} = C_0 - C_0X'(XC_0X' + V)^{-1}XC_0 \tag{1.9}$$

(see Problem 1.1). Using the above identity, it can be shown that the posterior expectation m_n can be expressed as

$$m_n = m_0 + C_0X'(XC_0X' + V)^{-1}(y - Xm_0) \tag{1.10}$$

(see Problem 1.2). Note that $Xm_0 = \mathrm{E}(Y|\beta, X)$ is the prior point forecast of Y. So, the above expression writes the Bayes estimate of β as the prior guess m_0 corrected by a term that takes into account the forecast error $(y - Xm_0)$. The (matrix) weight $C_0X'(XC_0X' + V)^{-1}$ of the term $y - Xm_0$ specifies the extent to which unexpectedly small or large observations translate into an adjustment of the point estimate m_0. Loosely speaking, it is the weight given to experimental evidence. In the context of dynamic linear models this weight is called gain matrix.

Inference on the covariance matrix

Suppose now that β is known and we are interested in inference on the covariance matrix V. Analogously to the case of inference on the parameters of the Gaussian univariate model, it is convenient to work with the precision matrix $\Phi = V^{-1}$. In order to determine a conjugate prior for Φ, note that we can write the likelihood (1.8) as

$$\pi(y|\beta, \Phi, X) \propto |\Phi|^{1/2} \exp\left\{-\frac{1}{2}(y - X\beta)'\Phi(y - X\beta)\right\}$$
$$= |\Phi|^{1/2} \exp\left\{-\frac{1}{2}\mathrm{tr}((y - X\beta)(y - X\beta)'\Phi)\right\},$$

where $\mathrm{tr}(A)$ denotes the trace of a matrix A, since $(y - X\beta)'\Phi(y - X\beta) = \mathrm{tr}((y - X\beta)'\Phi(y - X\beta))$ (the argument being a scalar) and recalling that $\mathrm{tr}(AB) = \mathrm{tr}(BA)$. We see that, as a function of Φ, the likelihood is proportional to a Wishart density with parameters $(n/2+1, 1/2(y - X\beta)(y - X\beta)')$ (see Appendix A). So, a conjugate prior for the precision Φ is Wishart

$$\Phi \sim \mathcal{W}(\nu_0, S_0).$$

The posterior is Wishart with updated parameters,

$$\Phi|Y, X, \beta \sim \mathcal{W}(\nu_n, S_n),$$

and it can be easily checked that

$$\nu_n = \nu_0 + \frac{1}{2}$$
$$S_n = \frac{1}{2}(y - X\beta)(y - X\beta)' + S_0.$$

It is often convenient to express the prior hyperparameters as

$$\nu_0 = \frac{\delta + n - 1}{2}; \qquad S_0 = \frac{1}{2}V_0,$$

(Lindley; 1978), so that $\mathrm{E}(V) = V_0/(\delta - 2)$ for $\delta > 2$ and the posterior expectation can be written as a weighted average between the prior guess and the sample covariance,

$$\mathrm{E}(V|y) = \frac{\delta - 2}{\delta + n - 2} \cdot \mathrm{E}(V) + \frac{n}{\delta + n - 2} \cdot \frac{(y - X\beta)(y - X\beta)'}{n},$$

with weights depending on δ.

Inference on (β, V)

If both β and V are random, analytical computations may become complicated; a simple case is when V has the form $V = \sigma^2 D$, where σ^2 is a random variable and the $n \times n$ matrix D is known; e.g., $D = I_n$. Let $\phi = \sigma^{-2}$. A conjugate prior for (β, ϕ) is a Normal-Gamma, with parameters $(\beta_0, N_0^{-1}, a, b)$

$$\pi(\beta, \phi) \propto \phi^{a-1} \exp(-b\phi)\, \phi^{\frac{p}{2}} \exp\left\{-\frac{\phi}{2}(\beta - \beta_0)'N_0(\beta - \beta_0)\right\}$$

that is

$$\begin{aligned}\beta|\phi &\sim \mathcal{N}(\beta_0, (\phi N_0)^{-1})\\ \phi &\sim \mathcal{G}(a,b).\end{aligned}$$

Note that, conditionally on ϕ, β has covariance matrix $(\phi N_0)^{-1} = \sigma^2 \tilde{C}_0$ where we let $\tilde{C}_0 = N_0^{-1}$, a symmetric $(p \times p)$ positive-definite matrix that "rescales" the observation variance σ^2.

It can be shown (see Problem 1.3) that the posterior is a Normal-Gamma with parameters

$$\begin{aligned}\beta_n &= \beta_0 + \tilde{C}_0 X'(X\tilde{C}_0 X' + D)^{-1}(y - X\beta_0),\\ \tilde{C}_n &= \tilde{C}_0 - \tilde{C}_0 X'(X\tilde{C}_0 X' + D)^{-1} X\tilde{C}_0\\ a_n &= a + \frac{n}{2}\\ b_n &= b + \frac{1}{2}(\beta_0' \tilde{C}_0^{-1}\beta_0 + y'D^{-1}y - \beta_n'\tilde{C}_n\beta_n).\end{aligned} \tag{1.11}$$

Furthermore, we can simplify the expression of b_n; in particular, it can be shown that

$$b_n = b + \frac{1}{2}(y - X\beta_0)'(D + X\tilde{C}_0 X')^{-1}(y - X\beta_0). \tag{1.12}$$

These formulae have again the estimation-error correction structure that we have underlined in the simple Gaussian model, see (1.3a), and in the regression model with known covariance, compare with (1.10).

1.6 Markov chain Monte Carlo methods

In Bayesian inference, it is very often the case that the posterior distribution of the parameters, denoted here by ψ, is analytically intractable. By this we mean that it is impossible to derive in closed form summaries of the posterior, such as its mean and variance, or the marginal distribution of a particular parameter. In fact, most of the times the posterior density is only known up to a normalizing factor. To overcome this limitation, the standard practice is to resort to simulation methods. For example, if one could draw $\psi_1, \ldots, \psi_N$ i.i.d. from the posterior distribution π, then, using the standard Monte Carlo method, the mean of any function $g(\psi)$ having finite posterior expectation can be approximated by a sample average:

$$\mathrm{E}_\pi(g(\psi)) \approx N^{-1} \sum_{j=1}^{N} g(\psi_j). \tag{1.13}$$

Unfortunately, independent samples from the posterior are not always easy to obtain. However, (1.13) holds more generally for some types of dependent samples. In particular, it holds for certain Markov chains. Monte Carlo methods based on simulating random variables from a Markov chain, called Markov chain Monte Carlo (MCMC) methods, are nowadays the standard way of performing the numerical analysis required by Bayesian data analysis. In the next subsections we review the main general methods that are commonly employed to simulate a Markov chain such that (1.13) holds for a specific π. For details we refer the reader to Gelman et al. (2004) and also, at a higher level, to Robert and Casella (2004) or the excellent article by Tierney (1994).

For an irreducible, aperiodic and recurrent Markov chain $(\psi_t)_{t\geq 1}$, having invariant distribution π, it can be shown that for every[2] initial value ψ_1, the distribution of ψ_t tends to π as t increases to infinity. Therefore, for M sufficiently large, $\psi_{M+1}, \ldots, \psi_{M+N}$ are all approximately distributed according to π and, jointly, they have statistical properties similar to those enjoyed by an independent sample from π. In particular, the law of large numbers, expressed by (1.13), holds, so that one has the approximation:

$$\mathrm{E}_\pi(g(\psi)) \approx N^{-1} \sum_{j=1}^{N} g(\psi_{M+j}). \tag{1.14}$$

We note, in passing, that if the Markov chain is only irreducible and recurrent, but has period $d > 1$, (1.14) still holds, even if in this case the distribution of ψ_t depends on where the chain started, no matter how large t is. In practice it is important to determine how large M should be, i.e., how many iterations of a simulated Markov chain are to be considered *burn-in* and discarded in the calculation of ergodic averages like (1.14).

Another issue is the assessment of the accuracy of an ergodic average as an estimator of the corresponding expected value. When the ψ_j's are simulated from a Markov chain, the usual formula for estimating the variance of a sample mean in the i.i.d. case no longer holds. For simplicity, suppose that the burn-in part of the chain has already been discarded, so that we can safely assume that ψ_1 is distributed according to π and $(\psi_t)_{t\geq 1}$ is a stationary Markov chain. Let $\bar{g}_N$ denote the right-hand side of (1.14). It can be shown that, for N large,

$$\mathrm{Var}(\bar{g}_N) \approx N^{-1}\mathrm{Var}(g(\psi_1))\tau(g),$$

where $\tau(g) = \sum_{t=-\infty}^{+\infty} \rho_t$ and $\rho_t = \mathrm{corr}(g(\psi_s), g(\psi_{s+t}))$. An estimate of the term $\mathrm{Var}(g(\psi_1))$ is provided by the sample variance of $g(\psi_1), \ldots, g(\psi_N)$. In order to estimate $\tau(g)$, Sokal (1989) suggests to truncate the summation and plug in empirical correlations for theoretical correlations:

$$\hat{\tau}_n = \sum_{|t|\leq n} \hat{\rho}_t,$$

with $n = \min\{k : k \geq 3\hat{\tau}_k\}$.

[2] We omit here some measure-theoretic details, trying to convey only the main ideas. For rigorous results the reader should consult the suggested references.

In the remainder of the section, we briefly present the most popular MCMC algorithms for simulating from a given distribution π.

1.6.1 Gibbs sampler

Suppose that the unknown parameter is multidimensional, so the posterior distribution is multivariate. In this case we can write $\psi = (\psi^{(1)}, \psi^{(2)}, \ldots, \psi^{(k)})$. Let $\pi(\psi) = \pi(\psi^{(1)}, \ldots, \psi^{(k)})$ be the target density. The Gibbs sampler starts from an arbitrary point $\psi_0 = (\psi_0^{(1)}, \ldots, \psi_0^{(k)})$ in the parameter space and "updates" one component at a time by drawing $\psi^{(i)}$, $i = 1, \ldots, k$, from the relevant conditional distribution, according to the scheme in Algorithm 1.1. An important point, one that is often used in practical applications of the

```
0. Initialize the starting point ψ_0 = (ψ_0^(1), ..., ψ_0^(k));
1. for j = 1, ..., N:
  1.1) generate ψ_j^(1) from π(ψ^(1)|ψ^(2) = ψ_{j-1}^(2), ..., ψ^(k) = ψ_{j-1}^(k));
  1.2) generate ψ_j^(2) from π(ψ^(2)|ψ^(1) = ψ_j^(1), ψ^(3) = ψ_{j-1}^(3) ..., ψ^(k) = ψ_{j-1}^(k));
       ⋮
  1.k) generate ψ_j^(k) from π(ψ^(k)|ψ^(1) = ψ_j^(1), ..., ψ^(k-1) = ψ_j^(k-1)).
```

Algorithm 1.1: The Gibbs sampler

Gibbs sampler, is that the basic algorithm just described still works when one or more of the components $\psi^{(i)}$ is itself multidimensional. In this case the Gibbs sampler updates in turn "blocks" of components of ψ, drawing from their conditional distribution, given all the remaining components.

1.6.2 Metropolis–Hastings algorithm

A very flexible method to generate a Markov chain having a prescribed invariant distribution is provided by Metropolis–Hastings algorithm (Metropolis et al.; 1953; Hastings; 1970). The method is very general, since it allows us to generate the next state of the chain from an essentially arbitrary distribution: the invariance of the target distribution is then enforced by an accept/reject step. This is how the algorithm works. Suppose that the chain is currently at ψ. Then a *proposal* $\tilde{\psi}$ is generated from a density $q(\psi, \cdot)$. Here q is a density in its second argument (the "$\cdot$"), but it is parametrized by the first argument. In practice this means that the proposal density may depend on the current state ψ. The proposal $\tilde{\psi}$ is accepted as the new state of the chain with probability

$$\alpha(\psi, \tilde{\psi}) = \min\left\{1, \frac{\pi(\tilde{\psi})q(\tilde{\psi}, \psi)}{\pi(\psi)q(\psi, \tilde{\psi})}\right\}. \tag{1.15}$$

If the proposal is rejected, the chain stays in the current state ψ. Algorithm 1.2 details the steps involved, assuming the chain starts at an arbitrary value ψ_0. The choice of the proposal density is an important practical issue. A proposal

```
0. Initialize the starting point ψ_0;
1. for j = 1,...,N:
  1.1) generate ψ̃_j from q(ψ_{j-1},·);
  1.2) compute α = α(ψ_{j-1}, ψ̃_j) according to (1.15);
  1.3) generate an independent random variable U_j ~ Be(α);
  1.4) if U_j = 1 set ψ_j = ψ̃_j, otherwise set ψ_j = ψ_{j-1}.
```

Algorithm 1.2: Metropolis–Hastings algorithm

leading to a high rejection rate will result in a "sticky" Markov chain, in which the state will tend to stay constant for many iterations. Ergodic averages like (1.14) provide in such a situation poor approximations, unless N is extremely large. On the other hand, a high acceptance rate is not a guarantee, *per se*, of good behavior of the chain. Consider, for example, a uniform proposal on $(\psi - a, \psi + a)$, where a is a very small positive number, and ψ is the current state. In this case $q(\psi, \tilde{\psi})$ is symmetric in its arguments, and hence it cancels out in α. Moreover, since the proposal $\tilde{\psi}$ will be close to ψ, in most cases one will have $\pi(\tilde{\psi}) \approx \pi(\psi)$ and $\alpha \approx 1$. However, the resulting simulated chain will move very slowly through its state space, exhibiting a strong positive autocorrelation, which in turn implies that in order to obtain good approximations via (1.14), one has to take N very large. Generally speaking, one shoud try to devise a proposal that is a good approximation—possibly local, in a neighborhood of the current state—of the target distribution. In the next section we illustrate a general method to construct such a proposal.

The Gibbs sampler and Metropolis–Hastings algorithm are by no means competing approaches to Markov chain simulation: in fact, they can be combined and used together. When taking a Gibbs sampling approach, it may be unfeasible, or simply not practical, to sample from one or more conditional distributions. Suppose for example that $\pi(\psi^{(1)}|\psi^{(2)})$ does not have a standard form and is therefore difficult to simulate from. In this case one can, instead of generating $\psi^{(1)}$ from $\pi(\psi^{(1)}|\psi^{(2)})$, update $\psi^{(1)}$ using a Metropolis–Hastings step. It can be shown that this does not alter the invariant distribution of the Markov chain.

1.6.3 Adaptive rejection Metropolis sampling

Rejection sampling is a simple algorithm that allows one to generate a random variable from a target distribution π by drawing from a different proposal distribution f and then accepting with a specific probability. Suppose that there is a constant C such that $\pi(\psi) \leq Cf(\psi)$ for every ψ and define

$r(\psi) = \pi(\psi)/Cf(\psi)$, so that $0 \leq r(\psi) \leq 1$. Generate two independent random variables U and V, with U uniformly distributed on $(0,1)$ and $V \sim f$. If $U \leq r(V)$ set $\psi = V$, otherwise repeat the process. In other words, draw V from $f(\psi)$ and accept V as a draw from $\pi(\psi)$ with probability $r(V)$. In case of rejection, restart the process. It can be shown that if the support of π is included in the support of f, the algorithm terminates in a finite time, i.e., one eventually generates a V that is accepted. To see that the resulting draw has the correct distribution, consider that the proposed V is accepted only if $U \leq r(V)$, so that the distribution of an accepted V is not just f, but f conditional on the event $\{U \leq r(V)\}$. Denoting by Π the cumulative distribution function of the target distribution π, one has:

$$\begin{aligned}
\mathrm{P}(V \leq v, U \leq r(V)) &= \int_{-\infty}^{v} \mathrm{P}(U \leq r(V) | V = \zeta) f(\zeta)\, \mathrm{d}\zeta \\
&= \int_{-\infty}^{v} \mathrm{P}(U \leq r(\zeta)) f(\zeta)\, \mathrm{d}\zeta = \int_{-\infty}^{v} r(\zeta) f(\zeta)\, \mathrm{d}\zeta \\
&= \int_{-\infty}^{v} \frac{\pi(\zeta)}{Cf(\zeta)} f(\zeta)\, \mathrm{d}\zeta = \frac{1}{C}\Pi(v).
\end{aligned}$$

Letting v go to $+\infty$, one obtains $\mathrm{P}(U \leq r(V)) = C^{-1}$. Therefore,

$$\mathrm{P}(V \leq v | U \leq r(V)) = \frac{\mathrm{P}(V \leq v, U \leq r(V))}{\mathrm{P}(U \leq r(V))} = \Pi(v).$$

The most favorable situations, in terms of acceptance probability, are obtained when the proposal distribution is close to the target: in this case C can be taken close to one and the acceptance probability $r(\cdot)$ will also be close to one. It is worth noting the analogy with Metropolis–Hastings algorithm. In both methods one generates a proposal from an instrumental density, and then accepts the proposal with a specific probability. However, while in rejection sampling one keeps on generating proposals until a candidate is accepted, so that, repeating the process, one can generate a sequence of independent draws exactly from the target distribution, in the Metropolis–Hastings algorithm the simulated random variables are in general dependent and are distributed according to the target only in the limit.

If π is univariate, log-concave[3], and it has bounded support, it is possible to construct a continuous piecewise linear envelope for $\log \pi$, see Figure 1.1, which corresponds to a piecewise exponential envelope for π. Appropriately normalized, this results in a piecewise exponential proposal density, which is easy to sample from using standard random number generators. Moreover, due to the interplay between C and the normalizing constant of the piecewise

[3] A function g is concave if it is defined in an interval (a,b) and $g(\alpha x + (1-\alpha)y) \geq \alpha g(x) + (1-\alpha)g(y)$ for every $\alpha \in (0,1)$ and $x, y \in (a,b)$. π is log-concave if $\log \pi(\psi)$ is a concave function.

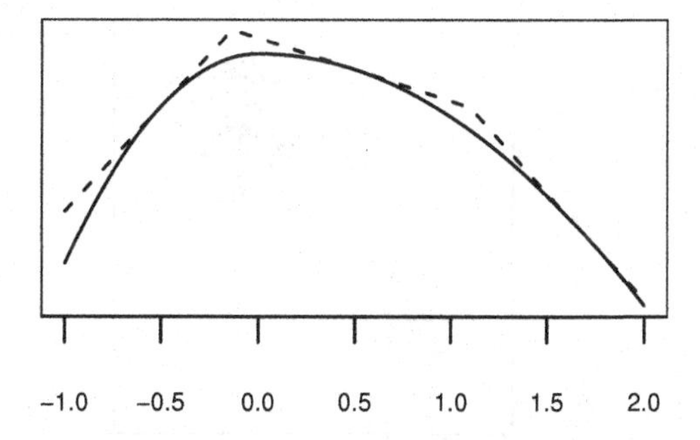

Fig. 1.1. Target log density with a piecewise linear envelope

exponential density, the target density π needs only to be known up to a normalizing factor. Clearly, the more points one uses in constructing the envelope to the target log density, the closer the proposal density will be to the target, and the sooner a proposal V will be accepted. This suggests an adaptive version of the method, according to which every time a proposal V is rejected, one refines the piecewise linear envelope using the point $(V, \log \pi(V))$, so that the next proposal will be drawn from a density that is closer to π. This algorithm is called adaptive rejection sampling in Gilks and Wild (1992). If the univariate target π is not log-concave, one can combine adaptive rejection sampling with the Metropolis–Hastings algorithm to obtain a Markov chain having π as invariant distribution. The details can be found in Gilks et al. (1995), where the algorithm is termed adaptive rejection Metropolis sampling (ARMS).

Within an MCMC setting, the univariate ARMS algorithm described above can be adapted to work also for a multivariate target distribution using the following simple device. Suppose that the chain is currently at $\psi \in \mathbb{R}^k$. Generate a uniformly distributed unit vector $u \in \mathbb{R}^k$. Then apply ARMS to the univariate density proportional to

$$t \longmapsto \pi(\psi + tu).$$

Up to a normalizing factor, this is the conditional target density, given that the new draw belongs to the straight line through the current ψ and having direction u. In R the function `arms`, originally written as part of the package *HI* (see Petris and Tardella; 2003) and now included in package `dlm`, performs this kind of multivariate version of ARMS. The function needs the arguments `y.start`, `myldens`, `indFunc`, and `n.sample` for the starting point, a function that evaluates the target logdensity, a function that evaluates the support of the density, and the number of draws to be simulated, respectively. It has also the additional argument ... that is passed on to `myldens` and `indFunc`. This is useful when the logdensity and the support depend on additional parameters. Figure 1.2 shows the plot of 500 simulated points from a mixture of two bivariate normal densities with unit variances and independent components

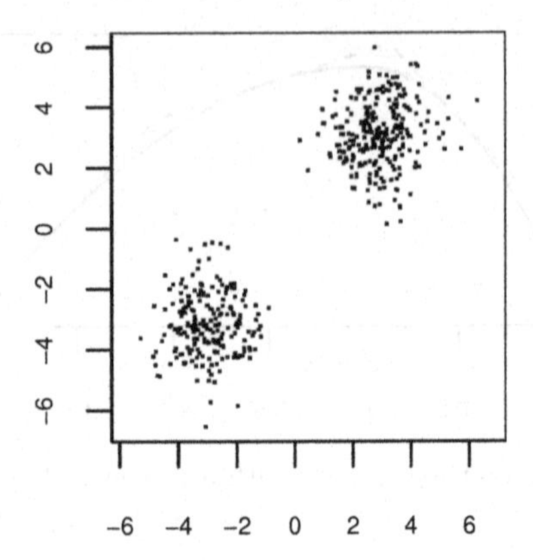

Fig. 1.2. Sample from a mixture of two bivariate normal distributions

and means $(-3, -3)$, $(3, 3)$, respectively. The code below was used to generate the sample.

R code

```
> bimodal <- function(x) log(prod(dnorm(x, mean = 3)) +
+                              prod(dnorm(x, mean = -3)))
> supp <- function(x) all(x > (-10)) * all(x < 10)
> y <- arms( c(-2, 2), bimodal, supp, 500 )
```

Note that for this target an ordinary Gibbs sampler would very likely get stuck in one of the two modes. This suggests that when one suspects a multivariate posterior distribution to be multimodal, it may be wise to include ARMS in a MCMC, and not to rely solely on a simple Gibbs sampler.

In addition to Markov chain Monte Carlo methods, which are in widespread use in every field of application of Bayesian statistics, there are other stochastic numerical methods that can be applied to some classes of models in order to compute posterior summaries. In particular, for state space models an alternative to MCMC that has become fairly popular in recent years—especially for nonlinear and non-Gaussian models—is provided by sequential Monte Carlo methods. Since this is a rather advanced topic, we postpone its treatment to Chapter 5.

Problems

1.1. Verify the identity (1.9).

1.2. Verify the identity (1.10).

1.3. Consider the linear regression model discussed in Section 1.5, with $V = \sigma^2 D$ for a known matrix D. Verify that the posterior density for the parameters $(\beta, \phi = \sigma^{-1})$, with a Normal-Gamma prior, in Normal-Gamma, with parameters given by (1.11). Then, verify the identity (1.12).

1.4. *(Shrinkage estimation)*. Consider random variables $Y_1, \ldots, Y_n$ such that

$$Y_1, \ldots, Y_n | \theta_1, \ldots, \theta_n \sim \prod_{t=1}^{n} \mathcal{N}(y_t | \theta_t, \sigma^2),$$

where σ^2 is known.

(a) Verify that, if $\theta_1, \ldots, \theta_n$ are i.i.d. $\sim \mathcal{N}(m, \tau^2)$, then the Y_t are independent. Compute the posterior density $p(\theta_1, \ldots, \theta_n | y_1, \ldots, y_n)$. With quadratic loss, the Bayesian estimate of θ_t is $\mathrm{E}(\theta_t | y_1, \ldots, y_n)$. Comment the expression of $\mathrm{E}(\theta_t | y_1, \ldots, y_n)$ that you found. What is the posterior variance, $V(\theta_t | y_1, \ldots, y_n)$?
(b) Now suppose that $\theta_1, \ldots, \theta_n$ are conditionally i.i.d. given λ, with common distribution $\mathcal{N}(\lambda, \sigma_w^2)$, and $\lambda \sim N(m, \tau^2)$, where m, σ_w^2, τ^2 are known. Compute the posterior density $p(\theta_1, \ldots, \theta_n | y_1, \ldots, y_n)$. Comment the expressions of $\mathrm{E}(\theta_t | y_1, \ldots, y_n)$ and of $V(\theta_t | y_1, \ldots, y_n)$ that you found.

1.5. Let $Y_1, \ldots, Y_n$ be i.i.d. random variables conditionally on θ, with $Y_i | \theta \sim \mathcal{N}(\theta, \sigma^2)$ with σ^2 known. Suppose that

$$\theta \sim \sum_{j=1}^{k} p_j \mathcal{N}(\mu_j, \tau_j^2).$$

Given $Y_1 = y_1, \ldots, Y_n = y_n$, compute the posterior distribution of θ, and the predictive distribution of Y_{n+1}.

1.6. Consider the linear model $y = X\beta + \epsilon$, $\epsilon \sim \mathcal{N}(0, V)$, where β and V are unknown. Suppose that they have independent priors, $\beta \sim \mathcal{N}(m_0, C_0)$ and $\Phi = V^{-1} \sim \mathcal{W}(\nu_0, S_0)$. Write a Gibbs sampler to approximate the joint posterior distribution of (β, Φ).

2

Dynamic linear models

In this chapter we discuss the basic notions about state space models and their use in time series analysis. The dynamic linear model is presented as a special case of a general state space model, being linear and Gaussian. For dynamic linear models, estimation and forecasting can be obtained recursively by the well-known Kalman filter.

2.1 Introduction

In recent years there has been an increasing interest in the application of state space models in time series analysis; see, for example, Harvey (1989), West and Harrison (1997), Durbin and Koopman (2001), the recent overviews by Künsch (2001) and Migon et al. (2005), and the references therein. State space models consider a time series as the output of a dynamic system perturbed by random disturbances. They allow a natural interpretation of a time series as the combination of several components, such as trend, seasonal or regressive components. At the same time, they have an elegant and powerful probabilistic structure, offering a flexible framework for a very wide range of applications. Computations can be implemented by recursive algorithms. The problems of estimation and forecasting are solved by recursively computing the conditional distribution of the quantities of interest, given the available information. In this sense, they are quite naturally treated within a Bayesian framework.

State space models can be used to model univariate or multivariate time series, also in the presence of non-stationarity, structural changes, and irregular patterns. In order to develop a feeling for the possible applications of state space models in time series analysis, consider for example the data plotted in Figure 2.1. This time series appears fairly predictable, since it repeats quite regularly its behavior over time: we see a trend and a rather regular seasonal component, with a slightly increasing variability. For data of this kind, we would probably be happy with a fairly simple time series model, with a trend

G. Petris et al., *Dynamic Linear Models with R*, Use R, DOI: 10.1007/b135794_2,

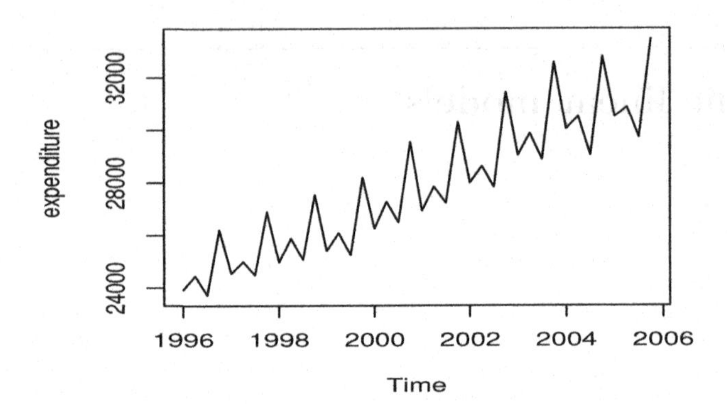

Fig. 2.1. Family food expenditure, quarterly data (1996Q1 to 2005Q4). Data available from `http://con.istat.it`

and a seasonal component. In fact, basic time series analysis relies on the possibility of finding a reasonable regularity in the behavior of the phenomenon under study: forecasting future behavior is clearly easier if the series tends to repeat a regular pattern over time. Things get more complex for time series

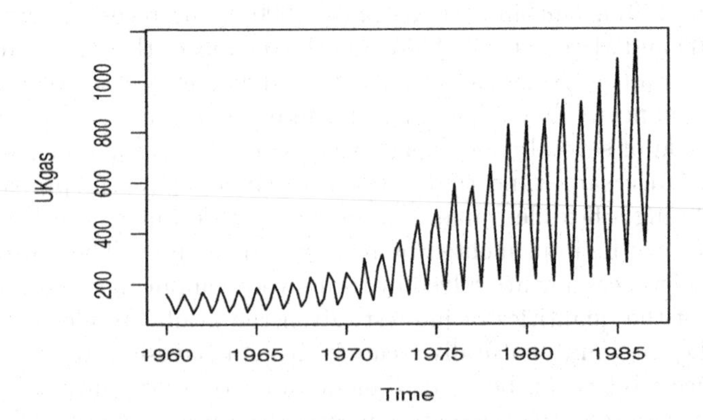

Fig. 2.2. Quarterly UK gas consumption from 1960Q1 to 1986Q4, in millions of therms

such as the ones plotted in Figures 2.2-2.4. Figure 2.2 shows the quarterly UK gas consumption from 1960 to 1986 (the data are available in R as `UKgas`). We clearly see a change in the seasonal component. Figure 2.3 shows a well-studied

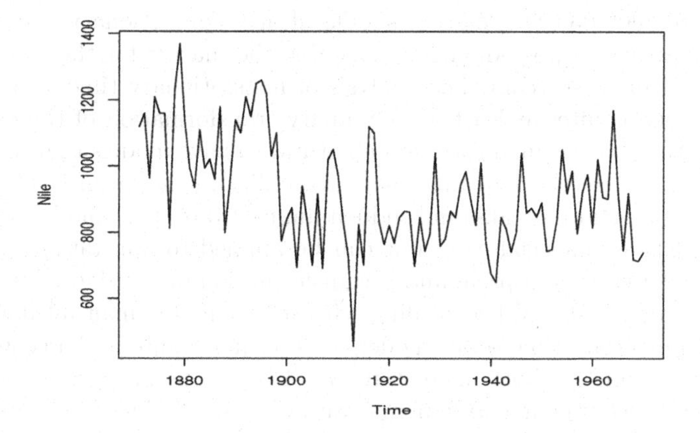

Fig. 2.3. Measurements of the annual flow of the river Nile at Ashwan, 1871-1970

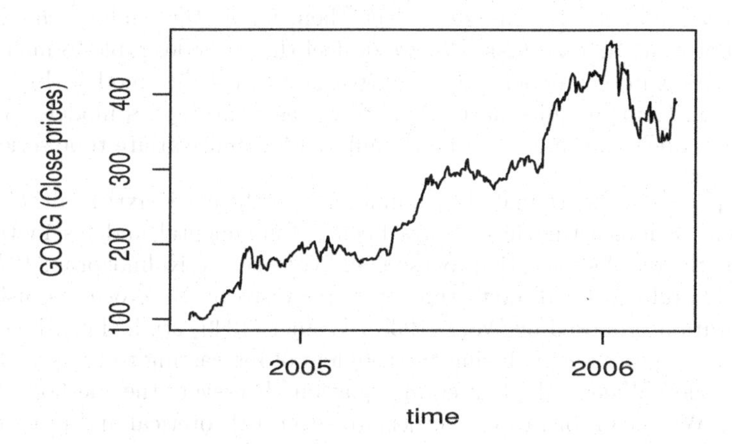

Fig. 2.4. Daily prices for Google Inc. (GOOG)

data set: the measurements of the annual flow of the river Nile at Ashwan from 1871 to 1970. The series shows level shifts. We know that the construction of the first dam of Ashwan started in 1898; the second big dam was completed in 1971: if you have ever seen these huge dams, you can easily understand the enormous changes that they caused on the Nile flow and in the vast surrounding area. Thus, we begin to feel the need for more flexible time series models, which do not assume a regular pattern and stability of the underlying system, but can include change points or structural breaks. Possibly more irregular is the series plotted in Figure 2.4, showing daily prices of Google[1] (close prices, 2004-08-19 to 2006-03-31). This series looks clearly nonstationary and in fact quite irregular: indeed, we know how unstable the market for the new economy has been in those years. The analysis of nonstationary time series with ARMA models requires at least a preliminary transformation of the data to get stationarity; but we might feel more natural to have models that allow us to analyze more directly data that show instability in the mean level and in the variance, structural breaks, and sudden jumps. State space models include ARMA models as a special case, but can be applied to nonstationary time series without requiring a preliminary transformation of the data. But there is a further basic issue. When dealing with economic or financial data, for example, a univariate time series model is often quite limited. An economist might want to gain a deeper understanding of the economic system, looking for example at relevant macroeconomic variables that influence the variable of specific interest. For the financial example of Figure 2.4, a univariate series model might be satisfying for high frequency data (the data in Figure 2.4 are daily prices), quickly *adapting* to irregularities, structural breaks or jumps; however, it will be hardly capable of *predicting* sudden changes without a further effort in a deeper and broader study of the economic and socio-political variables that influence the markets. Even then, forecasting sudden changes is clearly not at all an easy task! But we do feel that it is desirable to include regression terms in our model or use multivariate time series models. Including regression terms is quite natural in state space time series models. And state space models can in general be formulated for multivariate time series.

State space models originated in engineering in the early sixties, although the problem of forecasting has always been a fundamental and fascinating issue in the theory of stochastic processes and time series. Kolmogorov (1941) studied this problem for discrete time stationary stochastic processes, using a representation proposed by Wold (1938). Wiener (1949) studied continuous time stochastic processes, reducing the problem of forecasting to the solution of the so-called Wiener–Hopf integral equation. However, the methods for solving the Wiener problem were subject to several theoretical and practical limitations. A new look at the problem was given by Kalman (1960), using the Bode–Shannon representation of random processes and the "state transition" method of analsyis of dynamical systems. Kalman's solution, known as the Kalman filter (Kalman; 1960; Kalman and Bucy; 1963), applies to stationary and nonstationary random processes. These methods quickly gained popularity in other fields and were applied to a wide array of problems, from the determination of the orbits of the Voyager spacecraft to oceanographic problems, from agriculture to economics and speech recognition (see for instance the special issue of the IEEE Transactions on Automatic Control (1983) dedicated to applications of the Kalman filter). The importance of these methods

[1] Financial data can be easily downloaded in R using the function `get.hist.quote` in package `tseries`, or the function `priceIts` in package `its`.

was recognized by statisticians only later, although the idea of latent variables and recursive estimation can be found in the statistical literature at least as early as Thiele (1880) and Plackett (1950); see Lauritzen (1981). One reason for this delay is that the work on the Kalman filter was mostly published in the engineering literature. This means not only that the language of these works was not familiar to statisticians, but also that some issues that are crucial in applications in statistics and time series analysis were not sufficiently understood yet. Kalman himself, in his 1960 paper, underlines that the problem of obtaining the transition model, which is crucial in practical applications, was treated as a separate question and not solved. In the engineering literature, it was common practice to assume the structure of the dynamic system as known, except for the effects of random disturbances, the main problem being to find an optimal estimate of the state of the system, given the model. In time series analysis, the emphasis is somehow different. The physical interpretation of the underlying states of the dynamic system is often less evident than in engineering applications. What we have is the observable process, and even if it may be convenient to think of it as the output of a dynamic system, the problem of forecasting is often the most relevant. In this context, model building can be more difficult, and even when a state space representation is obtained, there are usually quantities or parameters in the model that are unknown and need to be estimated.

State space models appeared in the time series literature in the seventies (Akaike; 1974a; Harrison and Stevens; 1976) and became established during the eighties (Harvey; 1989; West and Harrison; 1997; Aoki; 1987). In the last decades they have become a focus of interest. This is due on one hand to the development of models well suited to time series analysis, but also to a wider range of applications, including, for instance, molecular biology or genetics, and on the other hand to the development of computational tools, such as modern Monte Carlo methods, for dealing with more complex nonlinear and non-Gaussian situations.

In the next sections we discuss the basic formulation of state space models and the structure of the recursive computations for estimation. Then, as a special case, we present the Kalman filter for Gaussian linear dynamic models.

2.2 A simple example

Before presenting the general formulation of state space models, it is useful to give an intuition of the basic ideas and of the recursive computations through a simple, introductory example. Let's think of the problem of determining the position θ of an object, based on some measurements $(Y_t : t = 1, 2, \ldots)$ affected by random errors. This problem is fairly intuitive, and dynamics can be incorporated into it quite naturally: in the static problem, the object does not move over time, but it is natural to extend the discussion to the case of a moving target. If you prefer, you may think of some economic problem, such as forecasting the sales of a good; in short-term forecasting, the observed sales are often modeled as measurements of the unobservable average sales level plus a random error; in turn, the average sales are supposed to be constant or randomly evolving over time (this is the so-called random walk plus noise model, see page 42).

We have already discussed Bayesian inference in the static problem in Chapter 1 (page 7). There, you were lost at sea, on a small island, and θ was your unknown position (univariate: distance from the coast, say). The observations were modeled as

$$Y_t = \theta + \epsilon_t, \qquad \epsilon_t \overset{iid}{\sim} \mathcal{N}(0, \sigma^2);$$

that is, given θ, the Y_t's are conditionally independent and identically distributed with a $\mathcal{N}(\theta, \sigma^2)$ distribution; in turn, θ has a Normal prior $\mathcal{N}(m_0, C_0)$. As we have seen in Chapter 1, the posterior for θ is still Gaussian, with updated parameters given by (1.2), or by (1.3) if we compute them sequentially, as new data become available.

To be concrete, let us suppose that your prior guess about the position θ is $m_0 = 1$, with variance $C_0 = 2$; the prior density is plotted in the first panel of Figure 2.5. Note that m_0 is also your point forecast for the observation: $\mathrm{E}(Y_1) = \mathrm{E}(\theta + \epsilon_1) = \mathrm{E}(\theta) = m_0 = 1$.

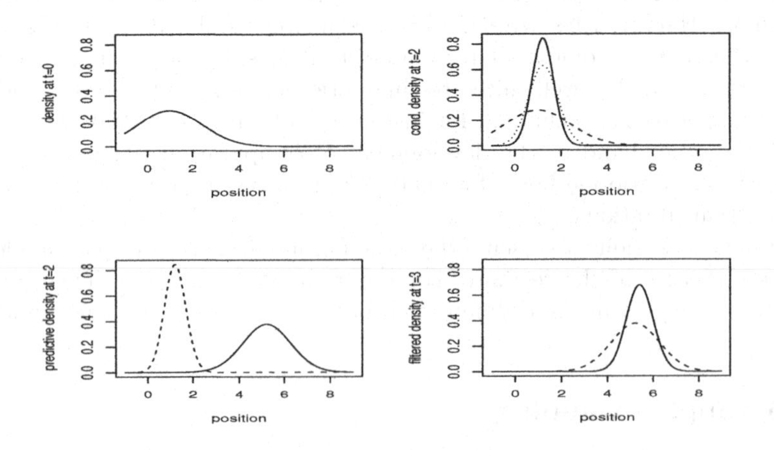

Fig. 2.5. Recursive updating of the density of θ_t

At time $t = 1$, we take a measurement, $Y_1 = 1.3$, say; from (1.3), the parameters of the posterior Normal density of θ are

$$m_1 = m_0 + \frac{C_0}{C_0 + \sigma^2}(Y_1 - m_0) = 1.24,$$

with precision $C_1^{-1} = \sigma^{-2} + C_0^{-1} = 0.4^{-1}$. We see that m_1 is obtained as our best guess at time zero, m_0, corrected by the forecast error $(Y_1 - m_0)$, weighted by a factor $K_1 = C_0/(C_0 + \sigma^2)$. The more precise the observation is, or the more vague our initial information was, the more we "trust the data": in the above formula, the smaller σ^2 is with respect to C_0, the bigger is the weight K_1 of the data-correction term in m_1. When a new observation, $Y_2 = 1.2$ say, becomes available at time $t = 2$, we can compute the density of $\theta | Y_{1:2}$, which is $\mathcal{N}(m_2, C_2)$, with $m_2 = 1.222$ and $C_2 = 0.222$, using again (1.3). The second panel in Figure 2.5 shows the updating from the prior density to the posterior density of θ, given $y_{1:2}$. We can proceed recursively in this manner as new data become available.

Let us introduce now a dynamic component to the problem. Suppose we know that at time $t = 2$ the object starts to move, so that its position changes between two consecutive measurements. Let us assume a motion of a simple form, say[2]

$$\theta_t = \theta_{t-1} + \nu + w_t, \qquad w_t \sim \mathcal{N}(0, \sigma_w^2). \tag{2.1}$$

where ν is a known nominal speed and w_t is a Gaussian random error with mean zero and known variance σ_w^2. Let, for example, $\nu = 4.5$ and $\sigma_w^2 = 0.9$. Thus, we have a process $(\theta_t : t = 1, 2, \ldots)$, which describes the unknown position of the target at successive time points. The observation equation is now

$$Y_t = \theta_t + \epsilon_t, \qquad \epsilon_t \overset{iid}{\sim} \mathcal{N}(0, \sigma^2), \tag{2.2}$$

and we assume that the sequences (θ_t) and (ϵ_t) are independent. To make inference about the unknown position θ_t, we proceed along the following steps.

Initial step. By the previous results, at time $t = 2$ we have

$$\theta_2 | y_{1:2} \sim \mathcal{N}(m_2 = 1.222, C_2 = 0.222).$$

Prediction step. At time $t = 2$, we can predict where the object will be at time $t = 3$, based on the dynamics (2.1). We easily find that

---

[2] Equation (2.1) can be thought of as a discretization of a motion law in continuous time, such as

$$d\theta_t = \nu dt + dW_t$$

where ν is the nominal speed and dW_t is an error term. For simplicity, we consider a discretization in small intervals of time (t_{i-1}, t_i), as follows:

$$\frac{\theta_{t_i} - \theta_{t_{i-1}}}{t_i - t_{i-1}} = \nu + w_{t_i},$$

that is

$$\theta_{t_i} = \theta_{t_{i-1}} + \nu(t_i - t_{i-1}) + w_{t_i}(t_i - t_{i-1}),$$

where we assume that the random error w_{t_i} has density $\mathcal{N}(0, \sigma_w^2)$. With a further simplification, we take unitary time intervals, $(t_i - t_{i-1}) = 1$, so that the above expression is rewritten as (2.1).

$$\theta_3|y_{1:2} \sim \mathcal{N}(a_3, R_3),$$

with

$$a_3 = \mathrm{E}(\theta_2 + \nu + w_3|y_{1:2}) = m_2 + \nu = 5.722$$

and variance

$$R_3 = \mathrm{Var}(\theta_2 + \nu + w_3|y_{1:2}) = C_2 + \sigma_w^2 = 1.122.$$

The third plot in Figure 2.5 illustrates the prediction step, from the conditional distribution of $\theta_2|y_{1:2}$ to the "predictive" distribution of $\theta_3|y_{1:2}$. Note that even if we were fairly confident about the position of the target at time $t = 2$, we become more uncertain about its position at time $t = 3$. This is the effect of the random error w_t in the dynamics of θ_t: the larger σ_w^2 is, the more uncertain we are about the position at the time of the next measurement. We can also predict the next observation Y_3, given $y_{1:2}$. Based on the observation equation (2.2), we easily find that

$$Y_3|y_{1:2} \sim \mathcal{N}(f_3, Q_3),$$

where

$$f_3 = \mathrm{E}(\theta_3 + \epsilon_3|y_{1:2}) = a_3 = 5.722$$

and

$$Q_3 = \mathrm{Var}(\theta_3 + \epsilon_3|y_{1:2}) = R_3 + \sigma^2 = 1.622.$$

The uncertainty about Y_3 depends on the measurement error (the term σ^2 in Q_3) as well as the uncertainty about the position at time $t = 3$ (expressed by R_3).

Estimation step (filtering). At time $t = 3$, the new observation $Y_3 = 5$ becomes available. Our point forecast of Y_3 was $f_3 = a_3 = 5.722$, so we have a forecast error $e_t = y_t - f_t = -0.722$. Intuitively, we have overestimated θ_3 and consequently Y_3; thus, our new estimate $\mathrm{E}(\theta_3|y_{1:3})$ of θ_3 will be smaller than $a_3 = \mathrm{E}(\theta_3|y_{1:2})$. For computing the posterior density of $\theta_3|y_{1:3}$, we use the Bayes formula, where the role of the prior is played by the density $\mathcal{N}(a_3, R_3)$ of θ_3 given $y_{1:2}$, and the likelihood is the density of Y_3 given (θ_3, y_1, y_2). Note that (2.2) implies that Y_3 is independent from the past observations given θ_3 (assuming independence among the error sequences), with

$$Y_3|\theta_3 \sim \mathcal{N}(\theta_3, \sigma^2).$$

Thus, by the Bayes formula (see (1.3)), we obtain

$$\theta_3|y_1, y_2, y_3 \sim \mathcal{N}(m_3, C_3),$$

where

$$m_3 = a_3 + \frac{R_3}{R_3 + \sigma^2}(y_3 - f_3) = 5.568$$

and

$$C_3 = \frac{\sigma^2 R_3}{\sigma^2 + R_3} = R_3 - \frac{R_3}{R_3 + \sigma^2} R_3 = 0.346.$$

We see again the estimation-correction structure of the updating mechanism in action. Our best estimate of θ_3 given the data $y_{1:3}$ is computed as our previous best estimate a_3, corrected by a fraction of the forecast error $e_3 = y_3 - f_3$, having weight $K_3 = R_3/(R_3 + \sigma^2)$. This weight is bigger the more uncertain we are about our forecast a_3 of θ_3 (that is, the larger R_3 is, which in turn depends on C_2 and σ_w^2) and the more precise the observation Y_3 is (i.e., the smaller σ^2 is). From these results we see that a crucial role in determining the effect of the data on estimation and forecasting is played by the magnitude of the system variance σ_w^2 relative to the observation variance σ^2, the so-called *signal-to-noise* ratio. The last plot in Figure 2.5 illustrates this estimation step. We can proceed repeating recursively the previous steps for updating our estimates and forecasts as new observations become available.

The previous simple example illustrates the basic aspects of dynamic linear models, which can be summarized as follows.

- The observable process $(Y_t : t = 1, 2, \ldots)$ is thought of as determined by a latent process $(\theta_t : t = 1, 2, \ldots)$, up to Gaussian random errors. If we knew the position of the object at successive time points, the Y_t's would be independent: what remains are only unpredictable measurement errors. Furthermore, the observation Y_t depends only on the position θ_t of the target at time t.
- The latent process (θ_t) has a fairly simple dynamics: θ_t does not depend on the entire past trajectory but only on the previous position θ_{t-1}, through a linear relationship, up to Gaussian random errors.
- Estimation and forecasting can be obtained sequentially, as new data become available.

The assumption of linearity and Gaussianity is specific to dynamic linear models, but the dependence structure of the processes (Y_t) and (θ_t) is part of the definition of a general state space model.

2.3 State space models

Consider a time series $(Y_t)_{t\geq 1}$. Specifying the joint finite-dimensional distributions of $(Y_1, \ldots, Y_t)$, for any $t \geq 1$, is not an easy task. In particular, in time series applications the assumptions of independence or exchangeability are seldom justified, since they would essentially make time irrelevant. Markovian dependence is arguably the simplest form of dependence among the Y_t's in which time has a definite role. We say that $(Y_t)_{t\geq 1}$ is a *Markov chain* if, for any $t > 1$,

$$\pi(y_t | y_{1:t-1}) = \pi(y_t | y_{t-1}).$$

This means that the information about Y_t carried by all the observations up to time $t-1$ is exactly the same as the information carried by y_{t-1} alone. Another way of saying the same thing is that Y_t and $Y_{1:t-2}$ are conditionally independent given y_{t-1}. For a Markov chain the finite-dimensional joint distributions can be written in the fairly simple form

$$\pi(y_{1:t}) = \pi(y_1) \cdot \prod_{j=2}^{t} \pi(y_j|y_{j-1}).$$

Assuming a Markovian structure for the observations is, however, not appropriate in many applications. State space models build on the relatively simple dependence structure of a Markov chain to define more complex models for the observations. In a state space model we assume that there is an unobservable Markov chain (θ_t), called the state process, and that Y_t is an imprecise measurement of θ_t. In engineering applications θ_t usually describes the state of a physically observable system that produced the output Y_t. On the other hand, in econometric applications θ_t is often a latent construct, which may, however, have a useful interpretation. In any case, one can think of (θ_t) as an auxiliary time series that facilitates the task of specifying the probability distribution of the observable time series (Y_t).

Formally, a state space model consists of an $\mathbb{R}^p$-valued time series $(\theta_t : t = 0, 1 \ldots)$ and an $\mathbb{R}^m$-valued time series $(Y_t : t = 1, 2 \ldots)$, satisfying the following assumptions.

(A.1) (θ_t) is a Markov chain.
(A.2) Conditionally on (θ_t), the Y_t's are independent and Y_t depends on θ_t only.

The consequence of (A.1)-(A.2) is that a state space model is completely specified by the initial distribution $\pi(\theta_0)$ and the conditional densities $\pi(\theta_t|\theta_{t-1})$ and $\pi(y_t|\theta_t)$, $t \geq 1$. In fact, for any $t > 0$,

$$\pi(\theta_{0:t}, y_{1:t}) = \pi(\theta_0) \cdot \prod_{j=1}^{t} \pi(\theta_j|\theta_{j-1})\pi(y_j|\theta_j). \tag{2.3}$$

From (2.3) one can derive, by conditioning or marginalization, any other distribution of interest. For example, the joint density of the observations $Y_{1:t}$ can be obtained by integrating out the θ_j's in (2.3); note however that in this way the simple product form of (2.3) is lost.

The information flow assumed by a state space model is represented in Figure 2.6. The graph in the figure is a special case of a directed acyclic graph (see Cowell et al.; 1999). The graphical representation of the model can be used to deduce conditional independence properties of the random variables occurring in a state space model. In fact, two sets of random variables, A and B, can be shown to be conditionally independent given a third set of variables, C, if and only if C separates A and B, i.e., if any path connecting

$$\begin{array}{ccccccccccccccc}
\theta_0 & \longrightarrow & \theta_1 & \longrightarrow & \theta_2 & \longrightarrow & \cdots & \longrightarrow & \theta_{t-1} & \longrightarrow & \theta_t & \longrightarrow & \theta_{t+1} & \longrightarrow & \cdots \\
 & & \downarrow & & \downarrow & & & & \downarrow & & \downarrow & & \downarrow & & \\
 & & Y_1 & & Y_2 & & & & Y_{t-1} & & Y_t & & Y_{t+1} & &
\end{array}$$

Fig. 2.6. Dependence structure for a state space model

one variable in A to one in B passes through C. Note that in the previous statement the arrows in Figure 2.6 have to be considered as undirected edges of the graph that can be transversed in both directions. For a proof, see Cowell et al. (1999, Section 5.3). As an example, we will use Figure 2.6 to show that Y_t and $(\theta_{0:t-1}, Y_{1:t-1})$ are conditionally independent given θ_t. The proof simply consists in observing that any path connecting Y_t with one of the previous Y_s $(s < t)$ or with one of the states θ_s, $s < t$, has to go through θ_t; hence, $\{\theta_t\}$ separates $\{\theta_{0:t-1}, Y_{1:t-1}\}$ and $\{Y_t\}$. It follows that

$$\pi(y_t|\theta_{0:t-1}, y_{1:t-1}) = \pi(y_t|\theta_t).$$

In a similar way, one can show that θ_t and $(\theta_{0:t-2}, Y_{1:t-1})$ are conditionally independent given θ_{t-1}, which can be expressed in terms of conditional distributions as

$$\pi(\theta_t|\theta_{0:t-1}, y_{1:t-1}) = \pi(\theta_t|\theta_{t-1}).$$

State space models in which the states are discrete-valued random variables are often called *hidden Markov models.*

2.4 Dynamic linear models.

The first, important class of state space models is given by Gaussian linear state space models, also called dynamic linear models. A *dynamic linear model* (DLM) is specified by a Normal prior distribution for the p-dimensional state vector at time $t = 0$,

$$\theta_0 \sim \mathcal{N}_p(m_0, C_0), \tag{2.4a}$$

together with a pair of equations for each time $t \geq 1$,

$$Y_t = F_t\theta_t + v_t, \qquad v_t \sim \mathcal{N}_m(0, V_t), \tag{2.4b}$$
$$\theta_t = G_t\theta_{t-1} + w_t, \qquad w_t \sim \mathcal{N}_p(0, W_t), \tag{2.4c}$$

where G_t and F_t are known matrices (of order $p\times p$ and $m\times p$ respectively) and $(v_t)_{t\geq1}$ and $(w_t)_{t\geq1}$ are two independent sequences of independent Gaussian random vectors with mean zero and known variance matrices $(V_t)_{t\geq1}$ and $(W_t)_{t\geq1}$, respectively. Equation (2.4b) is called the *observation equation*, while (2.4c) is the *state equation* or *system equation.* Furthermore, it is assumed that θ_0 is independent of (v_t) and (w_t). One can show that a DLM satisfies the assumptions (A.1) and (A.2) of the previous section, with $Y_t|\theta_t \sim \mathcal{N}(F_t\theta_t, V_t)$ and $\theta_t|\theta_{t-1} \sim \mathcal{N}(G_t\theta_{t-1}, W_t)$ (see Problems 2.1 and 2.2).

In contrast to (2.4), a general state space model can be specified by a prior distribution for θ_0, together with the observation and evolution equations

$$\begin{aligned} Y_t &= h_t(\theta_t, v_t), \\ \theta_t &= g_t(\theta_{t-1}, w_t) \end{aligned}$$

for arbitrary functions g_t and h_t. *Linear* state space models specify g_t and h_t as linear functions, and *Gaussian* linear models add the assumptions of Gaussian distributions. The assumption of Normality is sensible in many applications, and it can be justified by central limit theorem arguments. However, there are many important extensions, such as heavy tailed errors for modeling outliers, or the dynamic generalized linear model for treating discrete time series. The price to be paid when removing the assumption of Normality is additional computational difficulties.

We introduce here some examples of DLMs for time series analysis, which will be treated more extensively in Chapter 3. The simplest model for a univariate time series $(Y_t : t = 1, 2, \ldots)$ is the so-called *random walk plus noise* model, defined by

$$\begin{aligned} Y_t &= \mu_t + v_t, & v_t &\sim \mathcal{N}(0, V) \\ \mu_t &= \mu_{t-1} + w_t, & w_t &\sim \mathcal{N}(0, W), \end{aligned} \tag{2.5}$$

where the error sequences (v_t) and (w_t) are independent, both within them and between them. This is a DLM with $m = p = 1$, $\theta_t = \mu_t$ and $F_t = G_t = 1$. It is the model used in the introductory example in Section 2.2, when there is no speed in the dynamics ($\nu = 0$ in the state equation (2.1)). Intuitively, it is appropriate for time series showing no clear trend or seasonal variation: the observations (Y_t) are modeled as noisy observations of a level μ_t which, in turn, is subject to random changes over time, described by a random walk. This is why the model is also called *local level* model. If $W = 0$, we are back to the constant mean model. Note that the random walk (μ_t) is nonstationary. Indeed, DLMs can be used for modeling nonstationary time series. On the contrary, the usual ARMA models require a preliminary transformation of the data to achieve stationarity.

A slightly more elaborated model is the *linear growth model*, or local linear trend, which has the same observation equation as the local level model, but includes a time-varying slope in the dynamics for μ_t:

$$\begin{aligned} Y_t &= \mu_t + v_t, & v_t &\sim \mathcal{N}(0, V), \\ \mu_t &= \mu_{t-1} + \beta_{t-1} + w_{t,1}, & w_{t,1} &\sim \mathcal{N}(0, \sigma_\mu^2), \\ \beta_t &= \beta_{t-1} + w_{t,2}, & w_{t,2} &\sim \mathcal{N}(0, \sigma_\beta^2), \end{aligned} \tag{2.6}$$

with uncorrelated errors v_t, $w_{t,1}$ and $w_{t,2}$. This is a DLM with

$$\theta_t = \begin{bmatrix} \mu_t \\ \beta_t \end{bmatrix}, \quad G = \begin{bmatrix} 1 & 1 \\ 0 & 1 \end{bmatrix}, \quad W = \begin{bmatrix} \sigma_\mu^2 & 0 \\ 0 & \sigma_\beta^2 \end{bmatrix}, \quad F = \begin{bmatrix} 1 & 0 \end{bmatrix}.$$

The system variances σ_μ^2 and σ_β^2 are allowed to be zero. We have used this model in the introductory example of Section 2.2; there, we had a constant nominal speed in the dynamics, that is $\sigma_\beta^2 = 0$.

Note that in these examples the matrices G_t and F_t and the covariance matrices V_t and W_t are constant; in this case the model is said to be *time invariant*. We will see other examples in Chapter 3. In particular, the popular Gaussian ARMA models can be obtained as special cases of DLM; in fact, it can be shown that Gaussian ARMA and DLM models are equivalent in the time-invariant case (see Hannan and Deistler; 1988).

DLMs can be regarded as a generalization of the linear regression model, allowing for time varying regression coefficients. The simple, static linear regression model describes the relationship between a variable Y and a nonrandom explanatory variable x as

$$Y_t = \theta_1 + \theta_2 x_t + \epsilon_t, \quad \epsilon_t \overset{iid}{\sim} \mathcal{N}(0, \sigma^2).$$

Here we think of $(Y_t, x_t), t = 1, 2, \dots$ as observed over time. Allowing for time varying regression parameters, one can model nonlinearity of the functional relationship between x and y, structural changes in the process under study, omission of some variables. A simple dynamic linear regression model assumes

$$Y_t = \theta_{t,1} + \theta_{t,2} x_t + \epsilon_t, \quad \epsilon_t \sim \mathcal{N}(0, \sigma_t^2),$$

with a further equation for describing the system evolution

$$\theta_t = G_t \theta_{t-1} + w_t, \quad w_t \sim \mathcal{N}_2(0, W_t).$$

This is a DLM with $F_t = [1, x_t]$ and states $\theta_t = (\theta_{t,1}, \theta_{t,2})'$. As a particuar case, if $G_t = I$, the identity matrix, $\sigma_t^2 = \sigma^2$ and $w_t = 0$ for every t, we are back to the simple static linear regression model.

2.5 Dynamic linear models in package dlm

DLMs are represented in package dlm as named lists with a class attribute, which makes them into objects of class "`dlm`". Objects of class `dlm` can represent constant or time-varying DLMs. A constant DLM is completely specified once the matrices F, V, G, W, C_0, and the vector m_0 are given. In R, these components are stored in a `dlm` object as elements `FF`, `V`, `GG`, `W`, `C0`, and `m0`, respectively. Extractor and replacement functions are available to access and modify specific parts of the model in a user-friendly way. The package also provides several functions that create particular classes of DLMs from minimal input; we will illustrate those functions in Chapter 3, where we discuss model specification. A general univariate or multivariate DLM can be specified using the function *dlm*. This function creates a *dlm* object from its components, performing some sanity checks on the input, such as testing the dimensions of the matrices for consistency. The input may be given as a list with named arguments or as individual arguments. Here is how to use *dlm* to create a *dlm* object corresponding to the random walk plus noise model and to the linear growth model introduced on page 42. We assume that $V = 1.4$ and $\sigma^2 = 0.2$. Note that 1×1 matrices can safely be passed to *dlm* as scalars, i.e., numerical vectors of length one.

R code

```
  > rw <- dlm(m0 = 0, C0 = 10, FF = 1, V = 1.4, GG = 1, W = 0.2)
2 > unlist(rw)
    m0   C0   FF    V   GG    W
4  0.0 10.0  1.0  1.4  1.0  0.2
  > lg <- dlm(FF = matrix(c(1, 0), nr = 1),
6 +           V = 1.4,
  +           GG = matrix(c(1, 0, 1, 1), nr = 2),
8 +           W = diag(c(0, 0.2)),
  +           m0 = rep(0, 2),
10 +          C0 = 10 * diag(2))
  > lg
12 $FF
       [,1] [,2]
14 [1,]    1    0

16 $V
       [,1]
18 [1,]  1.4

20 $GG
       [,1] [,2]
22 [1,]    1    1
   [2,]    0    1
24
   $W
26     [,1] [,2]
   [1,]    0  0.0
28 [2,]    0  0.2

30 $m0
   [1] 0 0
32
   $C0
```

```
     [,1] [,2]
[1,]   10    0
[2,]    0   10

> is.dlm(lg)
[1] TRUE
```

Suppose now that one wants to change the observation variance in the linear growth model `lg` to $V = 0.8$ and the system variance W so as to have $\sigma^2 = 0.5$. This can be easily achieved as illustrated in the following code.

R code

```
> V(lg) <- 0.8
> W(lg)[2,2] <- 0.5
> V(lg)
[1] 0.8
> W(lg)
     [,1] [,2]
[1,]    0  0.0
[2,]    0  0.5
```

In a similar way we can modify or view the other components of the model, including the mean and variance of the state at time zero, m_0 and C_0.

Let us turn now on time-varying DLMs and how they are represented in R. Most often, in a time-invariant DLM, only a few entries (possibly none) of each matrix change over time, while the remaining are constant. Therefore, instead of storing the entire matrices F_t, V_t, G_t, W_t for all values of t that one wishes to consider, we opted to store a template of each of them, and save the time-varying entries in a separate matrix. This matrix is the component `X` of a `dlm` object. Taking this approach, one also needs to know to which entry of which matrix each column of `X` corresponds. To this aim one has to specify one or more of the components `JFF`, `JV`, `JGG`, and `JW`. Let us focus on the first one, `JFF`. This should be a matrix of the same dimension of `FF`, with integer entries: if `JFF[i,j]` is `k`, a positive integer, that means that the value of `FF[i,j]` at time `s` is `X[s,k]`. If, on the other hand, `JFF[i,j]` is zero then `FF[i,j]` is taken to be constant in time. `JV`, `JGG`, and `JW` are used in the same way, for `V`, `GG`, and `W`, respectively. Consider, for example, the dynamic regression model introduced on page 43. The only time-varying element is the $(1,2)$-entry of F_t; therefore, `X` will be a one-column matrix (although `X` is allowed to have extra, unused, columns). The following code shows how a dynamic regression model can be defined in R.

R code

```
> x <- rnorm(100) # covariates
> dlr <- dlm(FF = matrix(c(1, 0), nr = 1),
+            V = 1.3,
+            GG = diag(2),
+            W = diag(c(0.4, 0.2)),
+            m0 = rep(0, 2), C0 = 10 * diag(2),
+            JFF = matrix(c(0, 1), nr = 1),
+            X = x)
> dlr
$FF
     [,1] [,2]
[1,]    1    0

$V
     [,1]
[1,]  1.3

$GG
     [,1] [,2]
[1,]    1    0
[2,]    0    1

$W
     [,1] [,2]
[1,]  0.4  0.0
[2,]  0.0  0.2

$JFF
     [,1] [,2]
[1,]    0    1

$X
     [,1]
[1,] 0.4779
[2,] 0.5414
[3,] ...

$m0
[1] 0 0

$C0
     [,1] [,2]
[1,]   10    0
[2,]    0   10
```

Note that the dots on line 36 of the display above were produced by the `print` method function for objects of class `dlm`. If you want the entire `X` component to be printed, you need to extract it as `X(dlr)`, or use `print.default`. When modifying individual components of a `dlm` object, the user must ensure that the new components are compatible with the rest of the `dlm` object, as the replacement functions do not perform any check. This is a precise design choice, reflecting the fact that one may want to modify a `dlm` object one component at a time in such a way that, while the intermediate steps result in an invalid specification, the final result is a well-defined `dlm` object. For example, suppose one wants to use `rw` with a time series of length 30, and one wants to specify a time-varying observation variance as

$$V_t = \begin{cases} 0.75 & \text{if } t = 1, \ldots, 10, \\ 1.25 & \text{if } t = 11, \ldots, 30. \end{cases}$$

Assuming the researcher is satisfied with the constant system variance previously specified, she has to add to `rw` the two components `JV` and `X`. Adding `JV` first temporarily produces an invalid `dlm` object, which is then made into a valid one by the further addition of the `X` component. To stay on the safe side, one can make sure that a model obtained from another one by changing, adding, or removing components "by hand" is a valid `dlm` object by calling the function `dlm` on the modified model. In this case `is.dlm` is not useful, as it only looks at the class attribute of the object. The original value of `V` is still present in the new model but will never be used. For this reason `V(rw)` gives back the old value of `V`, at the same time warning the user that in `rw` the component `V` is now time-varying. The code below illustrates the previous discussion.

R code

```
> JV(rw) <- 1
> is.dlm(rw)
[1] TRUE
> dlm(rw)
Error in dlm(rw) : Component X must be provided for time-varying
models
> X(rw) <- rep(c(0.75, 1.25), c(10, 20))
> rw <- dlm(rw)
> V(rw)
     [,1]
[1,]  1.4
Warning message:
In V.dlm(rw) : Time varying V
```

2.6 Examples of nonlinear and non-Gaussian state space models

Specification and estimation of DLMs for time series analysis will be treated in Chapters 3 and 4. Here we briefly present some important classes of nonlinear and non-Gaussian state space models. Although in this book we will limit ourself to the linear Gaussian case, this section should give the reader an idea of the extensions that are possible in state space modeling when dropping those assumptions.

Exponential family state space models

Dynamic linear models can be generalized by removing the assumption of Gaussian distributions. This generalization is required for modeling discrete time series; for example, if Y_t represents the presence/absence of a characteristic in the problem under study over time, we would use a Bernoulli distribution; if Y_t are counts, we might use a Poisson model, etc. *Dynamic Generalized Linear Models* (West et al.; 1985) assume that the conditional distribution $\pi(y_t|\theta_t)$ of Y_t given θ_t is a member of the exponential family, with natural parameter $\eta_t = F_t\theta_t$. The state equation is as for Gaussian linear models, $\theta_t = G_t\theta_{t-1} + w_t$. Inference for generalized DLMs presents computational difficulties, which can, however, be solved by MCMC techniques.

Hidden Markov models

State space models in which the state θ_t is discrete are usually referred to as *hidden Markov models*. Hidden Markov models are used extensively in speech recognition (see for example Rabiner and Juang; 1993). In economics and finance, they are often used to model a time series with structural breaks. The dynamics of the series and the change points are thought as determined by a latent Markov chain (θ_t), with state space $\{\theta_1^*, \ldots, \theta_k^*\}$ and transition probabilities

$$\pi(i|j) = P(\theta_t = \theta_i^*|\theta_{t-1} = \theta_j^*).$$

Consequently, Y_t can come from a different distribution depending on the state of the chain at time t, in the sense that

$$Y_t|\{\theta_t = \theta_j^*\} \sim \pi(y_t|\theta_j^*), \qquad j = 1, \ldots, k.$$

Although state space models and hidden Markov models have evolved as separate subjects, their basic assumptions and recursive computations are closely related. MCMC methods for hidden Markov models have been developed, see for example Rydén and Titterington (1998), Kim and Nelson (1999), Cappé et al. (2005), and the references therein.

Stochastic volatility models

Stochastic volatility models are widely used in financial applications. Let Y_t be the log-return of an asset at time t (i.e., $Y_t = \log P_t/P_{t-1}$, where P_t is the asset price at time t). Under the assumption of efficient markets, the log-returns have null conditional mean: $\mathrm{E}(Y_{t+1}|y_{1:t}) = 0$. However, the conditional variance, called volatility, varies over time. There are two main classes of models for analyzing volatility of returns. The popular ARCH and GARCH models (Engle; 1982; Bollerslev; 1986) describe the volatility as a function of the past values of the returns. Stochastic volatility models, instead, consider the volatility as an exogenous random process. This leads to a state space model where the volatility is (part of) the state vector, see for example Shephard (1996). The simplest stochastic volatility model has the following form:

$$\begin{aligned} Y_t &= \exp\left\{\frac{1}{2}\theta_t\right\} w_t, \qquad & w_t &\sim \mathcal{N}(0,1), \\ \theta_t &= \eta + \phi\theta_{t-1} + v_t, \qquad & v_t &\sim \mathcal{N}(0,\sigma^2), \end{aligned}$$

that is, θ_t follows an autoregressive model of order one. These models are nonlinear and non-Gaussian, and computations are usually more demanding than for ARCH and GARCH models; however, MCMC approximations are available (Jacquier et al.; 1994). On the other hand, stochastic volatility models seem easier to generalize to the case of returns of a collection of assets, while for multivariate ARCH and GARCH models the number of parameters quickly becomes too large. Let $Y_t = (Y_{t,1}, \ldots, Y_{t,m})$ be the log-returns for m assets. A simple multivariate stochastic volatility model might assume that

$$Y_{t,i} = \exp\left(z_t + x_{t,i}\right) v_{t,i}, \qquad i = 1, \ldots, m,$$

where z_t describes a common market volatility factor and the $x_{t,i}$'s are individual volatilities. The state vector is $\theta_t = (z_t, x_{t,1}, \ldots, x_{t,m})'$, and a simple state equation might assume that the components of θ_t are independent AR(1) processes.

2.7 State estimation and forecasting

The great flexibility of state space models is one reason for their extensive application in an enormous range of applied problems. Of course, as in any statistical application, a crucial and often difficult step is a careful model specification. In many problems, the statistician and the experts together can build a state space model where the states have an intuitive meaning, and expert knowledge can be used to specify the transition probabilities in the state equation, determine the dimension of the state space, etc. However, often the model building can be a major difficulty: there might be no clear identification of physically interpretable states, or the state space representation could be non unique, or the state space is too big and poorly identifiable, or the model is too complicated. We will discuss some issues about model building for time series analysis with DLMs in Chapter 3. Here, to get started, we consider the model as given; that is, we assume that the densities $\pi(y_t|\theta_t)$ and $\pi(\theta_t|\theta_{t-1})$ have been specified, and we present the basic recursions for estimation and forecasting. In Chapter 4, we will let these densities depend on unknown parameters ψ and discuss their estimation.

For a given state space model, the main tasks are to make inference on the unobserved states or predict future observations based on a part of the observation sequence. Estimation and forecasting are solved by computing the conditional distributions of the quantities of interest, given the available information.

To estimate the state vector we compute the conditional densities $\pi(\theta_s|y_{1:t})$. We distinguish between problems of *filtering* (when $s = t$), *state prediction* ($s > t$) and *smoothing* ($s < t$). It is worth underlining the difference between filtering and smoothing. In the filtering problem, the data are supposed to arrive sequentially in time. This is the case in many applied problems: think for example of the problem of tracking a moving object, or of financial applications where one has to estimate, day by day, the term structure of interest rates, updating the current estimates as new data are observed on the markets the following day. In these cases, we want a procedure to estimate the current value of the state vector, based on the observations up to time t ("now"), and to update our estimates and forecasts as new data become available at time $t+1$. To solve the filtering problem, we compute the conditional density $\pi(\theta_t|y_{1:t})$. In a DLM, the Kalman filter provides the formulae for updating our current inference on the state vector as new data become available, that is for passing from the filtering density $\pi(\theta_t|y_{1:t})$ to $\pi(\theta_{t+1}|y_{1:t+1})$.

The problem of smoothing, or retrospective analysis, consists instead in estimating the state sequence at times $1, \ldots, t$, given the data $y_1, \ldots, y_t$. In many applications, one has observations on a time series for a certain period, and wants to retrospectively study the behavior of the system underlying the observations. For example, in economic studies, the researcher might have the time series of consumption, or of the gross domestic product of a country, for a certain number of years, and she might be interested in retrospectively understanding the socio-economic behavior of the system. The smoothing problem is solved by computing the conditional distribution of $\theta_{1:t}$ given $y_{1:t}$. As for filtering, smoothing can be implemented as a recursive algorithm.

As a matter of fact, in time series analysis forecasting is often the main task; the state estimation is then just a step for predicting the value of future observations. For one-step-ahead forecasting, that is, predicting the next observation Y_{t+1} based on the data $y_{1:t}$, one first estimates the next value θ_{t+1} of the state vector, and then, based on this estimate, one computes the forecast for Y_{t+1}. The one-step-ahead state predictive density is $\pi(\theta_{t+1}|y_{1:t})$ and it is based on the filtering density of θ_t. From this, one obtains the one-step-ahead predictive density $\pi(y_{t+1}|y_{1:t})$.

One might be interested in looking a bit further ahead, estimating the evolution of the system, represented by the state vector θ_{t+k} for some $k \geq 1$, and making k-steps-ahead forecasts for Y_{t+k}. The state prediction is solved by computing the k-steps-ahead state predictive density $\pi(\theta_{t+k}|y_{1:t})$. Based on this density, one can compute the k-steps-ahead predictive density $\pi(y_{t+k}|y_{1:t})$ for the future observation at time $t+k$. Of course, forecasts become more and more uncertain as the time horizon $t+k$ gets farther away in the future, but note that we can anyway quantify the uncertainty through a probability density, namely the predictive density of Y_{t+1} given $y_{1:t}$. We will show how to compute the predictive densities in a recursive fashion. In particular, the conditional mean $\mathrm{E}(Y_{t+1}|y_{1:t})$ provides an optimal one-step-ahead point forecast of the value of Y_{t+1}, minimizing the conditional expected square prediction error. As a function of k, $\mathrm{E}(Y_{t+k}|y_{1:t})$ is usually called the *forecast function.*

2.7.1 Filtering

We first describe the recursive steps needed to compute the filtering densities $\pi(\theta_t|y_{1:t})$ in general state space models. Even if we will not make extensive use of these formulae, it is useful to look now at the general recursions to better understand the role of the conditional independence assumptions that have been introduced. Then we move to the DLM case, for which the filtering problem is solved by the well-known Kalman filter.

One of the advantages of state space models is that, due to the Markovian structure of the state dynamics (A.1) and the assumptions on the conditional independence for the observables (A.2), the filtered and predictive densities can be computed using a recursive algorithm. As we have seen in the introductory example of Section 2.2, starting from $\theta_0 \sim \pi(\theta_0)$ one can recursively compute, for $t = 1, 2, \ldots$:

(i) the one-step-ahead predictive distribution for θ_t given $y_{1:t-1}$, based on the filtering density $\pi(\theta_{t-1}|y_{1:t-1})$ and the conditional distribution of θ_t given θ_{t-1} specified by the model;

(ii) the one-step-ahead predictive distribution for the next observation;

(iii) the filtering distribution $\pi(\theta_t|y_{1:t})$, using the Bayes rule with $\pi(\theta_t|y_{1:t-1})$ as the prior distribution and likelihood $\pi(y_t|\theta_t)$.

The following proposition contains a formal presentation of the filtering recursions for a general state space model.

Proposition 2.1 (Filtering recursions). *For a general state space model defined by* (A.1)-(A.2) *(p.40), the following statements hold.*

(i) The one-step-ahead predictive density for the states can be computed from the filtered density $\pi(\theta_{t-1}|y_{1:t-1})$ *according to*

$$\pi(\theta_t|y_{1:t-1}) = \int \pi(\theta_t|\theta_{t-1})\pi(\theta_{t-1}|y_{1:t-1})\, d\theta_{t-1}. \qquad (2.7a)$$

(ii) The one-step-ahead predictive density for the observations can be computed from the predictive density for the states as

$$\pi(y_t|y_{1:t-1}) = \int \pi(y_t|\theta_t)\pi(\theta_t|y_{1:t-1})\, d\theta_t. \tag{2.7b}$$

(iii) The filtering density can be computed from the above densities as

$$\pi(\theta_t|y_{1:t}) = \frac{\pi(y_t|\theta_t)\pi(\theta_t|y_{1:t-1})}{\pi(y_t|y_{1:t-1})}. \tag{2.7c}$$

Proof. The proof relies heavily on the conditional independence properties of the model, which can be deduced from the graph in Figure 2.6.

To prove (i), note that θ_t is conditionally independent of $Y_{1:t-1}$, given θ_{t-1}. Therefore,

$$\begin{aligned}\pi(\theta_t|y_{1:t-1}) &= \int \pi(\theta_{t-1}, \theta_t|y_{1:t-1})\, d\theta_{t-1} \\ &= \int \pi(\theta_t|\theta_{t-1}, y_{1:t-1})\pi(\theta_{t-1}|y_{1:t-1})\, d\theta_{t-1} \\ &= \int \pi(\theta_t|\theta_{t-1})\pi(\theta_{t-1}|y_{1:t-1})\, d\theta_{t-1}.\end{aligned}$$

To prove (ii), note that Y_t is conditionally independent of $Y_{1:t-1}$ given θ_t. Therefore,

$$\begin{aligned}\pi(y_t|y_{1:t-1}) &= \int \pi(y_t, \theta_t|y_{1:t-1})\, d\theta_t \\ &= \int \pi(y_t|\theta_t, y_{1:t-1})\pi(\theta_t|y_{1:t-1})\, d\theta_t \\ &= \int \pi(y_t|\theta_t)\pi(\theta_t|y_{1:t-1})\, d\theta_t.\end{aligned}$$

Part (iii) follows from Bayes' rule and the conditional independence of Y_t and $Y_{1:t-1}$ given θ_t:

$$\pi(\theta_t|y_{1:t}) = \frac{\pi(\theta_t|y_{1:t-1})\, \pi(y_t|\theta_t, y_{1:t-1})}{\pi(y_t|y_{1:t-1})} = \frac{\pi(\theta_t|y_{1:t-1})\, \pi(y_t|\theta_t)}{\pi(y_t|y_{1:t-1})}.$$

□

From the one-step-ahead predictive distribution provided by the previous proposition, k-steps ahead predictive distributions for the state and for the observation can be computed recursively according to the formulae

$$\pi(\theta_{t+k}|y_{1:t}) = \int \pi(\theta_{t+k}|\theta_{t+k-1})\, \pi(\theta_{t+k-1}|y_{1:t})\, d\theta_{t+k-1}$$

and

$$\pi(y_{t+k}|y_{1:t}) = \int \pi(y_{t+k}|\theta_{t+k})\, \pi(\theta_{t+k}|y_{1:t})\, \mathrm{d}\theta_{t+k}.$$

Incidentally, these recursions also show that $\pi(\theta_t|y_{1:t})$ summarizes the information contained in the past observations $y_{1:t}$, which is sufficient for predicting Y_{t+k}, for any $k > 0$.

2.7.2 Kalman filter for dynamic linear models

The previous results solve in principle the filtering and the forecasting problems; however, in general the actual computation of the relevant conditional distributions is not at all an easy task. DLMs are one important case where the general recursions simplify considerably. In this case, using standard results about the multivariate Gaussian distribution, it is easily proved that the random vector $(\theta_0, \theta_1, \ldots, \theta_t, Y_1, \ldots, Y_t)$ has a Gaussian distribution for any $t \geq 1$. It follows that the marginal and conditional distributions are also Gaussian. Since all the relevant distributions are Gaussian, they are completely determined by their means and variances. The solution of the filtering problem for DLMs is given by the celebrated Kalman filter.

Proposition 2.2 (Kalman filter). *Consider the DLM specified by* (2.4) *(p.41). Let*

$$\theta_{t-1}|y_{1:t-1} \sim \mathcal{N}(m_{t-1}, C_{t-1}).$$

Then the following statements hold.

(i) The one-step-ahead predictive distribution of θ_t given $y_{1:t-1}$ is Gaussian, with parameters

$$\begin{aligned} a_t &= \mathrm{E}(\theta_t|y_{1:t-1}) = G_t m_{t-1}, \\ R_t &= \mathrm{Var}(\theta_t|y_{1:t-1}) = G_t C_{t-1} G_t' + W_t. \end{aligned} \tag{2.8a}$$

(ii) The one-step-ahead predictive distribution of Y_t given $y_{1:t-1}$ is Gaussian, with parameters

$$\begin{aligned} f_t &= \mathrm{E}(Y_t|y_{1:t-1}) = F_t a_t, \\ Q_t &= \mathrm{Var}(Y_t|y_{1:t-1}) = F_t R_t F_t' + V_t. \end{aligned} \tag{2.8b}$$

(iii) The filtering distribution of θ_t given $y_{1:t}$ is Gaussian, with parameters

$$\begin{aligned} m_t &= \mathrm{E}(\theta_t|y_{1:t}) = a_t + R_t F_t' Q_t^{-1} e_t, \\ C_t &= \mathrm{Var}(\theta_t|y_{1:t}) = R_t - R_t F_t' Q_t^{-1} F_t R_t, \end{aligned} \tag{2.8c}$$

where $e_t = Y_t - f_t$ is the forecast error.

Proof. The random vector $(\theta_0, \theta_1, \ldots, \theta_t, Y_1, \ldots, Y_t)$ has joint distribution given by (2.3), where the marginal and conditional distributions involved are Gaussian. From standard results on the multivariate Normal distribution (see Appendix A), it follows that the joint distribution of $(\theta_0, \theta_1, \ldots, \theta_t, Y_1, \ldots, Y_t)$ is Gaussian, for any $t \geq 1$. Consequently, the distribution of any subvector is also Gaussian, as is the conditional distribution of some components given some other components. Therefore the predictive distributions and the filtering distributions are Gaussian, and it suffices to compute their means and variances.

To prove (i), let $\theta_t | y_{1:t-1} \sim \mathcal{N}(a_t, R_t)$. Using (2.4c), a_t and R_t can be obtained as follows:

$$\begin{aligned} a_t &= \mathrm{E}(\theta_t | y_{1:t-1}) = \mathrm{E}(\mathrm{E}(\theta_t | \theta_{t-1}, y_{1:t-1}) | y_{1:t-1}) \\ &= \mathrm{E}(G_t \theta_{t-1} | y_{1:t}) = G_t m_{t-1} \end{aligned}$$

and

$$\begin{aligned} R_t &= \mathrm{Var}(\theta_t | y_{1:t-1}) \\ &= \mathrm{E}(\mathrm{Var}(\theta_t | \theta_{t-1}, y_{1:t-1}) | y_{1:t-1}) + \mathrm{Var}(\mathrm{E}(\theta_t | \theta_{t-1}, y_{1:t-1}) | y_{1:t-1}) \\ &= W_t + G_t C_{t-1} G_t'. \end{aligned}$$

To prove (ii), let $Y_t | y_{1:t-1} \sim \mathcal{N}(f_t, Q_t)$. Using (2.4b), f_t and Q_t can be obtained as follows:

$$f_t = \mathrm{E}(Y_t | y_{1:t-1}) = \mathrm{E}(\mathrm{E}(Y_t | \theta_t, y_{1:t-1}) | y_{1:t-1}) = \mathrm{E}(F_t \theta_t | y_{1:t-1}) = F_t a_t$$

and

$$\begin{aligned} Q_t &= \mathrm{Var}(Y_t | y_{1:t-1}) \\ &= \mathrm{E}(\mathrm{Var}(Y_t | \theta_t, y_{1:t-1}) | y_{1:t-1}) + \mathrm{Var}(\mathrm{E}(Y_t | \theta_t, y_{1:t-1}) | y_{1:t-1}) \\ &= V_t + F_t R_t F_t'. \end{aligned}$$

Let us prove (iii) next. We can adapt Proposition 2.1(iii) to the present special case. There, we showed that, in order to compute the filtering distribution at time t, we have to apply the Bayes formula to combine the prior $\pi(\theta_t | y_{1:t-1})$ and the likelihood $\pi(y_t | \theta_t)$. In the DLM case all the distributions are Gaussian and the problem is the same as the Bayesian inference problem for the linear model

$$Y_t = F_t \theta_t + v_t, \qquad v_t \sim \mathcal{N}(0, V_t),$$

with a regression parameter θ_t following a conjugate Gaussian prior $\mathcal{N}(a_t, R_t)$. (Here V_t is known.) From the results in Section 1.5 we have that

$$\theta_t | y_{1:t} \sim \mathcal{N}(m_t, C_t),$$

where, by (1.10),

$$m_t = a_t + R_t F_t' Q_t^{-1}(Y_t - F_t a_t)$$

and, by (1.9),

$$C_t = R_t - R_t F_t' Q_t^{-1} F_t R_t.$$

□

The Kalman filter allows us to compute the predictive and filtering distributions recursively, starting from $\theta_0 \sim \mathcal{N}(m_0, C_0)$, then computing $\pi(\theta_1|y_1)$, and proceeding recursively as new data become available.

The conditional distribution of $\theta_t|y_{1:t}$ solves the filtering problem. However, in many cases one is interested in a point estimate. As we have discussed in Section 1.3, the Bayesian point estimate of θ_t given the information $y_{1:t}$, with respect to the quadratic loss function $L(\theta_t, a) = (\theta_t - a)'H(\theta_t - a)$, is the conditional expected value $m_t = \mathrm{E}(\theta_t|y_{1:t})$. This is the optimal estimate since it minimizes the conditional expected loss $\mathrm{E}((\theta_t - a)'H(\theta_t - a)|y_{1:t-1})$ with respect to a. If $H = I_p$, the minimum expected loss is the conditional variance matrix $\mathrm{Var}(\theta_t|y_{1:t})$.

As we noted in the introductory example in Section 2.2, the expression of m_t has the intuitive estimation-correction form "filter mean equals the prediction mean a_t plus a correction depending on how much the new observation differs from its prediction". The weight of the correction term is given by the gain matrix

$$K_t = R_t F_t' Q_t^{-1}.$$

Thus, the weight of current data point Y_t depends on the observation variance V_t (through Q_t) and on $R_t = \mathrm{Var}(\theta_t|y_{1:t-1}) = G_t C_{t-1} G_t' + W_t$.

As an example, consider the local level model (2.5). The Kalman filter gives

$$\begin{aligned}
\mu_t|y_{1:t-1} &\sim \mathcal{N}(m_{t-1}, R_t = C_{t-1} + W),\\
Y_t|y_{1:t-1} &\sim \mathcal{N}(f_t = m_{t-1}, Q_t = R_t + V),\\
\mu_t|y_{1:t} &\sim \mathcal{N}(m_t = m_{t-1} + K_t e_t, C_t = K_t V),
\end{aligned}$$

where $K_t = R_t/Q_t$ and $e_t = Y_t - f_t$. It is worth underlining that the behavior of the process (Y_t) is greatly influenced by the ratio between the two error variances, $r = W/V$, which is usually called the *signal-to-noise* ratio (a good exercise for seeing this is to simulate some trajectories of (Y_t), for different values of V and W). This is reflected in the structure of the estimation and forecasting mechanism. Note that $m_t = K_t y_t + (1 - K_t)m_{t-1}$, a weighted average of y_t and m_{t-1}. The weight $K_t = R_t/Q_t = (C_{t-1}+W)/(C_{t-1}+W+V)$ of the current observation y_t is also called *adaptive coefficient*, and it satisfies

$0 < K_t < 1$. For any given C_0, if the signal-to-noise r is small, K_t is small and y_t receives little weight. If, at the opposite extreme, $V = 0$, we have $K_t = 1$ and $m_t = y_t$, that is, the one-step-ahead forecast is given by the most recent data point. A practical illustration of how different relative magnitudes of W and V affect the mean of the filtered distribution and the one-step-ahead forecasts is given on pages 57 and 67.

The evaluation of the posterior variances C_t (and consequently also of R_t and Q_t) using the iterative updating formulae contained in Proposition 2.2, as simple as it may appear, suffers from numerical instability that may lead to nonsymmetric and even negative definite calculated variance matrices. Alternative, stabler, algorithms have been developed to overcome this issue. Apparently, the most widely used, at least in the Statistics literature, is the square root filter, which provides formulae for the sequential update of a square root[3] of C_t. References for the square root filter are Morf and Kailath (1975) and Anderson and Moore (1979, Ch. 6)

In our work we have found that occasionally, in particular when the observational noise has a small variance, even the square root filter incurs numerical stability problems, leading to negative definite calculated variances. A more robust algorithm is the one based on sequentially updating the singular value decomposition[4] (SVD) of C_t. The details of the algorithm can be found in Oshman and Bar-Itzhack (1986) and Wang et al. (1992). Strictly speaking, the SVD-based filter can be seen as a square root filter: in fact if $A = UD^2U'$ is the SVD of a variance matrix, then DU' is a square root of A. However, compared to the standard square root filtering algorithms, the SVD-based one is typically more stable (see the references for further discussion).

The Kalman filter is performed in package dlm by the function `dlmFilter`. The arguments are the data, `y`, in the form of a numerical vector, matrix, or time series, and the model, `mod`, an object of class `dlm` or a list that can be coerced to a `dlm` object. For the reasons of numerical stability mentioned above, the calculations are performed on the SVD of the variance matrices C_t and R_t. Accordingly, the output provides, for each t, an orthogonal matrix $U_{C,t}$ and a vector $D_{C,t}$ such that $C_t = U_{C,t}\,\mathrm{diag}(D^2_{C,t})\,U'_{C,t}$, and similarly for R_t.

The output produced by `dlmFilter`, a list with class attribute "`dlmFiltered`," includes, in addition to the original data and the model (components `y` and `mod`), the means of the predictive and filtered distributions (components `a` and `m`) and the SVD of the variances of the predictive and filtered distributions (components `U.R`, `D.R`, `U.C`, and `D.C`). For convenience, the component `f` of the output list provides the user with one-step-ahead forecasts. The component `U.C` is a list of matrices, the $U_{C,t}$ above, while `D.C`

[3] We define a square root of variance matrix A to be any square matrix N such that $A = N'N$.

[4] See Appendix B for a definition.

is a matrix containing, stored by row, the vectors $D_{C,t}$ of the SVD of the C_t's. Similarly for `U.R` and `D.R`. The utility function `dlmSvd2var` can be used to reconstruct the variances from their SVD. In the display below we use a random walk plus noise model with the Nile data (Figure 2.3). The variances $V = 15100$ and $W = 1468$ are maximum likelihood estimates. To set up the model we use, instead of `dlm`, the more convenient `dlmModPoly`, which will be discussed in Chapter 3.

R code

```
> NilePoly <- dlmModPoly(order = 1, dV = 15100, dW = 1468)
> unlist(NilePoly)
      m0       C0       FF        V       GG        W
       0 10000000        1    15100        1     1468
> NileFilt <- dlmFilter(Nile, NilePoly)
> str(NileFilt, 1)
List of 9
 $ y  : Time-Series [1:100] from 1871 to 1970: 1120 1160 ...
 $ mod:List of 10
  ..- attr(*, "class")= chr "dlm"
 $ m  : Time-Series [1:101] from 1870 to 1970:    0 1118 ...
 $ U.C:List of 101
 $ D.C: num [1:101, 1] 3162  123 ...
 $ a  : Time-Series [1:100] from 1871 to 1970:    0 1118 ...
 $ U.R:List of 100
 $ D.R: num [1:100, 1] 3163  129 ...
 $ f  : Time-Series [1:100] from 1871 to 1970:    0 1118 ...
 - attr(*, "class")= chr "dlmFiltered"
> n <- length(Nile)
> attach(NileFilt)
> dlmSvd2var(U.C[[n + 1]], D.C[n + 1, ])
         [,1]
[1,] 4031.035
```

The last number in the display is the variance of the filtering distribution of the 100-th state vector. Note that m_0 and C_0 are included in the output, which is the reason why `U.C` has one element more than `U.R`, and `m` and `U.D` one row more than `a` and `D.R`.

As we already noted on page 55, the relative magnitude of W and V is an important factor that enters the gain matrix, which, in turn, determines how sensitive the state prior-to-posterior updating is to unexpected observations. To illustrate the role of the signal-to-noise ratio W/V in the local level model, we use two models, with a significantly different signal-to-noise ratio, to estimate the true level of the Nile River. The filtered values for the two models can then be compared.

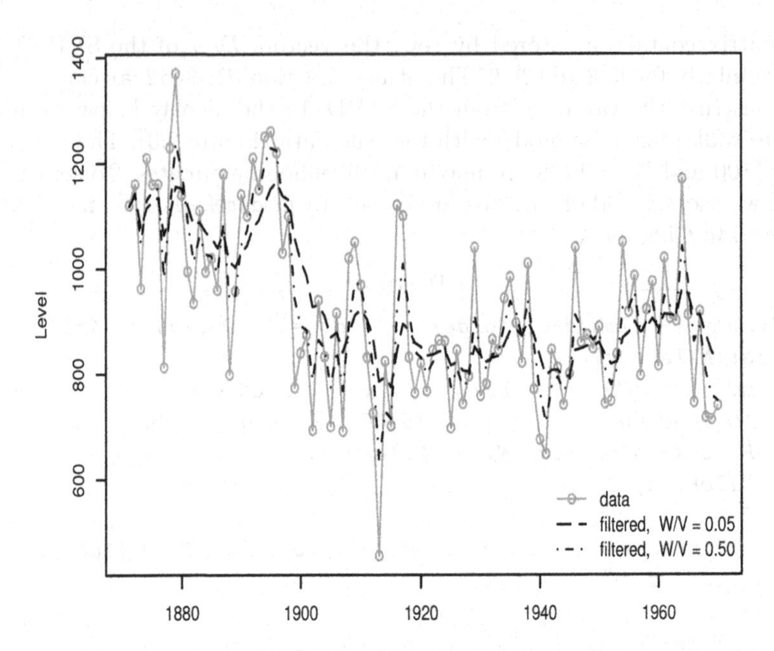

Fig. 2.7. Filtered values of the Nile River level for two different signal-to-noise ratios

R code

```
> plot(Nile, type='o', col = c("darkgrey"),
+       xlab = "", ylab = "Level")
> mod1 <- dlmModPoly(order = 1, dV = 15100, dW = 755)
> NileFilt1 <- dlmFilter(Nile, mod1)
> lines(dropFirst(NileFilt1$m), lty = "longdash")
> mod2 <- dlmModPoly(order = 1, dV = 15100, dW = 7550)
> NileFilt2 <- dlmFilter(Nile, mod2)
> lines(dropFirst(NileFilt2$m), lty = "dotdash")
> leg <- c("data", paste("filtered,  W/V =",
+                        format(c(W(mod1) / V(mod1),
+                                 W(mod2) / V(mod2)))))
> legend("bottomright", legend = leg,
+        col=c("darkgrey", "black", "black"),
+        lty = c("solid", "longdash", "dotdash"),
+        pch = c(1, NA, NA), bty = "n")
```

Figure 2.7 displays the filtered levels resulting from the two models. It is appearent that for model 2, which has a signal-to-noise ratio ten times larger than model 1, the filtered values tend to follow more closely the data.

2.7.3 Filtering with missing observations

In applied data analysis it is not infrequent to have to deal with a time series containing one or more missing observations. In multivariate time series, missing observations can be of two different types: totally missing and partially missing observations. The first type is the one that occurs when the observation vector at some time t is not available. In the second case only some of the components of the observation vector are not available. This may happen for example when considering a daily time series of closing prices of a set of stock indices in several countries: if day t is a holiday in country A but not country B, then for that day the closing price for the index of country A is not even defined, i.e., it is missing, while the closing price of the index of country B is normally recorded. Clearly, for a univariate time series an observation is either missing or not missing. Luckily, the structure of state space models is such that missing observations can be easily accomodated in the filtering recursion. We will first consider the case of totally missing observations. Following R convention, we will consider a missing observation as one having the special value `NA`. If the observation at time t is missing, then $y_t = \texttt{NA}$ and y_t does not carry any information, so that

$$\pi(\theta_t|y_{1:t}) = \pi(\theta_t|y_{1:t-1}). \tag{2.9}$$

This means that in this case the filtering distribution at time t is just the one-step-ahead predictive distribution at time $t-1$. Operationally, in the filtering recursion (Proposition 2.1) one has to replace (2.7c) with (2.9). In particular, for a DLM, since $\theta_t|y_{1:t-1} \sim \mathcal{N}(a_t, R_t)$, all one needs to do is to set $m_t = a_t$ and $C_t = R_t$. From time $t+1$ the standard filtering recursion resumes as usual, provided y_{t+1} is nonmissing. Note that formally, in a DLM, having $y_t = \texttt{NA}$ is the same as setting $F_t = 0$ or $V_t = \infty$. In the first case y_t is not linked to θ_t in any way, in the second the observation is so noisy as to be totally unreliable in providing meaningful information about θ_t. Either way leads to a gain matrix $K_t = 0$ and consequently $m_t = a_t$ and $C_t = R_t$.

Consider now a state space model with m-dimensional observation vectors, $m > 1$. Suppose that some, but not all, of the components of y_t are missing. The vector y_t in this case provides some information about θ_t, but all this information is contained in the nonmissing components. Let $\tilde{y}_t$ be the vector comprising only the nonmissing components of y_t. Then in the filtering recursion (2.7), $\pi(y_t|\theta_t)$ should be replaced by $\pi(\tilde{y}_t|\theta_t)$ and $\pi(y_t|y_{1:t-1})$ by $\pi(\tilde{y}_t|y_{1:t-1})$. Let us take a closer look at the DLM case. Denote by $\tilde{m}_t$ the dimension of $\tilde{y}_t$ and consider the $\tilde{m}_t$ by m matrix M_t obtained by removing from an m by m identity matrix the rows corresponding to the missing components of y_t, so that $\tilde{y}_t = M_t y_t$. The fact that we observed $\tilde{y}_t$ instead of y_t implies that in updating the prior $\mathcal{N}(a_t, R_t)$ to the posterior $\mathcal{N}(m_t, C_t)$, the correct observation equation to consider is

$$\tilde{y}_t = \tilde{F}_t\theta_t + \tilde{v}_t \qquad \tilde{v}_t \sim \mathcal{N}(0, \tilde{V}_t),$$

with $\tilde{F}_t = M_t F_t$ and $\tilde{V}_t = M_t V_t M_t'$. In practice, this implies that when computing the Kalman filter (Proposition 2.2), one has simply to replace F_t and V_t with $\tilde{F}_t$ and $\tilde{V}_t$ in (2.8b) and (2.8c).

The function `dlmFilter` accepts data containing `NA`'s, computing the moments of the correct filtering distributions.

2.7.4 Smoothing

One of the attractive features of state space models is that estimation and forecasting can be applied sequentially, as new data become available. However, in time series analysis one often has observations on Y_t for a certain period, $t = 1, \ldots, T$, and wants to retrospectively reconstruct the behavior of the system, to study the socio-economic construct or physical phenomenon underlying the observations. In this case, one can use a backward-recursive algorithm to compute the conditional distributions of θ_t given $y_{1:T}$, for any $t < T$, starting from the filtering distribution $\pi(\theta_T | y_{1:T})$ and estimating backward all the states' history. The result for general state space models is contained in the following proposition.

Proposition 2.3 (Smoothing recursion). *For a general state space model defined by* (A.1)-(A.2) *(p. 40), the following statements hold.*

(i) Conditional on $y_{1:T}$, the state sequence $(\theta_0, \ldots, \theta_T)$ has backward transition probabilities given by

$$\pi(\theta_t | \theta_{t+1}, y_{1:T}) = \frac{\pi(\theta_{t+1} | \theta_t) \pi(\theta_t | y_{1:t})}{\pi(\theta_{t+1} | y_{1:t})}.$$

(ii) The smoothing distributions of θ_t given $y_{1:T}$ can be computed according to the following backward recursion in t, starting from $\pi(\theta_T | y_{1:T})$:

$$\pi(\theta_t | y_{1:T}) = \pi(\theta_t | y_{1:t}) \int \frac{\pi(\theta_{t+1} | \theta_t)}{\pi(\theta_{t+1} | y_{1:t})} \pi(\theta_{t+1} | y_{1:T}) \, d\theta_{t+1}.$$

Proof. To prove (i), note that θ_t and $Y_{t+1:T}$ are conditionally independent given θ_{t+1}; moreover, θ_{t+1} and $Y_{1:T}$ are conditionally independent given θ_t. (Use the DAG in Figure 2.6 to show this.) Using the Bayes formula, one has

$$\begin{aligned}
\pi(\theta_t | \theta_{t+1}, y_{1:T}) &= \pi(\theta_t | \theta_{t+1}, y_{1:t}) \\
&= \frac{\pi(\theta_t | y_{1:t}) \pi(\theta_{t+1} | \theta_t, y_{1:t})}{\pi(\theta_{t+1} | y_{1:t})} \\
&= \frac{\pi(\theta_t | y_{1:t}) \pi(\theta_{t+1} | \theta_t)}{\pi(\theta_{t+1} | y_{1:t})}.
\end{aligned}$$

To prove (ii), marginalize $\pi(\theta_t, \theta_{t+1} | y_{1:T})$ with respect to θ_{t+1}:

$$\begin{aligned}\pi(\theta_t|y_{1:T}) &= \int \pi(\theta_t, \theta_{t+1}|y_{1:T})\, d\theta_{t+1} \\ &= \int \pi(\theta_{t+1}|y_{1:T})\pi(\theta_t|\theta_{t+1}, y_{1:T})\, d\theta_{t+1} \\ &= \int \pi(\theta_{t+1}|y_{1:T}) \frac{\pi(\theta_{t+1}|\theta_t)\pi(\theta_t|y_{1:t})}{\pi(\theta_{t+1}|y_{1:t})}\, d\theta_{t+1} \\ &= \pi(\theta_t|y_{1:t}) \int \pi(\theta_{t+1}|\theta_t) \frac{\pi(\theta_{t+1}|y_{1:T})}{\pi(\theta_{t+1}|y_{1:t})}\, d\theta_{t+1}.\end{aligned}$$

□

For a DLM, the smoothing recursion can be stated more explicitely in terms of means and variances of the smoothing distributions.

Proposition 2.4 (Kalman smoother). *For a DLM defined by (2.4), if $\theta_{t+1}|y_{1:T} \sim \mathcal{N}(s_{t+1}, S_{t+1})$, then $\theta_t|y_{1:T} \sim \mathcal{N}(s_t, S_t)$, where*

$$\begin{aligned}s_t &= m_t + C_t G'_{t+1} R^{-1}_{t+1}(s_{t+1} - a_{t+1}) \\ S_t &= C_t - C_t G'_{t+1} R^{-1}_{t+1}(R_{t+1} - S_{t+1}) R^{-1}_{t+1} G_{t+1} C_t.\end{aligned}$$

Proof. It follows from the properties of the multivariate Gaussian distribution that the conditional distribution of θ_t given $y_{1:T}$ is Gaussian; thus, it suffices to compute its mean and variance. We have

$$s_t = \mathrm{E}(\theta_t|y_{1:T}) = \mathrm{E}(\mathrm{E}(\theta_t|\theta_{t+1}, y_{1:T})|y_{1:T})$$

and

$$S_t = \mathrm{Var}(\theta_t|y_{1:T}) = \mathrm{Var}(\mathrm{E}(\theta_t|\theta_{t+1}, y_{1:T})|y_{1:T}) + \mathrm{E}(\mathrm{Var}(\theta_t|\theta_{t+1}, y_{1:T})|y_{1:T}).$$

As shown in the proof of Proposition 2.3, θ_t and $Y_{t+1:T}$ are conditionally independent given θ_{t+1}, so that $\pi(\theta_t|\theta_{t+1}, y_{1:T}) = \pi(\theta_t|\theta_{t+1}, y_{1:t})$. We can use the Bayes formula to compute this distribution. Note that the likelihood $\pi(\theta_{t+1}|\theta_t, y_{1:t}) = \pi(\theta_{t+1}|\theta_t)$ is expressed by the state equation (2.4c), that is,

$$\theta_{t+1}|\theta_t \sim \mathcal{N}(G_{t+1}\theta_t, W_{t+1}).$$

The prior is $\pi(\theta_t|y_{1:t})$, which is $\mathcal{N}(m_t, C_t)$. Using (1.10) and (1.9), we find that

$$\begin{aligned}\mathrm{E}(\theta_t|\theta_{t+1}, y_{1:t}) &= m_t + C_t G'_{t+1}(G_{t+1} C_t G'_{t+1} + W_{t+1})^{-1}(\theta_{t+1} - G_{t+1} m_t) \\ &= m_t + C_t G'_{t+1} R^{-1}_{t+1}(\theta_{t+1} - a_{t+1}) \\ \mathrm{Var}(\theta_t|\theta_{t+1}, y_{1:t}) &= C_t - C_t G'_{t+1} R^{-1}_{t+1} G_{t+1} C_t,\end{aligned}$$

from which it follows that

$$\begin{aligned}
s_t &= \mathrm{E}(\mathrm{E}(\theta_t|\theta_{t+1}, y_{1:t})|y_{1:T}) = m_t + C_t G'_{t+1} R^{-1}_{t+1}(s_{t+1} - a_{t+1}) \\
S_t &= \mathrm{Var}(\mathrm{E}(\theta_t|\theta_{t+1}, y_{1:t})|y_{1:T}) + \mathrm{E}(\mathrm{Var}(\theta_t|\theta_{t+1}, y_{1:t})|y_{1:T}) \\
&= C_t - C_t G'_{t+1} R^{-1}_{t+1} G_{t+1} C_t + C_t G'_{t+1} R^{-1}_{t+1} S_{t+1} R^{-1}_{t+1} G_{t+1} C_t \\
&= C_t - C_t G'_{t+1} R^{-1}_{t+1}(R_{t+1} - S_{t+1}) R^{-1}_{t+1} G_{t+1} C_t,
\end{aligned}$$

being $\mathrm{E}(\theta_{t+1}|y_{1:T}) = s_{t+1}$ and $\mathrm{Var}(\theta_{t+1}|y_{1:T}) = S_{t+1}$ by assumption. □

The Kalman smoother allows us to compute the distributions of $\theta_t|y_{1:T}$, starting from $t = T-1$, in which case $\theta_T|y_{1:T} \sim \mathcal{N}(s_T = m_T, S_T = C_T)$, and then proceeding backward to compute the distributions of $\theta_t|y_{1:T}$ for $t = T-2$, $t = T-3$, etc. Note that the smoothing recursion depends on the data only through the filtering and one-step-ahead predictive moments obtained using the Kalman filter. Therefore, if a time series contains missing observations, this should be accounted for when performing the filtering recursion, but no additional adjustment is required in the smoothing recursion.

About the numerical stability of the smoothing algorithm, the same caveat holds as for the filtering recursions. The formulae of Proposition 2.4 are subject to numerical instability, and more robust square root and SVD-based smoothers are available (see Zhang and Li; 1996). The function `dlmSmooth` performs the calculations in R, starting from an object of class `dlmFiltered`, typically the output produced by `dlmFilter`. Alternatively, the user can provide the data and the model, in which case `dlmFilter` is called internally. `dlmSmooth` returns a list with components `s`, the means of the smoothing distributions, and `U.S`, `D.S`, their variances, given in terms of their SVD. The following display illustrates the use of `dlmSmooth` on the Nile data.

R code

```
> NileSmooth <- dlmSmooth(NileFilt)
> str(NileSmooth, 1)
List of 3
 $ s  : Time-Series [1:101] from 1870 to 1970: 1111 1111 ...
 $ U.S:List of 101
 $ D.S: num [1:101, 1] 74.1 63.5 ...
> attach(NileSmooth)
> drop(dlmSvd2var(U.S[[n + 1]], D.S[n + 1,]))
[1] 4031.035
> drop(dlmSvd2var(U.C[[n + 1]], D.C[n + 1,]))
[1] 4031.035
> drop(dlmSvd2var(U.S[[n / 2 + 1]], D.S[n / 2 + 1,]))
[1] 2325.985
> drop(dlmSvd2var(U.C[[n / 2 + 1]], D.C[n / 2 + 1,]))
[1] 4031.035
```

In the display above, n is 100, the number of observations, so, accounting for time $t = 0$, n/2 + 1 corresponds to time 50. Observe that the smoothing and filtering variances are equal at the end of the observation period – time T (lines 9 and 11); but the smoothing variance at time 50 (line 13) is much smaller than the filtering variance at the same time (line 15). This is due to the fact that in the filtering distribution at time 50 we are conditioning on the first fifty observations only, while in the smoothing distribution the conditioning is with respect to all the one hundred observations available. Note also, incidentally, that the filtering variance at time 50 is the same as the filtering variance at time 100. It is the case for many constant models that the filtering variance, C_t, tends to a limiting value as t increases. In very informal terms, the explanation of this behavior is the following. In DLMs the learning process about the state of the system occurs in a dynamic environment, that is, one in which the state changes as one gains information about it. Therefore, in the updating of the filtering variance from time $t-1$ to time t, there are two conflicting processes going on: on one hand, the observation y_t brings new information about θ_{t-1}, but in the meanwhile the state of the system has changed to θ_t, with the additional uncertainty carried by w_t. This additional uncertainty is represented by the variance $W_t = W$, say. If C_0 is large – typically one does not have much confidence in his prior guess about the state – then the first observations are very informative and their impact on C_t is much more important than that of the dynamics of the state, resulting in an overall decrease of the filtering variance. However, as more data are collected, the impact of one additional observation on the information about the state of the system decreases and, at some point, it will be exactly balanced by the loss of information represented by the additional variance W. From that time on, C_t will essentially stay constant.

The display below illustrates how the variance of the smoothing distribution can be used to construct pointwise probability intervals for the state components – only one in this example. The plot produced by the code below is shown in Figure 2.8

R code

```
  > hwid <- qnorm(0.025, lower = FALSE) *
2 +     sqrt(unlist(dlmSvd2var(U.S, D.S)))
  > smooth <- cbind(s, as.vector(s) + hwid %o% c(-1, 1))
4 > plot(dropFirst(smooth), plot.type = "s", type = "l",
  +      lty = c(1, 5, 5), ylab = "Level", xlab = "",
6 +      ylim = range(Nile))
  > lines(Nile, type = "o", col = "darkgrey")
8 > legend("bottomleft", col = c("darkgrey", rep("black", 2)),
  +        lty = c(1, 1, 5), pch = c(1, NA, NA), bty = "n",
10 +       legend = c("data", "smoothed level",
  +        "95% probability limits"))
```

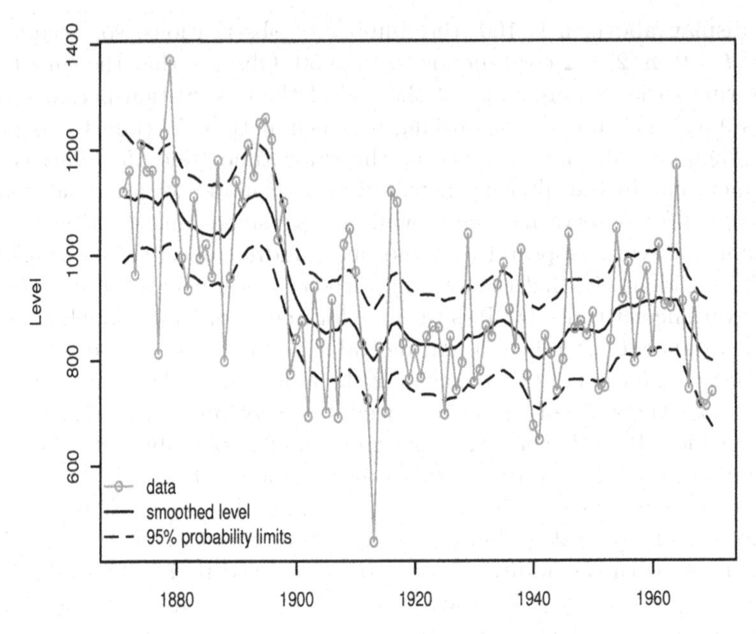

Fig. 2.8. Smoothed values of the Nile River level, with 95% probability limits

As an additional example, we consider a quarterly time series of consumer expenditure on durable goods in the UK, in 1958£, from the first quarter of 1957 to the last quarter of 1967[5]. A DLM including a local level plus a quarterly seasonal component was fitted to the data. This kind of model will be discussed in Chapter 3; here we focus on filtering and smoothing. In the model the state vector is 4-dimensional. Two of its components have a particularly relevant interpretation: the first one can be thought of as the true, deseasonalized, level of the series; the second is a dynamic seasonal component. According to the model, the observations are obtained by adding observational noise to the sum of the first and second component of the state vector, as can be deduced from the matrix *FF*. Figure 2.9 shows the data, together with the deseasonalized filtered and smoothed level. These values are just the first components of the series of filtered and smoothed state vectors. In addition to the level of the series, one can also estimate the seasonal component, which is just the second component of the smoothed or filtered state vector. Figure 2.10 shows the smoothed seasonal component. It is worth stressing that the model is dynamic, hence the seasonal component is allowed to vary as time goes by. This is clearly the case in the present example: from an alternating of positive and negative values at the beginning of the observation period, the series moves to a two-positive two-negative pattern in the second half. The display below shows how filtered and smoothed values have been obtained in R, as

[5] Source: Hyndman (n.d.).

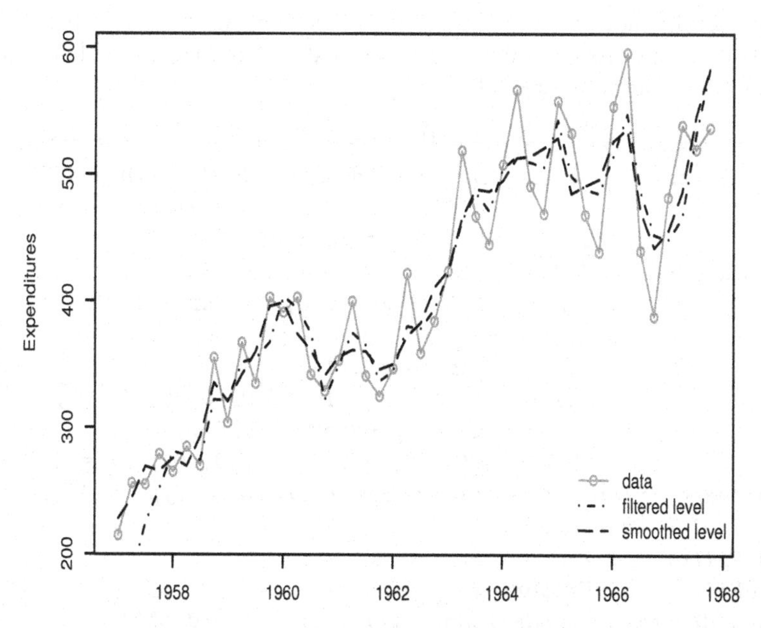

Fig. 2.9. Quarterly expenditure on durable goods, with filtered and smoothed level

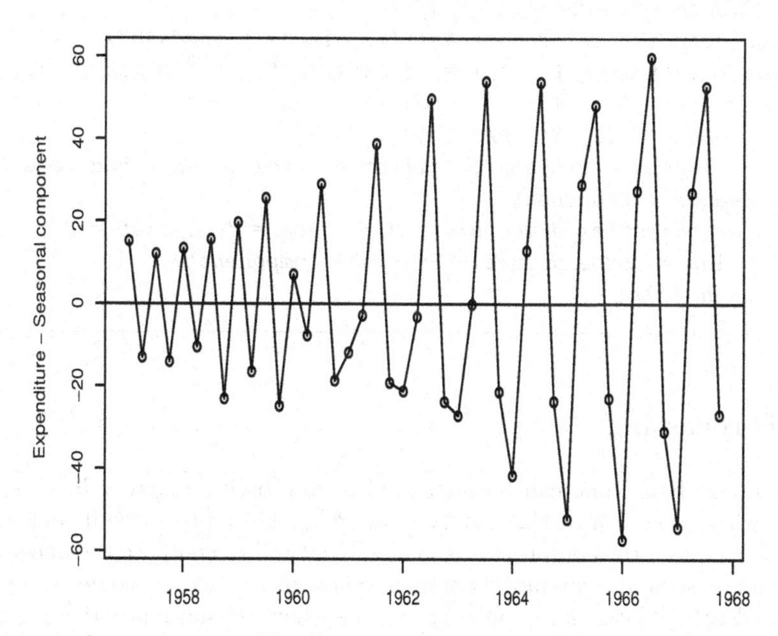

Fig. 2.10. Quarterly expenditure on durable goods: smoothed seasonal component

well as how the plots were created. The function *bdiag* is a utility function in package dlm that creates a block diagonal matrix from the individual blocks, or from a list containing the blocks.

R code

```
> expd <- ts(read.table("Datasets/qconsum.dat", skip = 4,
+                       colClasses = "numeric")[, 1],
+            start = c(1957, 1), frequency = 4)
> expd.dlm <- dlm(m0 = rep(0,4), C0 = 1e8 * diag(4),
+                 FF = matrix(c(1, 1, 0, 0), nr = 1),
+                 V = 1e-3,
+                 GG = bdiag(matrix(1),
+                 matrix(c(-1, -1, -1, 1, 0, 0, 0, 1, 0),
+                        nr = 3, byrow = TRUE)),
+                 W = diag(c(771.35, 86.48, 0, 0), nr = 4))
> plot(expd, xlab = "", ylab = "Expenditures", type = 'o',
+      col = "darkgrey")
> ### Filter
> expdFilt <- dlmFilter(expd, expd.dlm)
> lines(dropFirst(expdFilt$m[, 1]), lty = "dotdash")
> ### Smooth
> expdSmooth <- dlmSmooth(expdFilt)
> lines(dropFirst(expdSmooth$s[,1]), lty = "longdash")
> legend("bottomright", col = c("darkgrey", rep("black", 2)),
+        lty = c("solid", "dotdash", "longdash"),
+        pch = c(1, NA, NA), bty = "n",
+        legend = c("data", "filtered level", "smoothed level"))
> ### Seasonal component
> plot(dropFirst(expdSmooth$s[, 3]), type = 'o', xlab = "",
+      ylab = "Expenditure - Seasonal component")
> abline(h = 0)
```

2.8 Forecasting

With $y_{1:t}$ at hand, one can be interested in forecasting future values of the observations, Y_{t+k}, or of the state vectors, θ_{t+k}. For state space models, the recursive form of the computations makes it natural to compute the one-step-ahead forecasts and to update them sequentially as new data become available. This is clearly of interest in applied problems where the data do arrive sequentially, such as in day-by-day forecasting stock prices, or in tracking a moving target; but one-step-ahead forecasts are often also computed "in-sample", as a tool for checking the performance of the model.

For a DLM, the one-step-ahead predictive distributions, for states and observations, are obtained as a byproduct of the Kalman filter, as presented in Proposition 2.2.

In R, the one-step-ahead forecasts $f_t = \mathrm{E}(Y_t|y_{1:t-1})$ are provided in the output of the function **dlmFilter**. Since for each t the one-step-ahead forecast of the observation, f_t, is a linear function of the filtering mean m_{t-1}, the magnitude of the gain matrix plays the same role in determining how sensitive f_t is to an unexpected observation y_{t-1} as it did for m_{t-1}. In the case of the random walk plus noise model this is particularly evident, since in this case $f_t = m_{t-1}$. Figure 2.11, produced with the code below, contains the one-step-ahead forecasts obtained from the local level models with the different signal-to-noise ratios defined in the display on page 57.

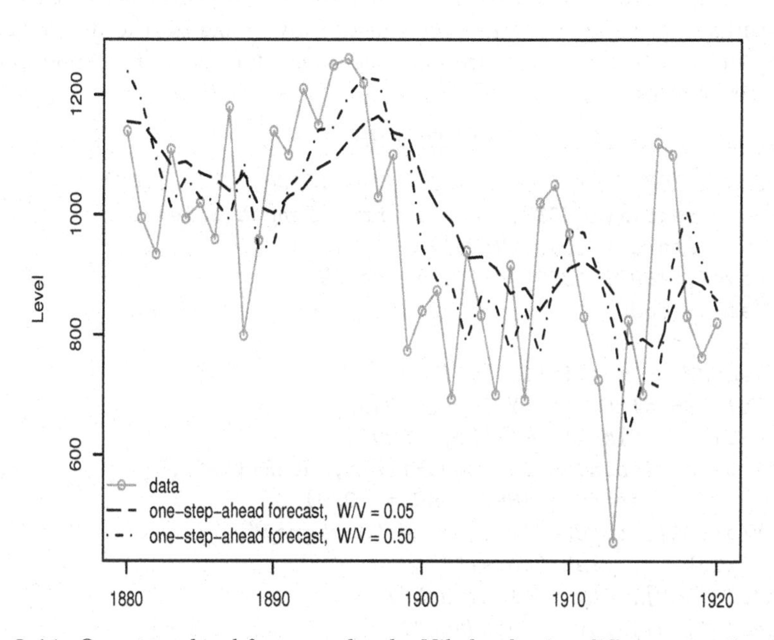

Fig. 2.11. One-step-ahead forecasts for the Nile level using different signal-to-noise ratios

R code

```
> a <- window(cbind(Nile, NileFilt1$f, NileFilt2$f),
+               start = 1880, end = 1920)
> plot(a[, 1], type = 'o', col = "darkgrey",
+       xlab = "", ylab = "Level")
> lines(a[, 2], lty = "longdash")
> lines(a[, 3], lty = "dotdash")
> leg <- c("data", paste("one-step-ahead forecast,  W/V =",
```

```
8  +                        format(c(W(mod1) / V(mod1),
   +                                 W(mod2) / V(mod2)))))
10 > legend("bottomleft", legend = leg,
   +        col = c("darkgrey", "black", "black"),
12 +        lty = c("solid", "longdash", "dotdash"),
   +        pch = c(1, NA, NA), bty = "n")
```

To further elaborate on the same example, we note that the signal-to-noise ratio need not be constant in time. The construction of the Ashwan dam in 1898, for instance, can be expected to produce a major change in the level of the Nile River. A simple way to incorporate this expected level shift in the model is to assume a system evolution variance W_t larger than usual (12 times larger in the display below) for that year and the following one. In this way the estimated true level of the river will quickly recognize the new regime, leading in turn to more accurate one-step-ahead forecasts. The code below illustrates this idea.

R code

```
   > mod0 <- dlmModPoly(order = 1, dV = 15100, dW = 1468)
2  > X <- ts(matrix(mod0$W, nc = 1, nr = length(Nile)),
   +         start = start(Nile))
4  > window(X, 1898, 1899) <- 12 * mod0$W
   > modDam <- mod0
6  > modDam$X <- X
   > modDam$JW <- matrix(1, 1, 1)
8  > damFilt <- dlmFilter(Nile, modDam)
   > mod0Filt <- dlmFilter(Nile, mod0)
10 > a <- window(cbind(Nile, mod0Filt$f, damFilt$f),
   +             start = 1880, end = 1920)
12 > plot(a[, 1], type = 'o', col = "darkgrey",
   +      xlab="", ylab="Level")
14 > lines(a[, 2], lty = "longdash")
   > lines(a[, 3], lty = "dotdash")
16 > abline(v=1898, lty=2)
   > leg <- c("data", paste("one-step-ahead forecast -",
18 +                        c("mod0", "modDam")))
   > legend("bottomleft", legend = leg,
20 +        col = c("darkgrey", "black", "black"),
   +        lty = c("solid", "longdash", "dotdash"),
22 +        pch = c(1, NA, NA), bty = "n")
```

Note (see Figure 2.12) how, using the modified model `modDam`, the forecast for the level of the river in 1900 is already around what the new river level actually is, while for the other model this happens only around 1907. On a

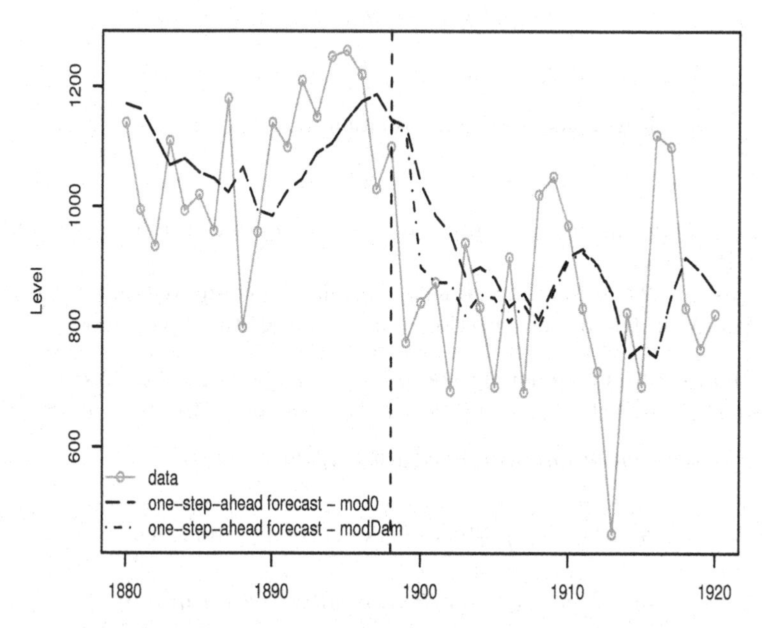

Fig. 2.12. One-step-ahead forecasts of Nile River level with and without change point

more technical note, it is instructive to note how we define the time varying model `modDam` by adding the components `X` and `JW` (lines 6 and 7) to the constant model `mod0`.

In many applications one is interested in looking a bit further in the future, and provide possible scenarios of the behavior of the series for k steps ahead. We present here the recursive formulae for the means and variances of the conditional distributions of states and observations at a future time $t+k$, given the data up to time t. In view of the Markovian nature of the model, the filtering distribution at time t acts like an initial distribution for the future evolution of the model. To be more precise, the joint distribution of present and future states $(\theta_{t+k})_{k\geq 0}$, and future observations $(Y_{t+k})_{k\geq 1}$ is that of a state space model having conditional distributions $\pi(\theta_{t+k}|\theta_{t+k-1})$ and $\pi(y_{t+k}|\theta_{t+k})$, and initial distribution $\pi(\theta_t|y_{1:t})$. The information about the future provided by the data is all contained in this distribution. For a DLM, in particular, since the data are only used to obtain m_t, the mean of $\pi(\theta_t|y_{1:t})$, it follows that m_t provides a summary of the data that is sufficient for predictive purposes. You can have a further intuition about that by looking at the DAG representing the dependence structure among the variables (Figure 2.6). We see that the path from $Y_{1:t}$ to Y_{t+k} is as in Figure 2.13, showing that the data $Y_{1:t}$ provide information about θ_t, which in turn gives information about the future state evolution up to θ_{t+k} and consequently on Y_{t+k}. Of course, as k

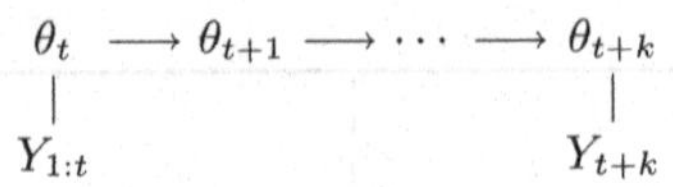

$$\begin{array}{ccccccc}\theta_t & \longrightarrow & \theta_{t+1} & \longrightarrow & \cdots & \longrightarrow & \theta_{t+k} \\ | & & & & & & | \\ Y_{1:t} & & & & & & Y_{t+k}\end{array}$$

Fig. 2.13. Flow of information from $Y_{1:t}$ to Y_{t+k}

gets larger, more uncertainty enters in the system, and the forecasts will be less and less precise.

Proposition 2.5 provides recursive formulae to compute the forecast distributions for states and observations for a general state space model.

Proposition 2.5 (Forecasting recursion). *For a general state space model defined by* (A.1)-(A.2) *(p.40), the following statements hold for any* $k > 0$.

(i) The k*-steps-ahead forecast distribution of the state is*

$$\pi(\theta_{t+k}|y_{1:t}) = \int \pi(\theta_{t+k}|\theta_{t+k-1})\pi(\theta_{t+k-1}|y_{1:t})\,\mathrm{d}\theta_{t+k-1}.$$

(ii) The k*-steps-ahead forecast distribution of the observation is*

$$\pi(y_{t+k}|y_{1:t}) = \int \pi(y_{t+k}|\theta_{t+k})\pi(\theta_{t+k}|y_{1:t})\,\mathrm{d}\theta_{t+k}.$$

Proof. Using the conditional independence properties of the model, we have:

$$\begin{aligned}\pi(\theta_{t+k}|y_{1:t}) &= \int \pi(\theta_{t+k}, \theta_{t+k-1}|y_{1:t})\,\mathrm{d}\theta_{t+k-1} \\ &= \int \pi(\theta_{t+k}|\theta_{t+k-1}, y_{1:t})\pi(\theta_{t+k-1}|y_{1:t})\,\mathrm{d}\theta_{t+k-1} \\ &= \int \pi(\theta_{t+k}|\theta_{t+k-1})\pi(\theta_{t+k-1}|y_{1:t})\,\mathrm{d}\theta_{t+k-1},\end{aligned}$$

which is (i). The proof of (ii) is again based on the conditional independence properties of the models. We have that

$$\begin{aligned}\pi(y_{t+k}|y_{1:t}) &= \int \pi(y_{t+k}, \theta_{t+k}|y_{1:t})\,\mathrm{d}\theta_{t+k} \\ &= \int \pi(y_{t+k}|\theta_{t+k}, y_{1:t})\pi(\theta_{t+k}|y_{1:t})\,\mathrm{d}\theta_{t+k} \\ &= \int \pi(y_{t+k}|\theta_{t+k})\pi(\theta_{t+k}|y_{1:t})\,\mathrm{d}\theta_{t+k},\end{aligned}$$

which is (ii). □

For DLMs, Proposition 2.5 takes a more specific form, since all the integrals can be computed explicitly. However, as is the case for filtering and smoothing, since all the forecast distributions are Gaussian, it is enough to compute their means and variances. Proposition 2.6 provides recursive formulae to compute them. We need to introduce some notation first. For $k \geq 1$, define

$$a_t(k) = \mathrm{E}(\theta_{t+k}|y_{1:t}), \tag{2.10a}$$
$$R_t(k) = \mathrm{Var}(\theta_{t+k}|y_{1:t}), \tag{2.10b}$$
$$f_t(k) = \mathrm{E}(Y_{t+k}|y_{1:t}), \tag{2.10c}$$
$$Q_t(k) = \mathrm{Var}(Y_{t+k}|y_{1:t}). \tag{2.10d}$$

Proposition 2.6. *For a DLM defined by (2.4), let $a_t(0) = m_t$ and $R_t(0) = C_t$. Then, for $k \geq 1$, the following statements hold.*

(i) The distribution of θ_{t+k} given $y_{1:t}$ is Gaussian, with

$$\begin{aligned} a_t(k) &= G_{t+k}a_{t,k-1}, \\ R_t(k) &= G_{t+k}R_{t,k-1}G'_{t+k} + W_{t+k}; \end{aligned}$$

(ii) The distribution of Y_{t+k} given $y_{1:t}$ is Gaussian, with

$$\begin{aligned} f_t(k) &= F_{t+k}a_t(k), \\ Q_t(k) &= F_{t+k}R_t(k)F'_{t+k} + V_t. \end{aligned}$$

Proof. As we have already noted, all conditional distributions are Gaussian. Therefore, we only need to prove the formulae giving the means and variances. We proceed by induction. The result holds for $k = 1$ in view of Proposition 2.2. For $k > 1$,

$$\begin{aligned} a_t(k) &= \mathrm{E}(\theta_{t+k}|y_{1:t}) = \mathrm{E}(\mathrm{E}(\theta_{t+k}|y_{1:t}, \theta_{t+k-1})|y_{1:t}) \\ &= \mathrm{E}(G_{t+k}\theta_{t+k-1}|y_{1:t}) = G_{t+k}a_{t,k-1}, \\ R_t(k) &= \mathrm{Var}(\theta_{t+k}|y_{1:t}) = \mathrm{Var}(\mathrm{E}(\theta_{t+k}|y_{1:t}, \theta_{t+k-1})|y_{1:t}) \\ &\quad + \mathrm{E}(\mathrm{Var}(\theta_{t+k}|y_{1:t}, \theta_{t+k-1})|y_{1:t}) \\ &= G_{t+k}R_{t,k-1}G'_{t+k} + W_{t+k}, \\ f_t(k) &= \mathrm{E}(Y_{t+k}|y_{1:t}) = \mathrm{E}(\mathrm{E}(Y_{t+k}|y_{1:t}, \theta_{t+k})|y_{1:t}) \\ &= \mathrm{E}(F_{t+k}\theta_{t+k}|y_{1:t}) = F_{t+k}a_t(k), \\ Q_t(k) &= \mathrm{Var}(Y_{t+k}|y_{1:t}) = \mathrm{Var}(\mathrm{E}(Y_{t+k}|y_{1:t}, \theta_{t+k})|y_{1:t}) \\ &\quad + \mathrm{E}(\mathrm{Var}(Y_{t+k}|y_{1:t}, \theta_{t+k})|y_{1:t}) \\ &= F_{t+k}R_t(k)F'_{t+k} + V_{t+k}, \end{aligned}$$

□

Note that the data only enter the predictive distributions through the mean of the filtering distribution at the time the last observation was taken. The function *dlmForecast* computes the means and variances of the predictive distributions of the observations and the states. Optionally, it can be used to draw a sample of future states and observations. The principal argument of *dlmForecast* is an object of class *dlmFiltered*. Alternatively, it can be a object of class *dlm* (or a list with the appropriate named components), where the components *m0* and *C0* are interpreted as being the mean and variance of the state vector at the end of the observation period, given the data, i.e., they are the mean and variance of the last (most recent) filtering distribution. The code below shows how to obtain predicted values of the expenditure series (Figure 2.9, p.65) for the three years following the last observation, together with a sample from their distribution. Figure 2.14 shows the forecasted and simulated future values of the series.

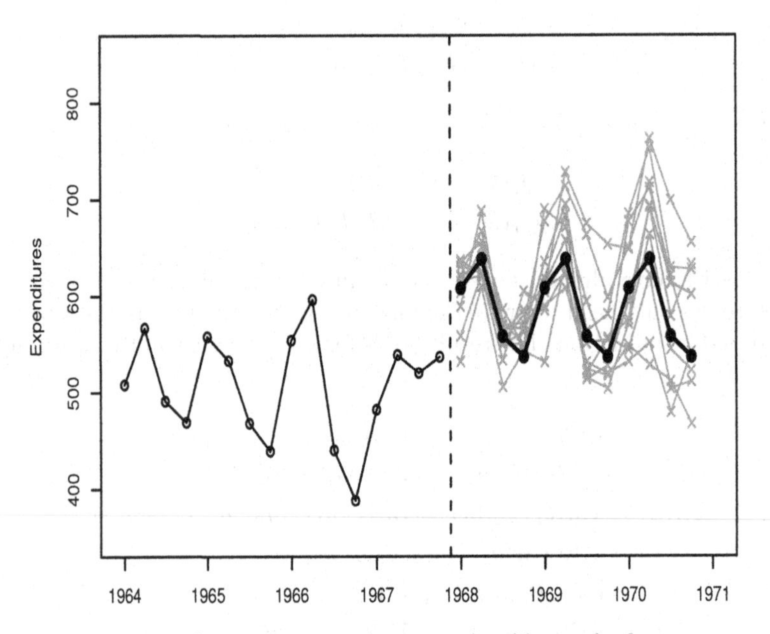

Fig. 2.14. Quarterly expenditure on durable goods: forecasts

R code

```
> set.seed(1)
> expdFore <- dlmForecast(expdFilt, nAhead = 12, sampleNew = 10)
> plot(window(expd, start = c(1964,1)), type = 'o',
+       xlim = c(1964,1971), ylim = c(350, 850),
+       xlab = "", ylab = "Expenditures")
> names(expdFore)
```

```
[1] "a"         "R"          "f"          "Q"
[5] "newStates" "newObs"
> attach(expdFore)
> invisible(lapply(newObs, function(x)
+                  lines(x, col = "darkgrey",
+                        type = 'o', pch = 4)))
> lines(f, type = 'o', lwd = 2, pch = 16)
> abline(v = mean(c(time(f)[1], time(expd)[length(expd)])),
+        lty = "dashed")
> detach()
```

2.9 The innovation process and model checking

As we have seen, for DLMs we can compute the one-step-ahead forecasts $f_t = \mathrm{E}(Y_y|y_{1:t-1})$, and we defined the forecast error as

$$e_t = Y_t - \mathrm{E}(Y_t|y_{1:t-1}) = Y_t - f_t.$$

The forecast errors can alternatively be written in terms of the one-step-ahead estimation errors as follows:

$$\begin{aligned} e_t &= Y_t - F_t a_t = F_t\theta_t + v_t - F_t a_t \\ &= F_t(\theta_t - a_t) + v_t. \end{aligned}$$

The sequence $(e_t)_{t\geq 1}$ of forecast errors enjoys some interesting properties, the most important of which are collected in the following proposition.

Proposition 2.7. *Let $(e_t)_{t\geq 1}$ be the sequence of forecast errors of a DLM. Then the following properties hold.*

(i) The expected value of e_t is zero.
(ii) The random vector e_t is uncorrelated with any function of $Y_1, \ldots, Y_{t-1}$.
(iii) For any $s < t$, e_t and Y_s are uncorrelated.
(iv) For any $s < t$, e_t and e_s are uncorrelated.
(v) e_t is a linear function of $Y_1, \ldots, Y_t$.
(vi) $(e_t)_{t\geq 1}$ is a Gaussian process.

Proof. (i) By taking iterated expected values,

$$\mathrm{E}(e_t) = \mathrm{E}(\mathrm{E}(Y_t - f_t|Y_{1:t-1})) = 0.$$

(ii) Let $Z = g(Y_1, \ldots, Y_{t-1})$. Then

$$\begin{aligned} \mathrm{Cov}(e_t, Z) &= \mathrm{E}(e_t Z) = \mathrm{E}(\mathrm{E}(e_t Z|Y_{1:t-1})) \\ &= \mathrm{E}(\mathrm{E}(e_t|Y_{1:t-1})Z) = 0. \end{aligned}$$

(iii) If the observations are univariate, this follows from (ii), taking $Z = Y_s$. Otherwise, apply (ii) to each component of Y_s.
(iv) This follows again from (ii), taking $Z = e_s$ if the observations are univariate. Otherwise, apply (ii) componentwise.
(v) Since $Y_1, \ldots, Y_t$ have a joint Gaussian distribution, $f_t = \mathrm{E}(Y_t|Y_{1:t-1})$ is a linear function of $Y_1, \ldots, Y_{t-1}$. Hence, e_t is a linear function of $Y_1, \ldots, Y_t$.
(vi) For any t, in view of (v), $(e_1, \ldots, e_t)$ is a linear transformation of $(Y_1, \ldots, Y_t)$, which has a joint Normal distribution. It follows that also $(e_1, \ldots, e_t)$ has a joint Normal distribution. Hence, since all finite-dimensional distributions are Gaussian, the process $(e_t)_{t \geq 1}$ is Gaussian.

□

The forecast errors e_t are also called *innovations*. The representation $Y_t = f_t + e_t$ justifies this terminology, since one can think of Y_t as the sum of a component, f_t, which is predictable from past observations, and another component, e_t, which is independent of the past and therefore contains the really new information provided by the observation Y_t.

Sometimes it may be convenient to work with the so-called *innovation form* of a DLM. This is obtained by choosing as new state variables the vectors $a_t = \mathrm{E}(\theta_t|y_{1:t-1})$. Then the observation equation is derived from $e_t = Y_t - f_t = Y_t - F_t a_t$:

$$Y_t = F_t a_t + e_t \tag{2.11a}$$

and, being $a_t = G_t m_{t-1}$, where m_{t-1} is given by the Kalman filter:

$$a_t = G_t m_{t-1} = G_t a_{t-1} + G_t R_{t-1} F'_{t-1} Q_{t-1}^{-1} e_t;$$

so, the new state equation is

$$a_t = G_t a_{t-1} + w_t^*, \tag{2.11b}$$

with $w_t^* = G_t R_{t-1} F'_{t-1} Q_{t-1}^{-1} e_t$. The system (2.11) is the innovation form of the DLM. Note that, in this form, the observation errors and the system errors are no longer independent, that is, the dynamics of the states is no longer independent of the observations. The main advantage is that in the innovation form all components of the state vector on which we cannot obtain any information from the observations are automatically removed. It is thus in some sense a minimal model.

When the observations are univariate, the sequence of standardized innovations, defined by $\tilde{e}_t = e_t/\sqrt{Q_t}$, is a Gaussian white noise, i.e., a sequence of independent identically distributed zero-mean normal random variables. This property can be exploited to check model assumptions: if the model is correct, the sequence $\tilde{e}_1, \ldots, \tilde{e}_t$ computed from the data should look like a sample of size t from a standard normal distribution. Many statistical tests, several of them readily available in R, can be carried out on the standardized innovations. Such tests fall into two broad categories: those aimed at checking

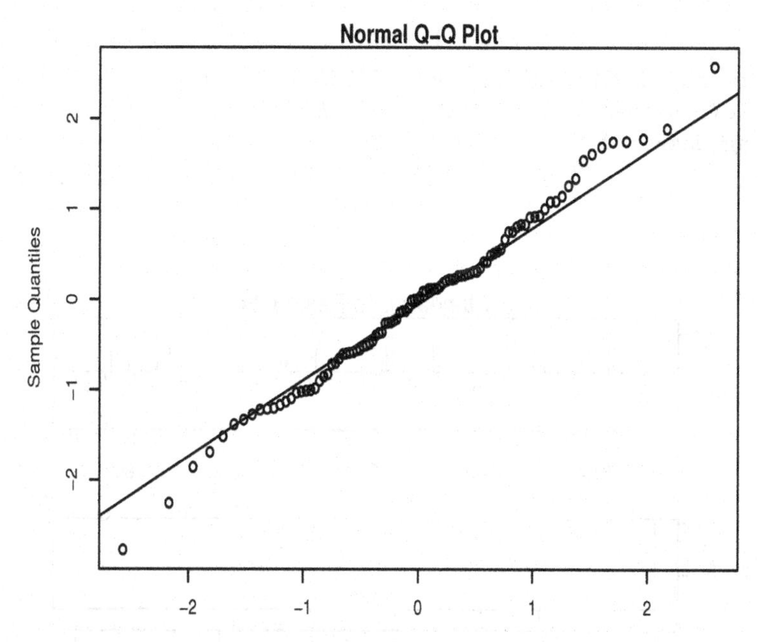

Fig. 2.15. Nile River: QQ-plot of standardized innovations

if the distribution of the $\tilde{e}_t$'s is standard normal, and those aimed at checking whether the $\tilde{e}_t$'s are uncorrelated. We will illustrate the use of some of these tests in Chapter 3. However, most of the time we take a more informal approach to model checking, based on the subjective assessment of selected diagnostic plots. The most useful are, in the opinion of the authors, a QQ-plot and a plot of the empirical autocorrelation function of the standardized innovations. The former can be used to assess normality, while the latter reveals departures from uncorrelatedness. A time series plot of the standardized innovations may prove useful in detecting outliers, change points, and other unexpected patterns.

In R, the standardized innovations can be extracted from an object of class `dlmFiltered` using the function `residuals`. Package dlm also provides a method function for `tsdiag` for objects of class `dlmFiltered`. This function, modeled after `tsdiag.Arima`, exctracts the standardized innovations and plots them, together with their empirical autocorrelation function and the p-values for Ljung-Box test statistics up to a specific lag (the default is 10). For the DLM `modDam` (p.68) used to model Nile River level data, Figure 2.9 shows a QQ-plot of the standardized innovations, while Figure 2.9 displays the plots produced by a call to `tsdiag`. The two figures were obtained with the code below.

R code

```
> qqnorm(residuals(damFilt, sd = FALSE))
> qqline(residuals(damFilt, sd = FALSE))
> tsdiag(damFilt)
```

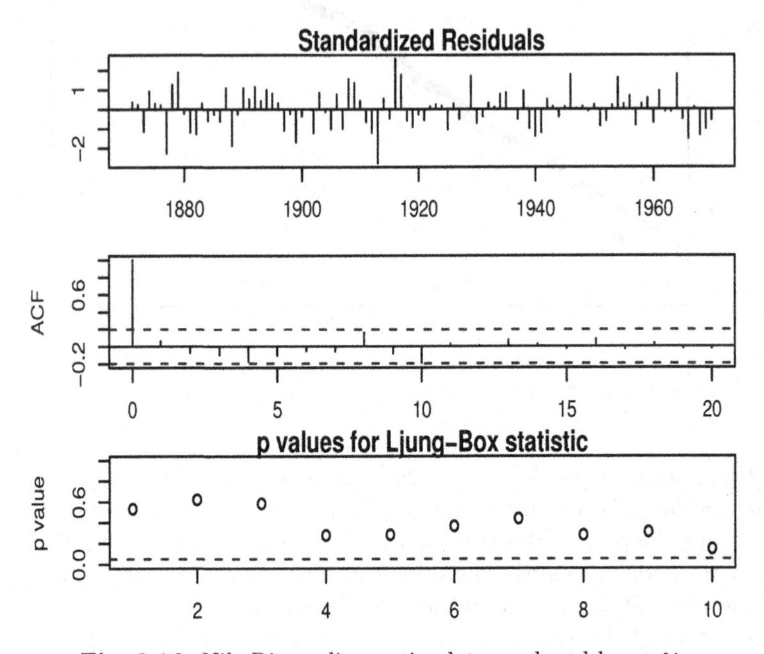

Fig. 2.16. Nile River: diagnostic plots produced by `tsdiag`

For multivariate observations we usually apply the same univariate graphical diagnostic tools component-wise to the innovation sequence. A further step would be to adopt the vector standardization $\tilde{e}_t = B_t e_t$, where B_t is a $p \times p$ matrix such that $B_t Q_t B_t' = I$. This makes the components of $\tilde{e}_t$ independent and identically distributed according to a standard normal distribution. Using this standardization, the sequence $\tilde{e}_{1,1}, \tilde{e}_{1,2} \ldots, \tilde{e}_{1,p}, \ldots, \tilde{e}_{t,p}$ should look like a sample of size tp from a univariate standard normal distribution. This approach, however, is not very popular in applied work and we will not employ it in this book.

2.10 Controllability and observability of time-invariant DLMs

In the engineering literature, DLMs are widely used in *control* problems; indeed, optimal control was one main objective in Kalman's contributions. See, for example, Kalman (1961), Kalman et al. (1963), and Kalman (1968). Here, the interest is in the state of the system, θ_t, which one wants to regulate by means of so-called *control variables* u_t. Problems of this nature are clearly of great relevance in many applied fields, besides engineering; for example, in economics, the monetary authority might want to regulate the *state* of macroeconomic variables, for example the inflation and the unemployment rates, by means of monetary instruments u_t under its control. A DLM including control variables will be referred to as a *controlled* DLM and will be written in the form

$$\begin{aligned} y_t &= F_t\theta_t + v_t, \\ \theta_t &= G_t\theta_{t-1} + H_t u_t + w_t \end{aligned}$$

where u_t is an r-dimensional vector of control variables, i.e., variables whose value can be regulated by the researcher, in order to obtain a desired level of the state θ_t, and H_t is a known $p \times r$ matrix; the usual assumptions are made for the stochastic errors v_t and w_t. Control problems have been first studied for deterministic systems (i.e., systems with no stochastic terms v_t and w_t); in most applications, however, a further difficulty is the presence of stochastic errors in the relationship between θ_t and y_t and in the state evolution. A comprehensive treatment of control problems is beyond the scope of this book; in this section we will only briefly recall some basic notions, limiting our attention to the case of a *time-invariant* controlled DLM, i.e., a controlled DLM where the matrices F_t, G_t, V_t, W_t, and H_t, are constant over time:

$$\begin{aligned} y_t &= F\theta_t + v_t, \\ \theta_t &= G\theta_{t-1} + Hu_t + w_t. \end{aligned}$$

Good references are Anderson and Moore (1979), Harvey (1989), Maybeck (1979), and Jazwinski (1970).

At a basic level, the goal of a control problem is to drive the state of a DLM from the initial value θ_0 to a target value θ^* in a finite time T, setting appropriately the control variables $u_1, \ldots, u_T$. Two issues immediately arise: the first is that the states of a DLM are not observed directly, so, in particular, θ_0 is not known exactly in general; the second is that, even if θ_0 were known, there is no guarantee that one can drive the system to the desired state θ^*. Let us take a closer look at the second problem first, considering the ideal case of a deterministic system equation, i.e., one in which $w_t = 0$ for every t. The system equation reduces in this case to

$$\theta_t = G\theta_{t-1} + Hu_t \tag{2.12}$$

Starting at θ_0 at time zero and applying (2.12) repeatedly, we have

$$\begin{aligned}
\theta_1 &= G\theta_0 + Hu_1, \\
\theta_2 &= G\theta_1 + Hu_2 = G^2\theta_0 + GHu_1 + Hu_2, \\
&\vdots \\
\theta_T &= G^T\theta_0 + \sum_{j=0}^{T-1} G^j H u_{T-j}.
\end{aligned}$$

Therefore, if we want the system to be in state θ^* at time T, we need to solve the equation $\theta_T = \theta^*$ with respect to the control variables $u_1, \ldots, u_T$. More explicitely, let $\mathcal{C}_T$ be the $p \times rT$ matrix defined by

$$\mathcal{C}_T = \left[G^{T-1}H \mid \cdots \mid GH \mid H\right].$$

Stacking the vectors $u_1, \ldots, u_T$, we obtain the following system of linear equations:

$$\mathcal{C}_T \begin{bmatrix} u_1 \\ \vdots \\ u_T \end{bmatrix} = \theta^* - G^T\theta_0. \tag{2.13}$$

If (2.13) has to have a solution for any arbitrary θ^* and θ_0, then $\mathcal{C}_T$ must be of rank p, and vice versa. In other words, a DLM with system equation (2.12) can be driven from an arbitrary initial state θ_0 to another arbitrary state θ^* in a finite time T through an appropriate choice of the control variables $u_1, \ldots, u_T$ if and only if $\mathcal{C}_T$ has full rank p. Moreover, using elementary linear algebra arguments, it can be shown that if $\mathcal{C}_T$ has rank p for some T, then $\mathcal{C}_p$ has rank p. For this reason the matrix $\mathcal{C}_p$ is called the *controllability* matrix of the DLM, and we will denote it $\mathcal{C}$, without subscript. A DLM is said to be *controllable* if its controllability matrix $\mathcal{C}$ has full rank p.

The definition of controllability given above can be transported to a standard time-invariant DLM with system equation

$$\theta_t = G\theta_{t-1} + w_t, \qquad w_t \sim \mathcal{N}(0, W). \tag{2.14}$$

After all, the only difference between (2.12) and (2.14) is that the control term Hu_t in the former is replaced by the system noise w_t in the latter. To carry the analogy one step further, we can write the noise as $w_t = B\eta_t$, where η_t is an r-dimensional random vector having independent standard normal components, and B is a full-rank $p \times r$ matrix. Note that $W = BB'$. When $r < p$, the rank of W is r and the possible values of w_t lie on an r-dimensional linear subspace of $\mathbb{R}^p$ – in this sense we can think of w_t as being essentially r-dimensional, and we can represent it via η_t. We define the *controllability* matrix of a DLM with system equation (2.14) to be

$$\mathcal{C} = \left[G^{p-1}B \mid \cdots \mid GB \mid B\right],$$

and the DLM to be *controllable* if its controllability matrix has full rank p.

Note that the decomposition $W = BB'$ does not identify B uniquely, since for any orthogonal matrix O of order r, the matrix $\tilde{B} = BO$ provides the representation $W = \tilde{B}\tilde{B}'$. However, the particular choice of B does not matter. In fact, one can also avoid computing the decomposition $W = BB'$ altogether. Note that the linear subspace of $\mathbb{R}^p$ spanned by the columns of B is the same as the one spanned by the columns of W. Hence, $\mathcal{C}$ and the matrix

$$\mathcal{C}^W = \left[G^{p-1}W \mid \cdots \mid GW \mid W\right]$$

have the same rank, although $\mathcal{C}^W$ has p^2 columns instead of rp.

As an example, consider an integrated random walk of order 2 (cf. p. 100), which is a DLM whose system equation is defined by the two matrices

$$\begin{aligned} G &= \begin{bmatrix} 1 & 1 \\ 0 & 1 \end{bmatrix}, \\ W &= \begin{bmatrix} 0 & 0 \\ 0 & \sigma_\beta^2 \end{bmatrix}, \end{aligned} \tag{2.15}$$

with $\sigma_\beta^2 > 0$. Here $p = 2$ and

$$\mathcal{C}^W = [GW \mid W] = \begin{bmatrix} 0 & \sigma_\beta^2 & 0 & 0 \\ 0 & \sigma_\beta^2 & 0 & \sigma_\beta^2 \end{bmatrix}.$$

Since $\mathcal{C}^W$ has rank 2, the DLM is controllable.

Clearly for a standard DLM, since the noise (w_t) cannot be set by the observer, the notion of controllability has a different interpretation than in the case of a controlled DLM. A controllable DLM with system equation (2.14) is one for which, by effect of the noise sequence (w_t), the state vector θ_t can reach any point in $\mathbb{R}^p$, no matter what the initial value of the state vector is. In other words, there are no inaccessible regions for the state of the system. In the general theory of Markov chains, this property is called *irreducibility* of the Markov chain (θ_t).

Let us turn now to the first issue raised at the beginning of the discussion, related to the observability of the states. Clearly, if the system or observation noises are nonzero, there is little hope of determining exactly the value of θ_t based solely on the observation y_t, or even on a finite number T of observations $y_{t:t+T-1}$. Therefore we will focus on the idealized situation of a time-invariant DLM in which we can set $V = 0$ and $W = 0$. The observation and system equation reduce to

$$\begin{aligned} y_t &= F\theta_t, \\ \theta_t &= G\theta_{t-1}. \end{aligned} \tag{2.16}$$

Applying repeatedly (2.16) we obtain

$$\begin{aligned} y_t &= F\theta_t, \\ y_{t+1} &= F\theta_{t+1} = FG\theta_t, \\ &\vdots \\ y_{t+T-1} &= FG^{T-1}\theta_t. \end{aligned}$$

Defining the matrix

$$\mathcal{O}_T = \begin{bmatrix} F \\ FG \\ \vdots \\ FG^{T-1} \end{bmatrix}$$

and stacking the observation vectors, the system above can be written as

$$\begin{bmatrix} y_t \\ \vdots \\ y_{t+T-1} \end{bmatrix} = \mathcal{O}_T \theta_t.$$

Therefore, the state θ_t can be determined from the data $y_{t:t+T-1}$ if and only if the previous system of linear equations has a unique solution (in θ_t). This is the case if and only if the $mT \times p$ matrix $\mathcal{O}_T$ has rank p. Also in this case, it can be shown that, if $\mathcal{O}_T$ has rank p for some T, then $\mathcal{O}_p$ has rank p. The matrix $\mathcal{O}_p$ is called the *observability* matrix of the given DLM and it will be denoted by $\mathcal{O}$, without subscript. A time-invariant DLM is said to be observable if its observability matrix $\mathcal{O}$ has full rank p.

Consider again, for example, the 2nd-order integrated random walk whose system equation is defined by (2.15). The observation matrix for this DLM is

$$F = \begin{bmatrix} 1 & 0 \end{bmatrix}.$$

Therefore the observability matrix is

$$\mathcal{O} = \begin{bmatrix} F \\ FG \end{bmatrix} = \begin{bmatrix} 1 & 0 \\ 1 & 1 \end{bmatrix}.$$

This matrix has rank 2, hence the DLM is observable.

In the next section we will link controllability and observability to the asymptotic behavior of the Kalman filter.

2.11 Filter stability

Consider a time-invariant DLM. As shown in Section 2.7, for any t we have that

$$\theta_t | y_{1:t-1} \sim \mathcal{N}_p(a_t, R_t),$$

where a_t and R_t are given by Proposition 2.2. Note that, if the matrices F, G, V and W are known, then the covariance matrix $R_t = \mathrm{Var}(\theta_t | y_{1:t-1})$ does not depend on the data, but only on the initial conditions m_0, C_0, on the system matrices F and G, and on the covariance matrices V and W. In this sense, the asymptotic behavior of R_t is intrinsic to the model, and it can be studied on the basis of the properties of the matrices F, G, V and W. In particular, one can study whether the conditional variance of θ_t given $y_{1:t-1}$ or $y_{1:t}$, tends to become stable for t increasing to infinity, forgetting the initial conditions m_0 and C_0.

Note that, by substituting the expressions of $m_{t-1}, C_{t-1}, f_{t-1}$ in the formulae given by (i) of Proposition 2.2 for a_t and R_t, the latter can be written in the form

$$a_t = (G - A_{t-1}F)a_{t-1} + A_{t-1}y_{t-1},$$

where we denoted by $A_{t-1} = GK_{t-1} = GR_{t-1}F'[V + FR_{t-1}F']^{-1}$ the gain matrix for the state forecast, and

$$R_t = GR_{t-1}G' - A_{t-1}FR_{t-1}G' + W. \tag{2.17}$$

The previous expression, when seen as an equation in the unknown matrix R_t, is called Riccati equation. Note that in (2.17), $A_t = A_t(R_{t-1})$. If there exists a constant positive semi-definite matrix R that satisfies

$$R = GRG' - GRF'[V + FRF']^{-1}FRG' + W \tag{2.18}$$

(which is called the *steady-state (or algebric) Riccati equation*), we say that the DLM has a *steady state* solution.

In the steady state,

$$\theta_t | y_{1:t-1} \sim \mathcal{N}_p(a_t, R),$$

where

$$a_t = (G - AF)a_{t-1} + Ay_t, \tag{2.19}$$

while $R = \mathrm{Var}(\theta_t | y_{1:t-1})$ is time-invariant. In this sense, R represents a bound, intrinsic to the system, to the information one can get in the state forecast. A sufficient condition for R_t to approach R as t increases can be given in terms of the eigenvalues of the matrix $G - AF$: the Kalman filter is asymptotically stable if all the eigenvalues of $G - AF$ are in modulus less than one.

Similarly, the filtering distribution is

$$\theta_t | y_{1:t} \sim \mathcal{N}_p(m_t, C),$$

where $m_t = a_t + K(y_t - Fa_{t-1})$ is recursively updated, while $C = R - KFR$, where $K = RF'[V + FRF']^{-1}$, is time-invariant, giving a bound to the information one can get in filtering.

Note that a solution of (2.18) – i.e., a steady state – does not always exist; and even when a solution is known to exist, it is not simple to show that it is unique nor that it is a positive semi-definite matrix. However, it can be proved (see Anderson and Moore; 1979) that, if the DLM is observable and controllable, then:

1. For any initial conditions m_0, C_0, we have $R_t \rightarrow R$ for $t \rightarrow \infty$, and R satisfies the algebraic Riccati equation (2.18);
2. All the eigenvalues of $G - AF$ are smaller than one in modulus, so the Kalman filter is asymptotically stable.

Problems

2.1. Show that

(i) w_t and $(Y_1, \ldots, Y_{t-1})$ are independent;
(ii) w_t and $(\theta_1, \ldots, \theta_{t-1})$ are independent;
(iii) v_t and $(Y_1, \ldots, Y_{t-1})$ are independent;
(iv) v_t and $(\theta_1, \ldots, \theta_t)$ are independent.

2.2. Show that a DLM satisfies the conditional independence assumptions A.1 and A.2 of state space models.

2.3. Give an alternative proof of Proposition 2.2, exploiting the independence properties of the error sequences (see Problem 2.1) and using the state equation directly:

$$\mathrm{E}(\theta_t | y_{1:t-1}) = \mathrm{E}(G_t \theta_{t-1} + w_t | y_{1:t-1}) = G_t m_{t-1}$$

$$\mathrm{Var}(\theta_t | y_{1:t-1}) = \mathrm{Var}(G_t \theta_{t-1} + w_t | y_{1:t-1}) = G_t C_{t-1} G_t' + W_t.$$

Analogously for (ii).

2.4. Give an alternative proof of Proposition 2.6 exploiting the independence properties of the error sequences (see Problem 2.1) and using the state equation directly:

$$\begin{aligned} a_t(k) &= \mathrm{E}(\theta_{t+k} | y_{1:t}) = \mathrm{E}(G_{t+k} \theta_{t+k-1} + w_{t+k} | y_{1:t}) = G_{t+k} a_{t,k-1}, \\ R_t(k) &= \mathrm{Var}(\theta_{t+k} | y_{1:t}) = \mathrm{Var}(G_{t+k} \theta_{t+k-1} + w_{t+k} | y_{1:t}) \\ &= G_{t+k} R_{t,k-1} G_{t+k}' + W_{t+k} \end{aligned}$$

and analogously, from the observation equation:

$$\begin{aligned} f_t(k) &= \mathrm{E}(Y_{t+k} | y_{1:t}) = \mathrm{E}(F_{t+k} \theta_{t+k} + v_{t+k} | y_{1:t}) = F_{t+k} a_t(k), \\ Q_t(k) &= \mathrm{Var}(Y_{t+k} | y_{1:t}) = \mathrm{Var}(F_{t+k} \theta_{t+k} + v_{t+k} | y_{1:t}) \\ &= F_{t+k} R_t(k) F_{t+k}' + V_{t+k}. \end{aligned}$$

2.5. Plot the following data:

$$(Y_t, t = 1, \ldots, 10) = (17, 16.6, 16.3, 16.1, 17.1, 16.9, 16.8, 17.4, 17.1, 17).$$

Consider the random walk plus noise model

$$\begin{aligned} Y_t &= \mu_t + v_t, & v_t &\sim N(0, 0.25), \\ \mu_t &= \mu_{t-1} + w_t, & w_t &\sim N(0, 25), \end{aligned}$$

with $V = 0.25$, $W = 25$, and $\mu_0 \sim N(17, 1)$.
(a) Compute the filtering states estimates.
(b) Compute the one-step ahead forecasts $f_t, t = 1, \ldots, 10$ and plot them,

together with the observations. Comment briefly.
(c) What is the effect of the observation variance V and of the system variance W on the forecasts? Repeat the exercise with different choices of V and W.
(d) Discuss the choice of the initial distribution.
(e) Compute the smoothing state estimates and plot them.

2.6. This requires maximum likelihood estimates (see Chapter 4). For the data and model of Problem 2.5, compute the maximum likelihood estimates of the variances V and W (since these must be positive, write them as $V = \exp(u_1)$, $W = \exp(u_2)$ and compute the MLE of the parameters (u_1, u_2)). Then repeat Problem 2.5, using the MLE of V and W.

2.7. Let $R_{t,h,k} = \text{Cov}(\theta_{t+h}, \theta_{t+k}|y_{1:t})$ and $Q_{t,h,k} = \text{Cov}(Y_{t+h}, Y_{t+k}|y_{1:t})$ for $h, k > 0$, so that $R_{t,k,k} = R_t(k)$ and $Q_{t,k,k} = Q_t(k)$, according to definition (2.10b) and (2.10d).

(i) Show that $R_{t,h,k}$ can be computed recursively via the formula:

$$R_{t,h,k} = G_{t+h} R_{t,h-1,k}, \qquad h > k.$$

(ii) Show that $Q_{t,h,k}$ is equal to $F_{t+h} R_{t,h,k} F'_{t+k}$.
(iii) Find explicit formulae for $R_{t,h,k}$ and $Q_{t,h,k}$ for the random walk plus noise model.

2.8. Derive the filter formulae for the DLM with intercepts:

$$v_t \sim \mathcal{N}(\delta_t, V_t), \; w_t \sim \mathcal{N}(\lambda_t, W_t).$$

3

Model specification

This chapter is devoted to the description of specific classes of DLMs that, alone or in combinations, are most often used to model univariate or multivariate time series. The additive structure of DLMs makes it easy to think of the observed series as originating from the sum of different components, a long term trend and a seasonal component, for example, possibly subject to an observational error. The basic models introduced in this chapter are in this view elementary building blocks in the hands of the modeler, who has to combine them in an appropriate way to analyze any specific data set. The focus of the chapter is the description of the basic models together with their properties; estimation of unknown parameters will be treated in the following chapter. For completeness we include in Section 3.1 a brief review of some traditional methods used for time series analysis. As we will see, those methods can be cast in a natural way in the DLM framework.

3.1 Classical tools for time series analysis

3.1.1 Empirical methods

There are several forecasting methods that, although originally proposed as empirical tools with no probabilistic interpretation, are quite popular and effective. Exponentially weighted moving average (EWMA) is a traditional method used to forecast a time series. It used to be very popular for forecasting sales and inventory level. Suppose one has observations $y_1, \ldots, y_t$ and one is interested in predicting y_{t+1}. If the series is non-seasonal and shows no systematic trend, a reasonable predictor can be obtained as a linear combination of the past observations in the following form:

$$\hat{y}_{t+1|t} = \lambda \sum_{j=0}^{t-1} (1-\lambda)^j y_{t-j} \qquad (0 \leq \lambda < 1). \tag{3.1}$$

G. Petris et al., *Dynamic Linear Models with R*, Use R, DOI: 10.1007/b135794_3,

For t large, the weights $(1-\lambda)^j\lambda$ sum approximately to one. From an operational point of view, (3.1) implies the following updating of the forecast at time $t-1$ when the new data point y_t becomes available:

$$\hat{y}_{t+1|t} = \lambda y_t + (1-\lambda)\hat{y}_{t|t-1},$$

starting from $\hat{y}_{2|1} = y_1$. This is also known as exponential smoothing or Holt point predictor. It can be rewritten as

$$\hat{y}_{t+1|t} = \hat{y}_{t|t-1} + \lambda(y_t - \hat{y}_{t|t-1}), \tag{3.2}$$

enlightening its "forecast-error correction" structure: the point forecast for y_{t+1} is equal to the previous forecast $\hat{y}_{t|t-1}$, corrected by the *forecast error* $e_t = (y_t - \hat{y}_{t|t-1})$ once we observe y_t. Notice the similarity between (3.2) and the state estimate updating recursion given by the Kalman filter for the local level model (see page 38). At time t, forecasts of future observations are taken to be equal to the forecast of y_{t+1}; in other words, $\hat{y}_{t+k|t} = \hat{y}_{t+1|t}$, $k = 1, 2, \ldots$, and the forecast function is constant.

Extensions of EWMA exist that allow for a linear forecast function. For example, the popular Holt linear predictor extends the simple exponential smoothing to nonseasonal time series that show a local linear trend, decomposing y_t as the sum of a local level and a local growth rate—or slope: $y_t = L_t + B_t$. Point forecasts are obtained by combining exponential smoothing forecasts of the level and the growth rate:

$$\hat{y}_{t+k|t} = \hat{L}_{t+1|t} + \hat{B}_{t+1|t}\, k,$$

where

$$\begin{aligned}
\hat{L}_{t+1|t} &= \lambda y_t + (1-\lambda)\hat{y}_{t|t-1} = \lambda y_t + (1-\lambda)(\hat{L}_{t|t-1} + \hat{B}_{t|t-1}) \\
\hat{B}_{t+1|t} &= \gamma(\hat{L}_{t+1|t} - \hat{L}_{t|t-1}) + (1-\gamma)\hat{B}_{t|t-1}.
\end{aligned}$$

The above recursive formulae can be rewritten as

$$\hat{L}_{t+1|t} = \hat{y}_{t|t-1} + \lambda e_t, \tag{3.3a}$$

$$\hat{B}_{t+1|t} = \hat{B}_{t|t-1} + \lambda\gamma e_t, \tag{3.3b}$$

where $e_t = y_t - \hat{y}_{t|t-1}$ is the forecast error. Further extensions to include a seasonal component are possible, such as the popular Holt and Winters forecasting methods; see, e.g., Hyndman et al. (2008).

Although of some practical utility, the empirical methods described in this subsection are not based on a probabilistic or statistical model for the observed series, which makes it impossible to assess uncertainty (about the forecasts, for example) using standard measures like confidence or probability intervals. As they stand, these methods can be used as exploratory tools. In fact, they can also be derived from an underlying DLM, which can in this case be used to provide a theoretical justification for the method and to derive probability intervals. For an in-depth treatment along these lines, the reader can consult Hyndman et al. (2008). Package forecast (Hyndman; 2008) contains functions that implement the methods described in the book.

3.1.2 ARIMA models

Among the most widely used models for time series analysis is the class of autoregressive moving average (ARMA) models, popularized by Box and Jenkins (see Box et al.; 2008). For nonnegative integers p and q, a univariate stationary ARMA(p,q) model is defined by the relation

$$Y_t = \mu + \sum_{j=1}^{p} \phi_j (Y_{t-j} - \mu) + \sum_{j=1}^{q} \psi_j \epsilon_{t-j} + \epsilon_t, \tag{3.4}$$

where (ϵ_t) is Gaussian white noise with variance σ_ϵ^2 and the parameters $\phi_1, \ldots, \phi_p$ satisfy a stationarity condition. To simplify the notation, we assume in what follows that $\mu = 0$. When the data appear to be nonstationary, one usually takes differences until stationarity is achieved, and then proceeds fitting an ARMA model to the differenced data. A model for a process whose dth difference follows an ARMA(p,q) model is called an autoregressive integrated moving average process of order (p, d, q), or ARIMA(p,d,q). The orders p, q can be chosen informally by looking at empirical autocorrelations and partial autocorrelations, or using a more formal model selection criterion like AIC or BIC. Univariate ARIMA models can be fit in R using the function `arima` (see Venables and Ripley (2002) for details on ARMA analysis in R).

ARMA models for m-dimensional vector observations are formally defined by the same formula (3.4), taking (ϵ_t) to be m-dimensional Gaussian white noise with variance Σ_ϵ and the parameters $\phi_1, \ldots, \phi_p$ and $\psi_1, \ldots, \psi_q$ to be $m \times m$ matrices satisfying appropriate stationarity restrictions. Although in principle as simple to define as in the univariate case, multivariate ARMA models are much harder to deal with than their univariate counterpart, in particular for what concerns identifiability issues and fitting procedures. The interested reader can find a thorough treatment of multivariate ARMA models in Reinsel (1997). Functions for the analysis of multivariate ARMA models in R can be found in the contributed package `dse1` (Gilbert; 2008).

It is possible to represent an ARIMA model, univariate or multivariate, as a DLM, as we will show in Sections 3.2.5 and 3.3.7. This may be useful for the evaluation of the likelihood function. However, in spite of the fact that formally an ARIMA model can be considered a DLM, the philosophy underlying the two classes of models is quite different: on the one hand, ARIMA models provide a black-box approach to data analysis, offering the possibility of forecasting future observations, but with a limited interpretability of the fitted model; on the other hand, the DLM framework encourages the analyst to think in terms of easily interpretable, albeit unobservable, processes—such as trend and seasonal components—that drive the observed time series. Forecasting the individual underlying components of the process, in addition to the observations, is also possible—and useful in many applications—within the DLM framework.

3.2 Univariate DLMs for time series analysis

As we have discussed in Chapter 2, the Kalman filter provides the formulae for estimation and prediction for a completely specified DLM, that is, a DLM where the matrices F_t, G_t and the covariance matrices V_t and W_t are known. In practice, however, specifying a model can be a difficult task. A general approach that works well in practice is to imagine a time series as obtained by combining simple elementary components, each one capturing a different feature of the series, such as trend, seasonality, and dependence on covariates (regression). Each component is represented by an individual DLM, and the different components are then combined in a unique DLM, producing a model for the given time series. To be precise, the components are combined in an additive fashion; series for which a multiplicative decomposition is more appropriate can be modeled using an additive decomposition after a log transformation. We detail below the additive decomposition technique in the univariate case, although the same approach carries over to multivariate time series with obvious modifications.

Consider a univariate series (Y_t). One may assume that the series can be written as the sum of *independent* components

$$Y_t = Y_{1,t} + \cdots + Y_{h,t}, \tag{3.5}$$

where $Y_{i,t}$ might represent a trend component, $Y_{2,t}$ a seasonal component, and so on. The ith component $Y_{i,t}$, $i = 1, \ldots, h$, might be described by a DLM as follows:

$$\begin{aligned} Y_{i,t} &= F_{i,t}\theta_{i,t} + v_{i,t}, & v_{i,t} &\sim \mathcal{N}(0, V_{i,t}), \\ \theta_{i,t} &= G_{i,t}\theta_{i,t-1} + w_{i,t}, & w_{i,t} &\sim \mathcal{N}(0, W_{i,t}), \end{aligned}$$

where the p_i-dimensional state vectors $\theta_{i,t}$ are distinct and the series $(Y_{i,t}, \theta_{i,t})$ and $(Y_{j,t}, \theta_{j,t})$ are mutually independent for all $i \neq j$. The component DLMs are then combined in order to obtain the DLM for (Y_t). By the assumption of independence of the components, it is easy to show that $Y_t = \sum_{i=1}^{h} Y_{i,t}$ is described by the DLM

$$\begin{aligned} Y_t &= F_t\theta_t + v_t, & v_t &\sim \mathcal{N}(0, V_t), \\ \theta_t &= G_t\theta_{t-1} + w_t, & w_t &\sim \mathcal{N}(0, W_t), \end{aligned}$$

where

$$\theta_t = \begin{bmatrix} \theta_{1,t} \\ \vdots \\ \theta_{h,t} \end{bmatrix}, \qquad F_t = [F_{1,t}|\cdots|F_{h,t}],$$

G_t and W_t are the block diagonal matrices

$$G_t = \begin{bmatrix} G_{1,t} & & \\ & \ddots & \\ & & G_{h,t} \end{bmatrix}, \qquad W_t = \begin{bmatrix} W_{1,t} & & \\ & \ddots & \\ & & W_{h,t} \end{bmatrix},$$

and $V_t = \sum_{i=1}^{j} V_{i,t}$. In this chapter, with the exception of Section 3.2.6, we will assume that all the matrices defining a DLM are known; the analysis will be extended to DLMs with unknown parameters in Chapters 4 and 5.

In R, dlm objects are created by the functions of the family `dlmMod*`, or by the general function `dlm`. DLMs having a common dimension of the observation vectors can be added together to produce another DLM. For example, `dlmModPoly(2) + dlmModSeas(4)` adds together a linear trend and a quarterly seasonal component. We start by introducing the families of DLMs that are commonly used as basic building blocks in the representation (3.5). In particular, Sections 3.2.1 and 3.2.2 cover trend and seasonal models, respectively. These two component models can be used to carry over to the DLM setting the classical decomposition "trend + seasonal component + noise" of a time series.

3.2.1 Trend models

Polynomial DLMs are the models most commonly used for describing the trend of a time series, where the trend is viewed as a smooth development of the series over time. At time t, the expected trend of the time series can be thought of as the expected behavior of Y_{t+k} for $k \geq 1$, given the information up to time t; in other words, the expected trend is the forecast function $f_t(k) = \mathrm{E}(Y_{t+k}|y_{1:t})$. A polynomial model of order n is a DLM with constant matrices $F_t = F$ and $G_t = G$, and a forecast function of the form

$$f_t(k) = E(Y_{t+k}|y_{1:t}) = a_{t,0} + a_{t,1}k + \cdots + a_{t,n-1}k^{n-1}, \quad k \geq 0\,, \tag{3.6}$$

where $a_{t,0}, \ldots, a_{t,n-1}$ are linear functions of $m_t = E(\theta_t|y_{1:t})$ and are independent of k. Thus, the forecast function is a polynomial of order $n-1$ in k (note that, as we will see, n is the dimension of the state vector and not the degree of the polynomial). Roughly speaking, any reasonable shape of the forecast function can be described or closely approximated by a polynomial, by choosing n sufficiently large. However, one usually thinks of the trend as a fairly smooth function of time, so that in practice small values of n are used. The most popular polynomial models are the random walk plus noise model, which is a polynomial model of order $n = 1$, and the linear growth model, that is a polynomial model of order $n = 2$.

The local level model

The random walk plus noise, or local level model, is defined by the two equations (2.5) of the previous chapter. As noted there, the behavior of the process

(Y_t) is greatly influenced by the signal-to-noise ratio $r = W/V$, the ratio between the two error variances. Figure 3.1 shows some simulated trajectories of (Y_t) and (μ_t) for different values of the ratio r (see Problem 3.1).

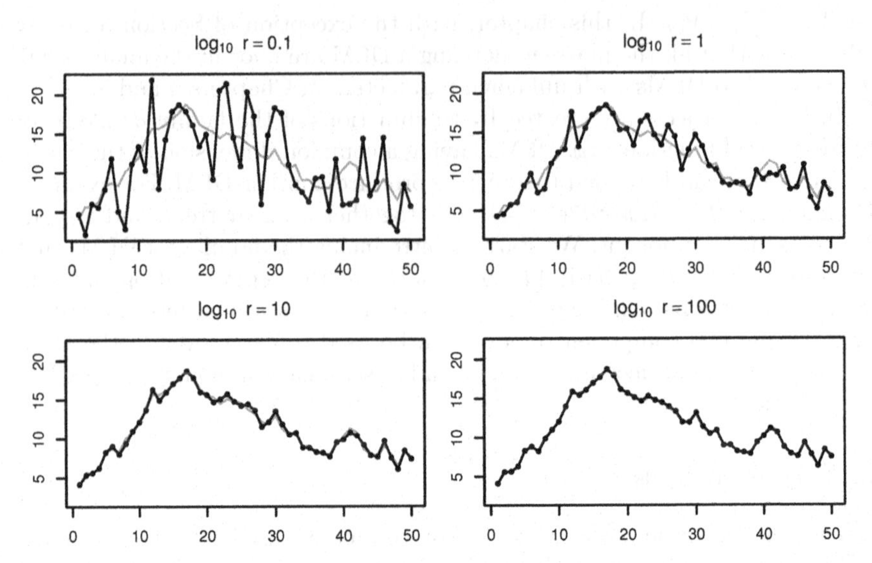

Fig. 3.1. Trajectories of the random walk plus noise, for different values of the signal-to-noise ratio. The trajectory of the state μ_t is shown in gray and is the same in the four plots

The k-steps-ahead predictive distribution for this simple model is

$$Y_{t+k}|y_{1:t} \sim N(m_t, Q_t(k)) \quad , \quad k \geq 1, \tag{3.7}$$

where $Q_t(k) = C_t + \sum_{j=1}^{k} W_{t+j} + V_{t+k} = C_t + kW + V$. We see that the forecast function $f_t(k) = E(Y_{t,k}|y_{1:t}) = m_t$ is constant (as a function of k). For this reason this model is also referred to as the *steady model.* The uncertainty on the future observations is summarized by the variance $Q_t(k) = C_t + kW + V$, and we clearly see that it increases as the time horizon $t+k$ gets further away.

The controllability and observability matrices of the model are

$$\mathcal{C} = \left[W^{1/2}\right],$$
$$\mathcal{O} = F = [1],$$

which are trivially of full rank, as long as $W > 0$. It follows from the results of Section 2.11 that the Kalman filter for this model is asymptotically stable, with R_t, C_t, and the gain matrix K_t converging to limiting values R, C, and K, respectively. It can be shown (see West and Harrison; 1997, Theorem 2.3) that

$$K = \frac{r}{2}\left(\sqrt{1+\frac{4}{r}}-1\right). \tag{3.8}$$

It follows that $C = KV$. This gives an upper bound to the precision attainable in estimating the current value of μ_t. Furthermore, we obtain a limit form of the one-step-ahead forecasts. From (3.7),

$$f_{t+1} = E(Y_{t+1}|y_{1:t}) = m_t = m_{t-1} + K_t(Y_t - m_{t-1}) = m_{t-1} + K_t e_t \,.$$

For large t, $K_t \approx K$ so that, asymptotically, the one-step-ahead forecast is given by

$$f_{t+1} = m_{t-1} + K e_t. \tag{3.9}$$

A forecast function of the kind (3.9) is used in many popular models for time series. It corresponds to Holt point predictor, see equation (3.2).

In can be shown that Holt point predictor is optimal if (Y_t) is an ARIMA$(0,1,1)$ process. In fact, the steady model has connections with the popular ARIMA(0,1,1) model. It can be shown (Problem 3.3) that, if Y_t is a random walk plus noise, then the first differences $Z_t = Y_t - Y_{t-1}$ are stationary, and have the same autocorrelation function as an MA(1) model. Furthermore, being $e_t = Y_t - m_{t-1}$ and $m_t = m_{t-1} + K_t e_t$, we have

$$\begin{aligned} Y_t - Y_{t-1} &= e_t + m_{t-1} - e_{t-1} - m_{t-2} \\ &= e_t + m_{t-1} - e_{t-1} - m_{t-1} + K_{t-1}e_{t-1} \\ &= e_t - (1 + K_{t-1})e_{t-1} \,. \end{aligned}$$

If t is large, so that $K_{t-1} \approx K$,

$$Y_t - Y_{t-1} \approx e_t - (1 - K)e_{t-1}.$$

Since the forecast errors are a white noise sequence (see Chapter 2, page 73), (Y_t) is asymptotically an ARIMA(0,1,1) process.

Example — Annual precipitation at Lake Superior

Figure 3.2 shows annual precipitation in inches at Lake Superior, from 1900 to 1986[1]. The series shows random fluctuations about a changing level over time, with no remarkable trend behavior; thus, a random walk plus noise model could be tentatively entertained. We suppose here that the evolution variance W and the observational variance V are known, and we assume that W is much smaller (0.121) than V (9.465) (so $r = 0.0128$). In R a local level model can be set up using the function `dlmModPoly` with first argument `order=1`.

Figure 3.3(a) shows the filtering estimates m_t of the underlying level of the series and Figure 3.3(b) shows the square root of the variances C_t. Recall that

[1] Source: Hyndman (n.d.).

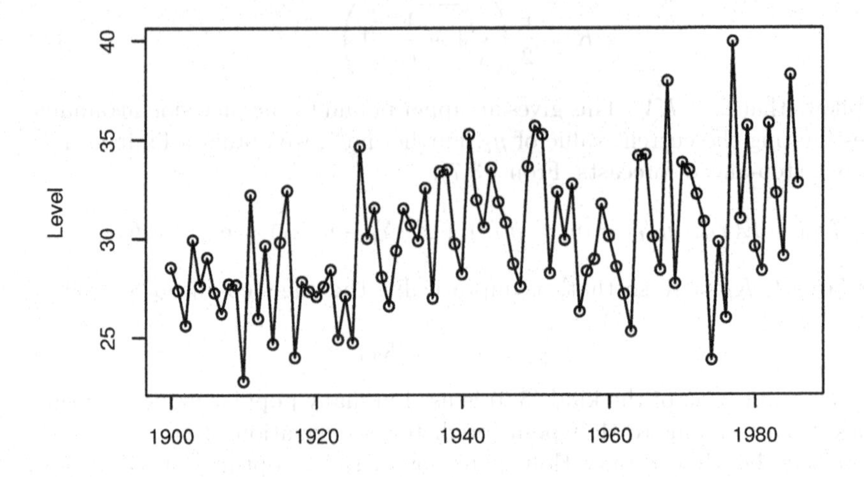

Fig. 3.2. Annual precipitation at Lake Superior

for the local level model C_t has a limiting value as t approaches infinity. The smoothed states s_t and square root of the variances S_t are plotted in Figures 3.3(c) and 3.3(d). The U-shaped behavior of the sequence of variances S_t reflects the intuitive fact that the states around the middle of the time interval spanned by the data are those that can be estimated more accurately.

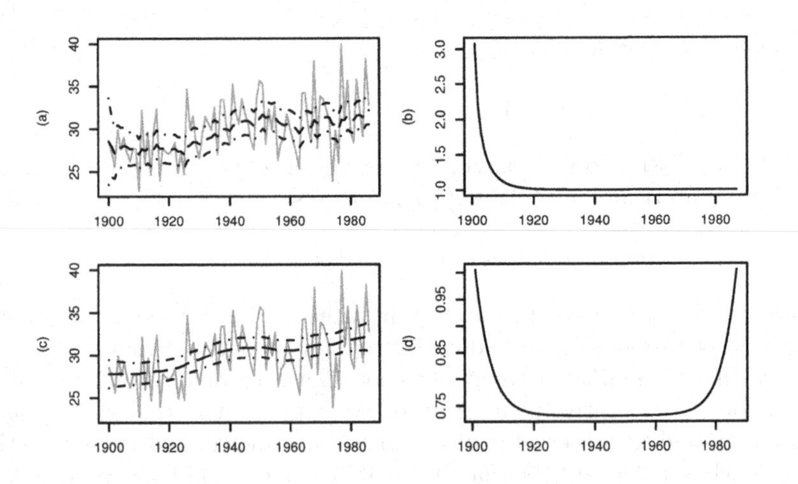

Fig. 3.3. (a): Filtered state estimates m_t with 90% confidence intervals; (b): Square root of filtering variances C_t; (c): Smoothed state estimates s_t with 90% confidence intervals; (d): Square root of smoothing variances S_t

The one-step-ahead forecasts, for the local level model, are $f_t = m_{t-1}$. The standardized one-step-ahead forecast errors, or standardized innovations, can be computed in R with a call to the *residuals* function, which has a method for dlmFiltered objects. The residuals can be inspected graphically (Figure 3.4(a)) to check for unusually large values or unexpected patterns—recall that the standardized innovation process has the distribution of a Gaussian white noise. Two additional very useful graphical tools to detect departures from the model assumptions are the plot of the empirical autocorrelation function (ACF) of the standardized innovations (Figure 3.4(b)) and their normal QQ-plot (Figure 3.4(c)). These can be drawn using the standard R functions *acf* and *qqnorm*. By looking at the plots, there does not seem to be any meaningful departure from the model assumptions.

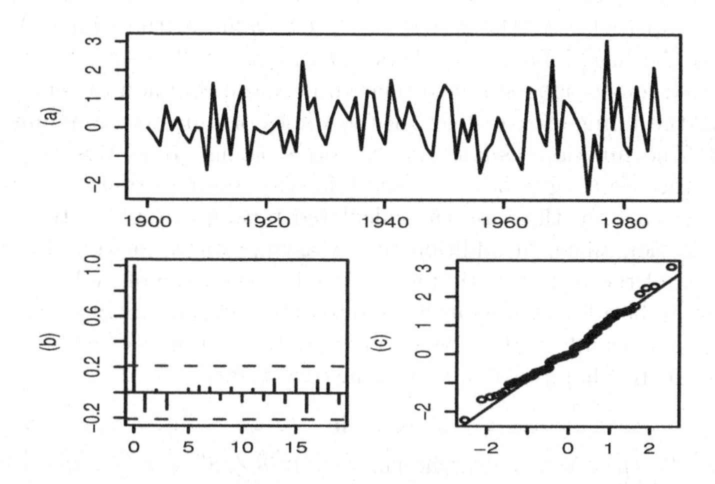

Fig. 3.4. (a): Standardized one-step-ahead forecast errors; (b): ACF of one-step-ahead forecast errors (c): Normal probability plot of standardized one-step-ahead forecast errors

Formal statistical tests may also be employed to assess model assumptions via the implied properties of the innovations. For example, the Shapiro–Wilk test can be used to test the standardized innovations for normality. It is available in R as *shapiro.test*. For the standardized innovations from the Lake Superior precipitation data, the p-value is 0.403, so the null hypothesis of normally distributed standardized innovations cannot be rejected. The Shapiro–Wilk normality test is commonly preferred to the Kolmogorov–Smirnov test, which is also available in R as *ks.test*, as being more powerful against a broad range of alternatives. R functions that perform other normality tests are available in contributed packages fBasics (Wuertz; 2008) and nortest (Gross; n.d.). For a thorough treatment of normality tests the reader is referred to D'Agostino and Stephens (1986). To test for lack of serial correlation one can use the Ljung and Box test (Ljung and Box; 1978), which is based on the first k sample autocorrelations, for a prespecified value of k. The test statistic is

$$Q(k) = n(n+2) \sum_{j=1}^{k} \hat{\rho}^2(j)/(n-j),$$

where n is the sample size and $\hat{\rho}(j)$ is the sample autocorrelation at lag j, defined by

$$\hat{\rho}(j) = \sum_{t=1}^{n-j} (\tilde{e}_t - \bar{\tilde{e}})(\tilde{e}_{t+j} - \bar{\tilde{e}}) \Big/ \sum_{t=1}^{n} (\tilde{e}_t - \bar{\tilde{e}})^2, \qquad j = 1, 2, \ldots.$$

What the Ljung–Box test effectively does is test for the absence of serial correlation up to lag k. Using $k = 20$, the p-value of the Ljung–Box test for the standardized innovations of the example is 0.813, confirming that the standardized innovations are uncorrelated. It is also common to compute the p-value of the Ljung–Box test for all the values of k up to a maximum, say 10 or 20. The function `tsdiag`, among other things, does this calculation and plots the resulting p-values versus k for the residuals of a fitted ARMA model. Of course, in this case the calculated p-values should only be taken as an indication since, in addition to the asymptotic approximation of the distribution of the test statistic for any fixed k, the issue of multiple testing would have to be addressed if one wanted to draw a conclusion in a formal way. The display below illlustrates how to obtain in R the standardized innovations and perform the Shapiro–Wilk and Ljung–Box tests.

R code

```
  > loc <- "http://www.robjhyndman.com/TSDL/roberts/plsuper.dat"
2 > lakeSup <- ts(read.table(loc, skip = 3,
  +                              colClasses = "numeric"),
4 +                 start = 1900)
  > modLSup <- dlmModPoly(1, dV = 9.465, dW = 0.121)
6 > lSupFilt <- dlmFilter(lakeSup, modLSup)
  > res <- residuals(lSupFilt, sd = FALSE)
8 > shapiro.test(res)

10          Shapiro-Wilk normality test


12 data:  res
   W = 0.9848, p-value = 0.4033

14
   > Box.test(res, lag = 20, type = "Ljung")

16
            Box-Ljung test

18
```

```
data:  res
X-squared = 14.3379, df = 20, p-value = 0.813

> sapply(1 : 20, function(i)
+       Box.test(res, lag = i, type = "Ljung-Box")$p.value)
 [1] 0.1552078 0.3565713 0.2980295 0.4508888 0.5829209 0.6718375
 [7] 0.7590090 0.8148123 0.8682010 0.8838797 0.9215812 0.9367660
[13] 0.9143456 0.9185912 0.8924318 0.7983241 0.7855680 0.7971489
[19] 0.8010898 0.8129607
```

Exponential smoothing one-step-ahead forecasts can be obtained using the function *HoltWinters*. The results for the annual precipitations in Lake Superior are plotted in Figure 3.2.1 (here the smoothing parameter is estimated as $\lambda = 0.09721$). We see that the steady model has, for large t, essentially the same forecast function as the simple exponential smoothing.

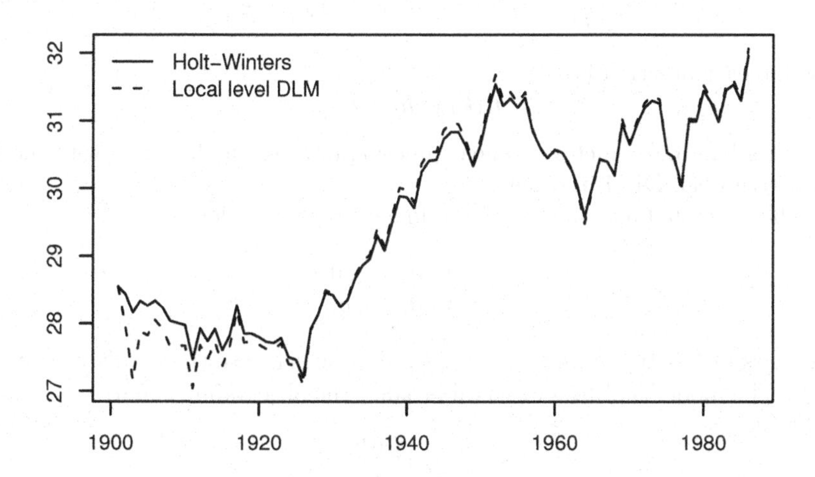

Fig. 3.5. One-step ahead forecasts

R code

```
> HWout <- HoltWinters(lakeSup, gamma = 0, beta = 0)
> plot(dropFirst(lSupFilt$f), lty = "dashed",
+      xlab = "", ylab = "")
> lines(HWout$fitted[, "level"])
> leg <- c("Holt-Winters", "Local level DLM")
> legend("topleft", legend = leg, bty = "n",
+        lty = c("solid", "dashed"))
```

Linear growth model

The linear growth, or local linear trend model, is defined by (2.6). The state vector is $\theta_t = (\mu_t, \beta_t)'$, where μ_t is usually interpreted as the local level and β_t as the local growth rate. The model assumes that the current level μ_t changes linearly through time and that the growth rate may also evolve. It is thus more flexible than a global linear trend model. A good exercise, also for this model, is to simulate trajectories of $(Y_t, t = 1, \ldots, T)$, for different values of V and W (see Problem 3.1).

Denoting $m_{t-1} = (\hat{\mu}_{t-1}, \hat{\beta}_{t-1})'$, the one-step-ahead point forecasts and the filtering state estimates are given by

$$a_t = G m_{t-1} = \begin{bmatrix} \hat{\mu}_{t-1} + \hat{\beta}_{t-1} \\ \hat{\beta}_{t-1} \end{bmatrix} \tag{3.10a}$$

$$f_t = F_t\, a_t = \hat{\mu}_{t-1} + \hat{\beta}_{t-1}, \tag{3.10b}$$

$$m_t = \begin{bmatrix} \hat{\mu}_t \\ \hat{\beta}_t \end{bmatrix} = a_t + K_t e_t = \begin{bmatrix} \hat{\mu}_{t-1} + \hat{\beta}_{t-1} + k_{t1} e_t \\ \hat{\beta}_{t-1} + k_{t2} e_t \end{bmatrix}. \tag{3.10c}$$

The forecast function is

$$f_t(k) = \hat{\mu}_t + k\hat{\beta}_t,$$

(see Problem 3.6) which is a linear function of k, so the linear growth model is a polynomial DLM of order 2.

The controllability matrix of the linear growth model is

$$\mathcal{C} = \begin{bmatrix} \sigma_\mu & \sigma_\beta & \sigma_\mu & 0 \\ 0 & \sigma_\beta & 0 & \sigma_\beta \end{bmatrix}.$$

The rank of $\mathcal{C}$ is two if and only if $\sigma_\beta > 0$, in which case the model is controllable. The model is always observable, since the observability matrix is always full-rank:

$$\mathcal{O} = \begin{bmatrix} 1 & 0 \\ 1 & 1 \end{bmatrix}.$$

Therefore, assuming that $\sigma_\beta > 0$, there are limiting values for R_t, C_t, and K_t. In particular, the Kalman gain K_t converges to a constant matrix $K = [k_1\, k_2]$ (see West and Harrison; 1997, Theorem 7.2). Therefore, the asymptotic updating formulae for the estimated state vector are given by

$$\begin{aligned} \hat{\mu}_t &= \hat{\mu}_{t-1} + \hat{\beta}_{t-1} + k_1\, e_t \\ \hat{\beta}_t &= \hat{\beta}_{t-1} + k_2\, e_t. \end{aligned} \tag{3.11}$$

Several popular point predictors' methods use expressions of the form (3.11), such as the Holt linear predictor (compare with (3.3)) and the Box and Jenkins' ARIMA(0,2,2) predictor (see West and Harrison; 1997, p. 221 for a discussion). In fact, the linear growth model is related to the ARIMA(0,2,2) process. It can be shown (Problem 3.5) that the second differences of (Y_t) are stationary and have the same autocorrelation function as an MA(2) model. Furthermore, we can write the second differences $z_t = Y_t - 2Y_{t-1} + Y_{t-2}$ as

$$z_t = e_t + (-2 + k_{1,t-1} + k_{2,t-1})e_{t-1} + (1 - k_{1,t-2})e_{t-2} \tag{3.12}$$

(see Problem 3.7). For large t, $k_{1,t} \approx k_1$ and $k_{2,t} \approx k_2$, so that the above expression reduces to

$$Y_t - 2Y_{t-1} + Y_{t-2} \approx e_t + \psi_1 e_{t-1} + \psi_2 e_{t-2}$$

where $\psi_1 = -2 + k_1 + k_2$ and $\psi_2 = 1 - k_1$, which is a MA(2) model. Thus, asymptotically, the series (Y_t) is an ARIMA(0,2,2) process.

Example — Spain annual investment

Consider Spain annual investments from 1960 to 2000[2], plotted in Figure 3.6. The time series shows a roughly linear increase, or decrease, in the level, with a slope changing every few years. In the near future it would not be unreasonable to predict the level of the series by linear extrapolation, i.e., using a linear forecast function. A linear growth model could be therefore

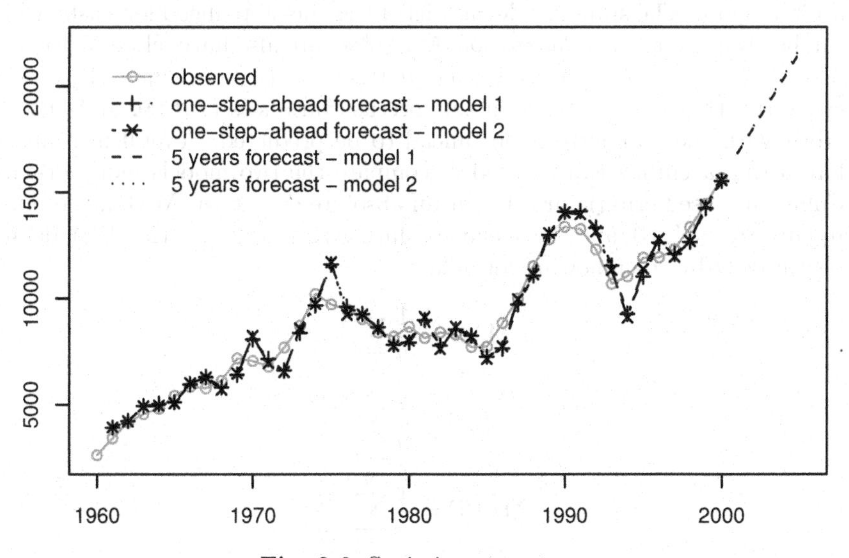

Fig. 3.6. Spain investments

appropriate for these data. We assume that the variances are known (they were actually estimated) and are as follows:

[2] Source: `http://www.fgn.unisg.ch/eumacro/macrodata`.

$$W = \text{diag}(102236,\ 321803), \qquad V = 10.$$

The function *dlmModPoly* with argument *order=2* (which is the default) can be used to set up the model in R, see display below. Visual inspection of a QQ-plot and ACF of the standardized innovations (not shown) do not raise any specific concern about the appropriateness of the model. An alternative model (an integrated random walk model, in fact, see page 100) that describes the data almost equally well, with one less parameter, is the linear growth model with the same V and

$$W = \text{diag}(0, 515939).$$

R code

```
> mod1 <- dlmModPoly(dV = 10, dW = c(102236, 321803))
> mod1Filt <- dlmFilter(invSpain, mod1)
> fut1 <- dlmForecast(mod1Filt, n = 5)
> mod2 <- dlmModPoly(dV = 10, dW = c(0, 515939))
> mod2Filt <- dlmFilter(invSpain, mod2)
> fut2 <- dlmForecast(mod2Filt, n = 5)
```

Figure 3.6 shows, together with the data, one-step-ahead forecasts and five years forecasts for the two models under consideration. It is clear that the forecasts, both in sample and out of sample, produced by the two models are very close. The standard deviations of the one-step-ahead forecasts, which can be obtained as *residuals(mod1Filt)$sd*, are also fairly close, 711 for the first model versus 718 for the second at time $t = 41$ (year 2000). The reader can verify that the difference in the forecast variances (*fut1$Q* and *fut2$Q*) grows with the number of steps ahead to be predicted. Several measures of forecasting accuracy can be used to compare the two models more formally. Commonly used criteria are the mean absolute deviation (MAD), the mean square error (MSE), and the mean absolute percentage error (MAPE), defined respectively by the following formulae:

$$\text{MAD} = \frac{1}{n}\sum_{t=1}^{n} |e_t|,$$

$$\text{MSE} = \frac{1}{n}\sum_{t=1}^{n} e_t^2,$$

$$\text{MAPE} = \frac{1}{n}\sum_{t=1}^{n} \frac{|e_t|}{y_t}.$$

For the two models under consideration and the Spain investment data, none of the two stands out as a clear winner, as the following display shows.

R code

```
> mean(abs(mod1Filt$f - invSpain))
[1] 623.5682
```

```
> mean(abs(mod2Filt$f - invSpain))
[1] 610.2621
> mean((mod1Filt$f - invSpain)^2)
[1] 655480.6
> mean((mod2Filt$f - invSpain)^2)
[1] 665296.7
> mean(abs(mod1Filt$f - invSpain) / invSpain)
[1] 0.08894788
> mean(abs(mod2Filt$f - invSpain) / invSpain)
[1] 0.08810524
```

An additional statistic that can be used to assess the forecasting performance of a specific model is Theil's U (Theil; 1966). This compares the MSE of the model with the MSE of the trivial "no-change" model that predicts the next observation to be the same as the current one. Formally, the definition is the following:

$$U = \sqrt{\frac{\sum_{t=2}^{n}(y_t - f_t)^2}{\sum_{t=2}^{n}(y_t - y_{t-1})^2}}.$$

A value of U less than one means that the entertained model produces better forecasts, on average, than the no-change model. For the Spain investment data the results are reported below.

R code

```
> sqrt(sum((mod1Filt$f - invSpain)[-(1:5)]^2) /
+      sum(diff(invSpain[-(1:4)])^2))
[1] 0.9245
> sqrt(sum((mod2Filt$f - invSpain)[-(1:5)]^2) /
+      sum(diff(invSpain[-(1:4)])^2))
[1] 0.9346
```

We left out the first five observations to give the Kalman filter the time to adapt to the data—remember that the prior variance C_0 is very large. In this example, both models perform better than the no-change model, with about a 7% reduction in the square root of the MSE.

nth order polynomial model

The local level and the linear growth models are special cases of the nth order polynomial model. The general nth order polynomial model has an n-dimensional state space and is described by the matrices

$$F = (1, 0, \ldots, 0) \tag{3.13a}$$

$$G = \begin{bmatrix} 1 & 1 & 0 & & \ldots & 0 \\ 0 & 1 & 1 & 0 & \ldots & 0 \\ \vdots & & & & \ddots & \vdots \\ 0 & & \ldots & 0 & 1 & 1 \\ 0 & & \ldots & & 0 & 1 \end{bmatrix} \tag{3.13b}$$

$$W = \mathrm{diag}(W_1, \ldots, W_n). \tag{3.13c}$$

In terms of its components, the model can be written in the form

$$\begin{cases} Y_t = \theta_{t,1} + v_t \\ \theta_{t,j} = \theta_{t-1,j} + \theta_{t-1,j+1} + w_{t,j} \quad j = 1, \ldots, n-1 \\ \theta_{t,n} = \theta_{t-1,n} + w_{t,n}. \end{cases} \tag{3.14}$$

So, for $j = 2, \ldots, n$, the jth component of the state vector at any time t represents, up to a random error, the increment of the $(j-1)$st component during the next time interval, while the first component represents the mean response, or the level of the series. The forecast function, $f_t(k)$, is a polynomial of degree $n-1$ in k (Problem 3.6).

The special case that is obtained by setting $W_1 = \cdots = W_{n-1} = 0$ is called *integrated random walk model.* The mean response function satisfies for this model the relation $\Delta^n \mu_t = \epsilon_t$ for some white noise sequence (ϵ_t). The form of the forecast function is again polynomial. With respect to the nth order polynomial model, the integrated random walk model has $n-1$ fewer parameters, which may improve the precision attainable in estimating unknown parameters. On the other hand, having only one degree of freedom in the system noise, it may be slower in adapting to random shocks to the state vector, which may reflect in a lower accuracy in the forecasts.

3.2.2 Seasonal factor models

In this section and the next we present two ways of modeling a time series that shows a cyclical behavior, or "seasonality": the seasonal factor model, covered below, and the Fourier-form seasonal model, treated in the following section.

Suppose that we have quarterly data $(Y_t, t = 1, 2, \ldots)$, for examples on the sales of a store, which show an annual cyclic behavior. Assume for simplicity that the series has zero mean: a non-zero mean, or a trend component, can be modeled separately, so for the moment we consider the series as purely seasonal. We might describe the series by introducing seasonal deviations from the mean, expressed by different coefficients α_i for the different quarters, $i = 1, \ldots, 4$. So, if Y_{t-1} refers to the first quarter of the year and Y_t to the second quarter, we assume

$$\begin{aligned} Y_{t-1} &= \alpha_1 + v_{t-1} \\ Y_t &= \alpha_2 + v_t \end{aligned} \tag{3.15}$$

and so on. This model can be written as a DLM as follows. Let $\theta_{t-1} = (\alpha_1, \alpha_4, \alpha_3, \alpha_2)'$ and $F_t = F = (1, 0, 0, 0)$. Then the observation equation of the DLM is given by

$$Y_{t-1} = F\theta_{t-1} + v_{t-1},$$

which corresponds to (3.15). The state equation must "rotate" the components of θ_{t-1} into the vector $\theta_t = (\alpha_2, \alpha_1, \alpha_4, \alpha_3)'$, so that $Y_t = F\theta_t + v_t = \alpha_2 + v_t$. The required permutation of the state vector can be obtained by a permutation matrix G so defined:

$$G = \begin{bmatrix} 0 & 0 & 0 & 1 \\ 1 & 0 & 0 & 0 \\ 0 & 1 & 0 & 0 \\ 0 & 0 & 1 & 0 \end{bmatrix}.$$

Then the state equation can be written as

$$\theta_t = G\theta_{t-1} + w_t = (\alpha_2, \alpha_1, \alpha_4, \alpha_3)' + w_t.$$

In the static seasonal model, w_t is degenerate on a vector of zeros (i.e., $W_t = 0$) More generally, the seasonal effects might change in time, so that W_t is nonzero and has to be carefully specified.

In general, a seasonal time series with period s can be modeled through an s-dimensional state vector θ_t of seasonal deviations, by specifying a DLM with $F = (1, 0, \ldots, 0)$ and G given by a s by s permutation matrix. Identifiability constraints have to be imposed on the seasonal factors $\alpha_1, \ldots, \alpha_s$. A common choice is to impose that they sum to zero, $\sum_{j=1}^{s} \alpha_j = 0$. The linear constraint on the s seasonal factors implies that there are effectively only $s-1$ free seasonal factors, and this suggests an alternative, more parsimonious representation that uses an $(s-1)$-dimensional state vector. In the previous example of quarterly data, one can consider $\theta_{t-1} = (\alpha_1, \alpha_4, \alpha_3)'$ and $\theta_t = (\alpha_2, \alpha_1, \alpha_4)'$, with $F = (1, 0, 0)$. To go from θ_{t-1} to θ_t, assuming for the moment a static model without system evolution errors and using the constraint $\sum_{i=1}^{4} \alpha_i = 0$, one has to apply the linear transformation given by the matrix

$$G = \begin{bmatrix} -1 & -1 & -1 \\ 1 & 0 & 0 \\ 0 & 1 & 0 \end{bmatrix}.$$

In general, for a seasonal model with period s, one can consider an $(s-1)$-dimensional state space, with $F = (1, 0, \ldots, 0)$ and

$$G = \begin{bmatrix} -1 & -1 & \dots & -1 & -1 \\ 1 & 0 & & 0 & 0 \\ 0 & 1 & & 0 & 0 \\ & & \ddots & & \\ 0 & 0 & & 1 & 0 \end{bmatrix}.$$

A dynamic variation in the seasonal components may be introduced via a system evolution error with variance $W = \text{diag}(\sigma_w^2, 0, \dots, 0)$.

A seasonal factor DLM can be specified in R using the function *dlmModSeas*. For example, a seasonal factor model for quarterly data, with $\sigma_w^2 = 4.2$ and observation variance $V = 3.5$ can be specified as follows.

R code

```
mod <- dlmModSeas(frequency = 4, dV = 3.5, dW = c(4.2, 0, 0))
```

3.2.3 Fourier form seasonal models

Any discrete-time periodic function having period s is characterized by the values it takes at time $t = 1, 2, \dots, s$: if g_t is such a function, with $g_t = \alpha_t$, $t = 1, \dots, s$, then after time s the values of g_t simply repeat themselves, and we have $g_{s+1} = \alpha_1$, $g_{s+2} = \alpha_2$ and so forth. Therefore, we can associate the periodic function g_t to the s-dimensional vector $\alpha = (\alpha_1, \dots, \alpha_s)'$. We can think of α as a linear combination of basis vectors:

$$\alpha = \sum_{j=1}^{s} \alpha_j u_j,$$

where u_j is the s-dimensional vector having jth component equal one and all the remaining components equal zero. The set $\{u_1, \dots, u_s\}$ is customarily referred to as the canonical basis of $\mathbb{R}^s$; clearly, any vector in $\mathbb{R}^s$ has a unique representation as a linear combination of basis vectors. For our purposes, however, this representation is not very useful, since it does not allow us to distinguish between smooth, less smooth, and not-so-smooth functions. Fortunately, there is an alternative basis of $\mathbb{R}^s$ that allows this finer distinction to be made. To begin with, suppose that s is even. In practice, this is the most commonly encountered case—think of quarterly or montly data. Define the *Fourier frequencies*

$$\omega_j = \frac{2\pi j}{s}, \qquad j = 0, 1, \dots, \frac{s}{2}$$

and consider the following s dimensional vectors:

$$
\begin{aligned}
e_0 &= (1, 1, \ldots, 1)' \\
c_1 &= (\cos \omega_1, \cos 2\omega_1, \ldots, \cos s\omega_1)' \\
s_1 &= (\sin \omega_1, \sin 2\omega_1, \ldots, \sin s\omega_1)' \\
&\vdots \\
c_j &= (\cos \omega_j, \cos 2\omega_j, \ldots, \cos s\omega_j)' \\
s_j &= (\sin \omega_j, \sin 2\omega_j, \ldots, \sin s\omega_j)' \\
&\vdots \\
c_{s/2} &= (\cos \omega_{s/2}, \cos 2\omega_{s/2}, \ldots, \cos s\omega_{s/2})'.
\end{aligned} \tag{3.16}
$$

Note that $c_{s/2} = (-1, 1, -1, \ldots, -1, 1)'$ and we do not consider $s_{s/2}$ since this would be a vector of zeros. The number of vectors defined in (3.16) is $1 + 2(\frac{s}{2} - 1) + 1 = s$ and, by using standard trigonometric identities, one can show that these vectors are orthogonal. In turn, this implies that every vector in $\mathbb{R}^s$ has a unique representation as a linear combination of $e_0, \ldots, c_{s/2}$, which we will write as

$$
\alpha = a_0 e_0 + \sum_{j=1}^{s/2-1} \left(a_j c_j + b_j s_j \right) + a_{s/2} c_{s/2}. \tag{3.17}
$$

There are two distinct advantages in using this basis of $\mathbb{R}^s$ over the canonical one. The first is that the extension of the basis vectors to periodic functions can be obtained naturally in terms of the trigonometric functions involved in their definition. Consider for example s_j: for any $1 \le t \le s$, its tth component is $s_j(t) = \sin(2\pi t j/s)$. To extend s_j to a periodic function we must define $s_j(t + ks) = s_j(t)$. But, since

$$
\sin \frac{2\pi(t + ks)j}{s} = \sin\left(\frac{2\pi t j}{s} + 2\pi k j \right) = \sin \frac{2\pi t j}{s},
$$

the extension of $s_j(t)$ to any integer t is achieved simply by plugging the argument t into the trigonometric expression defining s_j. The second and more substantial advantage of using this new basis composed by trigonometric functions is that the basis vectors now go from smoothest to roughest: at the two extremes we have the constant vector e_0, which corresponds to a constant periodic function, and the maximally oscillating $c_{s/2}$, which corresponds to a periodic function bouncing back and forth between -1 and 1. To understand what happens in between these two extremes, let us focus on the c_j's; similar considerations can be made regarding the s_j's. For a fixed j, as t goes from 1 to s, $2\pi t j/s$ increases monotonically from $2\pi j/s$ to $2\pi j$. Furthermore, the trigonometric functions sine and cosine are periodic, with period 2π. So, considering $j = 1$, the values $2\pi t/s$, $t = 1, \ldots, s$, consist of s equally spaced points in the interval $(0, 2\pi]$. For $j = 2$, $2\pi t j/s$ runs up to 4π, spanning two times the basic period 2π of the cosine function and therefore being twice as wiggling.

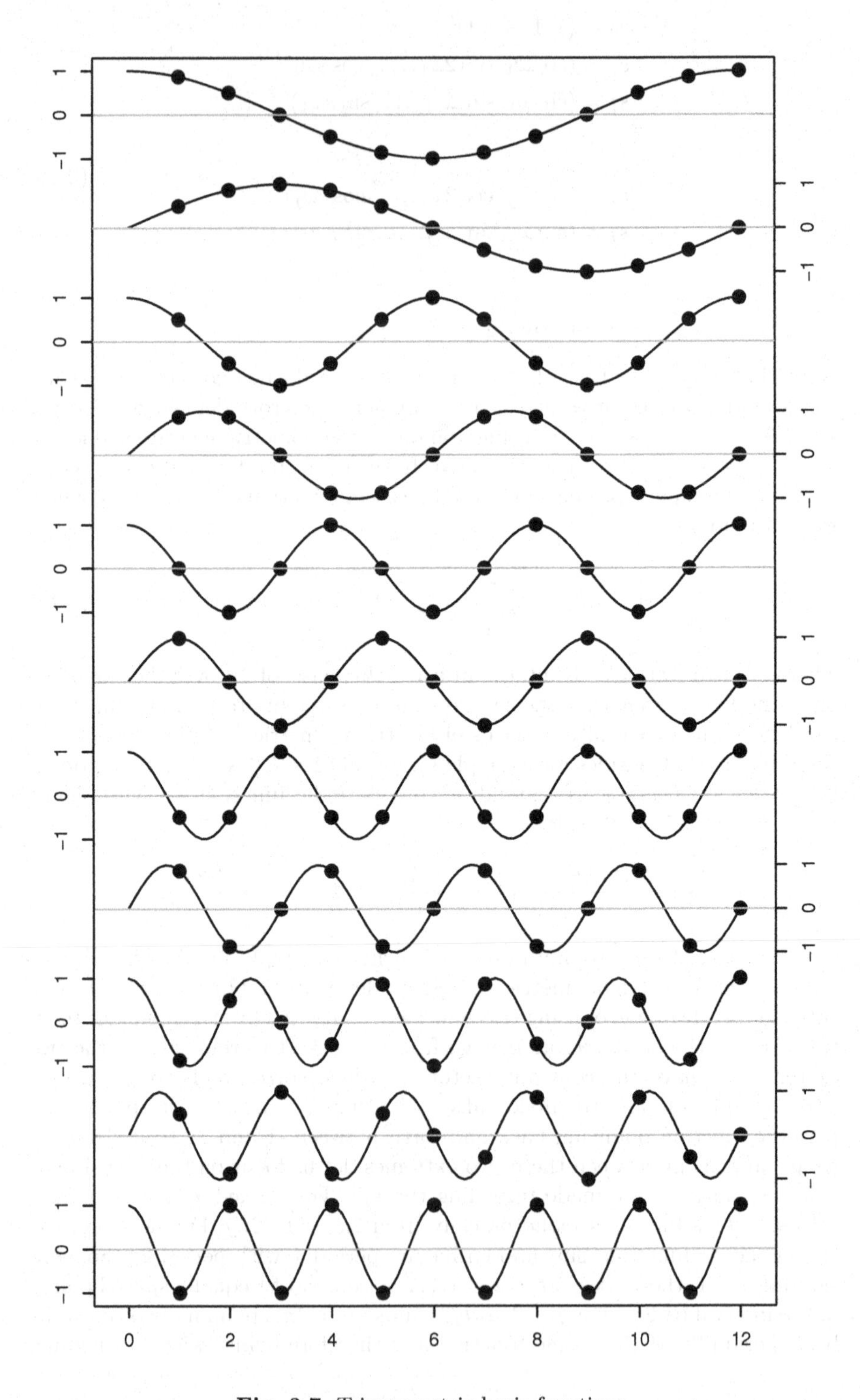

Fig. 3.7. Trigonometric basis functions

In general, for any j, the argument $2\pi tj/s$ spans the interval $(0, 2\pi j]$ and the corresponding cosine function c_j goes through a complete period exactly j times as t goes from 1 to s. Therefore, a higher value of j corresponds to a rougher c_j, in the sense that the function displays more frequent oscillations. To illustrate this feature for $s = 12$, Figure 3.7 shows, from top to bottom, the graphs of $c_1, s_1, \ldots, c_5, s_5, c_6$.

It is convenient to group together in the representation (3.17) the terms involving c_j and s_j, which are functions oscillating with the same frequency. For $j = 1, \ldots, s/2$, define the jth harmonic of g_t by

$$S_j(t) = a_j \cos(t\omega_j) + b_j \sin(t\omega_j), \tag{3.18}$$

where we take $b_{s/2} = 0$. Moreover, we will assume that $a_0 = 0$, since in the DLM framework—as in the classical time series decomposition—the mean is typically modelled separately from the seasonal component. Note that, in view of the orthogonality of the basis vectors, the sum of any harmonic over an entire period is zero. With these assumptions, we can write

$$g_t = \sum_{j=1}^{s/2} S_j(t). \tag{3.19}$$

The last step, in order to use in the context of DLMs the representation of seasonality in terms of harmonics discussed above, is to study, for a fixed j, the temporal dynamics of $S_j(t)$. The evolution of S_j from time t to time $t+1$ is given by

$$S_j(t) \longmapsto S_j(t+1) = a_j \cos\big((t+1)\omega_j\big) + b_j \sin\big((t+1)\omega_j\big).$$

If $j < s/2$, it is easy to realize that, from the knowledge of $S_j(t)$ alone, i.e., without knowing a_j and b_j individually, it is impossible to determine the value of $S_j(t+1)$. However, if in addition to $S_j(t)$ one also knows the conjugate harmonic

$$S_j^*(t) = -a_j \sin(t\omega_j) + b_j \cos(t\omega_j),$$

then one can explicitely compute $S_j(t+1)$ and also $S_j^*(t+1)$. In fact,

$$\begin{aligned} S_j(t+1) &= a_j \cos\big((t+1)\omega_j\big) + b_j \sin\big((t+1)\omega_j\big) \\ &= a_j \cos(t\omega_j + \omega_j) + b_j \sin(t\omega_j + \omega_j) \\ &= a_j \big(\cos(t\omega_j)\cos\omega_j - \sin(t\omega_j)\sin\omega_j\big) \\ &\quad + b_j\big(\sin(t\omega_j)\cos\omega_j + \cos(t\omega_j)\sin\omega_j\big) \\ &= \big(a_j\cos(t\omega_j) + b_j\sin(t\omega_j)\big)\cos\omega_j \\ &\quad + \big(-a_j\sin(t\omega_j) + b_j\cos(t\omega_j)\big)\sin\omega_j \\ &= S_j(t)\cos\omega_j + S_j^*(t)\sin\omega_j \end{aligned} \tag{3.20a}$$

and

$$
\begin{aligned}
S_j^*(t+1) &= -a_j \sin\left((t+1)\omega_j\right) + b_j \cos\left((t+1)\omega_j\right) \\
&= -a_j\big(\sin(t\omega_j)\cos\omega_j + \cos(t\omega_j)\sin\omega_j\big) \\
&\qquad + b_j\big(\cos(t\omega_j)\cos\omega_j - \sin(t\omega_j)\sin\omega_j\big) \\
&= \big(-a_j\sin(t\omega_j) + b_j\cos(t\omega_j)\big)\cos\omega_j \\
&\qquad - \big(a_j\cos(t\omega_j) + b_j\sin(t\omega_j)\big)\sin\omega_j \\
&= -S_j(t)\sin\omega_j + S_j^*(t)\cos\omega_j.
\end{aligned}
\tag{3.20b}
$$

The two equations (3.20) can be combined in the matrix equation

$$
\begin{bmatrix} S_j(t+1) \\ S_j^*(t+1) \end{bmatrix} = \begin{bmatrix} \cos\omega_j & \sin\omega_j \\ -\sin\omega_j & \cos\omega_j \end{bmatrix} \begin{bmatrix} S_j(t) \\ S_j^*(t) \end{bmatrix}.
$$

In this form, the jth harmonic fits naturally in the DLM framework by considering the bivariate state vector $\big(S_j(t), S_j^*(t)\big)'$ with evolution matrix

$$
H_j = \begin{bmatrix} \cos\omega_j & \sin\omega_j \\ -\sin\omega_j & \cos\omega_j \end{bmatrix}
$$

and observation matrix $F = [1\ 0]$. The case $j = s/2$ is even simpler, since

$$
S_{s/2}(t+1) = \cos\big((t+1)\pi\big) = -\cos(t\pi) = -S_{s/2}(t).
$$

That is, $S_{s/2}$ simply changes sign at every unit time increment. As a DLM, we can consider this to correspond to a univariate state vector with evolution matrix $H_{s/2} = [-1]$ and observation matrix $F = [1]$.

The DLM representations of the different harmonics can be combined to obtain back (3.19). To this aim, we can consider the state vector

$$
\theta_t = \big(S_1(t), S_1^*(t), \ldots, S_{\frac{s}{2}-1}(t), S_{\frac{s}{2}-1}^*(t), S_{\frac{s}{2}}(t)\big)', \qquad t = 0, 1, \ldots,
$$

together with the evolution matrix

$$
G = \text{blockdiag}(H_1, \ldots, H_{\frac{s}{2}})
$$

and observation matrix

$$
F = [1\ 0\ 1\ 0 \ldots 0\ 1].
$$

Setting to zero all the evolution and observation variances, the above definitions give a DLM representation of a periodic seasonal component. Non-zero evolution variances may be included to account for a stochastically evolving seasonal component. In this case, strictly speaking, the seasonal component will no longer be periodic. However, the forecast function will, see Problem 3.9. In any case,

$$
\theta_0 = \big(a_1, b_1, \ldots, a_{\frac{s}{2}-1}, b_{\frac{s}{2}-1}, a_{\frac{s}{2}}\big)'.
$$

By appropriately selecting θ_0, the representation above can give just any zero-mean periodic function of period s. In most applications, however, we want to

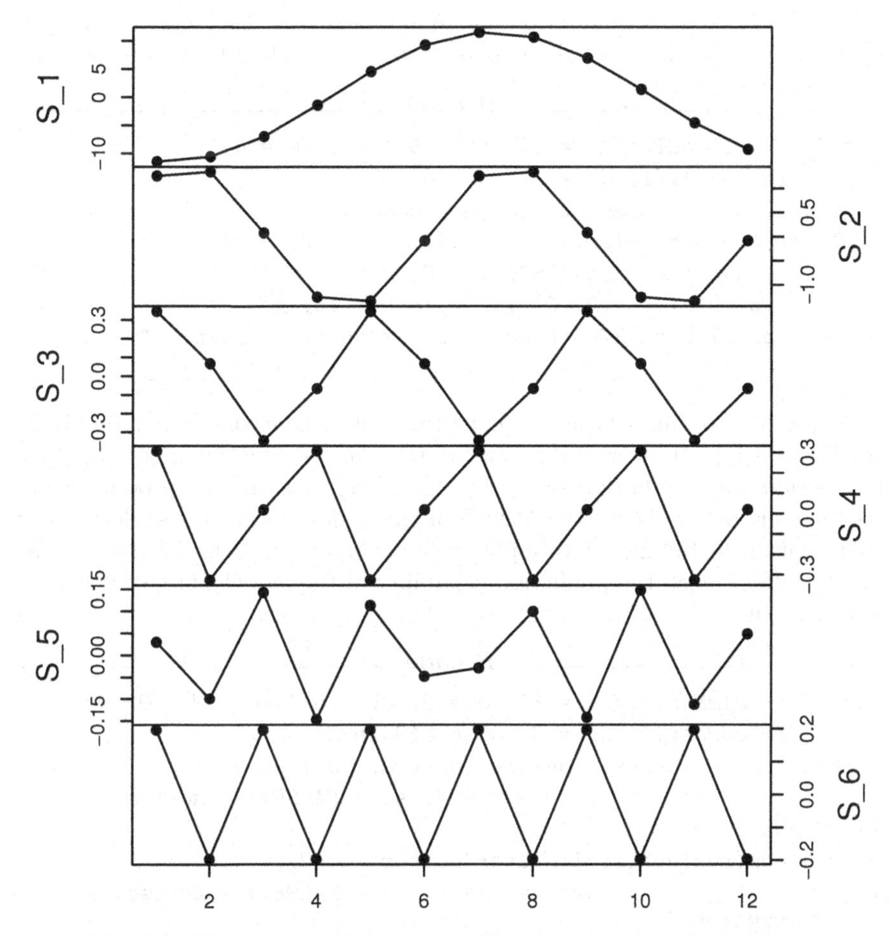

Fig. 3.8. Harmonics of average monthly temperature at Nottingham

model a fairly smooth seasonal component. This can be achieved by discarding a number of high-frequency harmonics and retaining only a few of the first harmonics, which are those that oscillate more slowly (cf. Figure 3.7). In this way we have a parsimonious representation of a periodic component, not available when using the seasonal factors of Section 3.2.2.

To create in R the representation of a Fourier-form seasonal DLM, we can use the function `dlmModTrig`, specifying the period via the argument `s`. In addition, we may specify the number of harmonics to retain in the DLM via the argument `q`. To illustrate how the Fourier form of a seasonal component can be used to obtain a more parsimonius model, consider the data set of average monthly temperatures at Nottingham, `nottem`. As a first model, we use a full 12-month static seasonal component added to a local level. The nonzero variances were estimated by maximum likelihood. Figure 3.8 shows the six harmonics of the seasonal component over a complete 12-month period.

R code

```
> mod1 <- dlmModTrig(s = 12, dV = 5.1118, dW = 0) +
+       dlmModPoly(1, dV = 0, dW = 81307e-3)
> smoothTem1 <- dlmSmooth(nottem, mod1)
> plot(ts(smoothTem1$s[2 : 13, c(1, 3, 5, 7, 9, 11)],
+           names = paste("S", 1 : 6, sep = "_")),
+       oma.multi = c(2, 0, 1, 0), pch = 16, nc = 1,
+       yax.flip = TRUE, type = 'o', xlab = "", main = "")
```

From the plot we can see that the harmonics after the second have a relatively small amplitude. In view of this, we can try a model containing only the first two harmonics, in addition to the local level representing the process mean. As the code below shows, the MAPE of this reduced model is slightly lower than that of the full model: dropping seven components from the state vector did not affect much the prediction capability of the model—in fact it helped improving it.

R code

```
> mod2 <- dlmModTrig(s = 12, q = 2, dV = 5.1420, dW = 0) +
+       dlmModPoly(1, dV = 0, dW = 81942e-3)
> mean(abs(residuals(dlmFilter(nottem, mod1),
+                          type = "raw", sd = FALSE)) / nottem)
[1] 0.08586188
> mean(abs(residuals(dlmFilter(nottem, mod2),
+                          type = "raw", sd = FALSE)) / nottem)
[1] 0.05789139
```

An explanation for the lower MAPE of the reduced model is that the higher-order harmonics are basically used to fit the noise in the data, and this fitted noise does not generalize well when it comes to making out-of-sample predictions, in particular one-step-ahead predictions.

So far in our discussion of periodic functions we have considered an even period. The case of an odd period can be treated in essentially the same way. The only difference is in the last harmonic which, instead of being a cosine function only, also has a sine component and has therefore the same form as the previous ones. More specifically, a zero-mean periodic function having odd period s has the representation

$$\alpha(t) = \sum_{j=1}^{(s-1)/2} S_j(t),$$

where the S_j are defined by (3.18). Each harmonic, including the last, has two degrees of freedom, expressed by the coefficients a_j and b_j. The DLM representation parallels the one outlined above in the even-period case, with the difference that the last diagonal block of the evolution matrix G is the 2×2 matrix

$$H_{\frac{s-1}{2}} = \begin{bmatrix} \cos\omega_{\frac{s-1}{2}} & \sin\omega_{\frac{s-1}{2}} \\ -\sin\omega_{\frac{s-1}{2}} & \cos\omega_{\frac{s-1}{2}} \end{bmatrix}$$

and the observation matrix is

$$F = [1\,0\,1\,0 \ldots 0\,1\,0].$$

3.2.4 General periodic components

The treatment of seasonal components given in the previous section is perfectly appropriate when the period of the process underlying the observations is a multiple of the time between consecutive observations. However, for many natural phenomena this is not the case. Consider, for example, the data set of monthly sunspots available in R as `sunspots`. A plot of the data reveals very clearly some kind of periodicity of about eleven years, but it would be naif to think that the period consists of an integer number of months. After all, a month is a totally arbitrary time scale to measure something happening on the Sun. In cases like this, it is useful to think of an underlying periodic countinuous-time process, $g(t)$, observed at discrete time intervals. A periodic function defined on the real line can be expressed as a sum of harmonics similar to (3.17) or (3.19). In fact, it can be proved that for a continuous periodic function $g(t)$ one has the representation

$$g(t) = a_0 + \sum_{j=1}^{\infty} \left(a_j \cos(t\omega_j) + b_j \sin(t\omega_j)\right), \qquad (3.21)$$

where $\omega_j = j\omega$ and ω is the so-called *fundamental frequency*. The fundamental frequency is related to τ, the period of $g(t)$, by the relationship $\tau\omega = 2\pi$. In fact, it is easy to see that, if $t_2 = t_1 + k\tau$ for some integer k, then

$$t_2\omega_j = t_2 j\omega = (t_1 + k\tau)j\omega = t_1 j\omega + kj\tau\omega = t_1\omega_j + kj \cdot 2\pi$$

and, therefore, $g(t_2) = g(t_1)$. Assuming, as we did in Section 3.2.3, that $a_0 = 0$, that is, that $g(t)$ has mean zero, and defining the jth harmonic of $g(t)$ to be

$$S_j(t) = a_j \cos(t\omega_j) + b_j \sin(t\omega_j),$$

we can rewrite (3.21) as

$$g(t) = \sum_{j=1}^{\infty} S_j(t). \qquad (3.22)$$

The discrete-time evolution of the jth harmonic from time t to time $t+1$ can be cast as a DLM in the same way as we did in Section 3.2.3. We just have to consider the bivariate state vector $\left(S_j(t), S_j^*(t)\right)'$, where $S_j^*(t)$ is the conjugate harmonic defined by

$$S_j^*(t) = -a_j \sin(t\omega_j) + b_j \cos(t\omega_j),$$

and the evolution matrix

$$H_j = \begin{bmatrix} \cos\omega_j & \sin\omega_j \\ -\sin\omega_j & \cos\omega_j \end{bmatrix}.$$

Then we have

$$\begin{bmatrix} S_j(t+1) \\ S_j^*(t+1) \end{bmatrix} = H_j \begin{bmatrix} S_j(t) \\ S_j^*(t) \end{bmatrix}.$$

A nonzero evolution error may be added to this evolution equation to account for a stochastically varying harmonic.

Clearly, in a DLM we cannot keep track of the infinitely many harmonics in the representation (3.22): we have to truncate the infinite series on the RHS to a finite sum of, say, q terms. As already seen in the previous section, the higher-order harmonics are those oscillating more rapidly. Thus, the decision of truncating the infinite sum in (3.22) can be interpreted as a subjective judgement about the degree of smoothness of the function $g(t)$. In practice it is not uncommon to model the periodic function $g(t)$ using only one or two harmonics.

To set up in R a general periodic DLM component, the function *dlmModTrig* introduced in Section 3.2.3 may be used. The user needs to specify the period via the argument `tau` and the number of harmonics, `q`. Alternatively, the fundamental frequency ω can be specified, via the argument `om`, instead of the period τ. The display below shows how to specify a two-harmonic stochastic periodic component added to a local level for the sunspots data, on the square root scale. The period $\tau = 130.51$ as well as the nonzero variances were estimated by maximum likelihood.

R code

```
> mod <- dlmModTrig(q = 2, tau = 130.51, dV = 0,
+                   dW = rep(c(1765e-2, 3102e-4), each = 2)) +
+      dlmModPoly(1, dV = 0.7452, dW = 0.1606)
```

The model can be used, for example, to smooth the data, extracting the level (the "signal") from the data. Figure 3.9, obtained with the code below, shows the data, the level, and the general stochastic periodic component. The level is the last (fifth) component of the state vector, while the periodic component is obtained by adding the two harmonics of the model (first and third components).

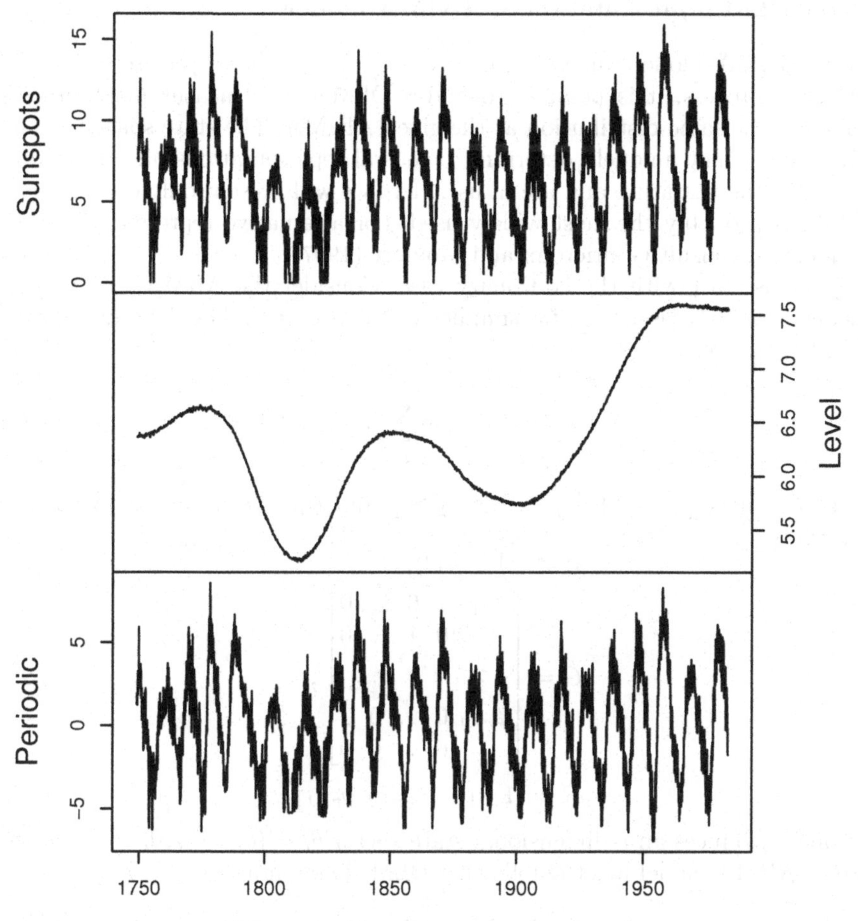

Fig. 3.9. Monthly sunspot numbers

R code

```
> sspots <- sqrt(sunspots)
> sspots.smooth <- dlmSmooth(sspots, mod)
> y <- cbind(sspots,
+            tcrossprod(dropFirst(sspots.smooth$s[, c(1, 3, 5)]),
+                       matrix(c(0, 0, 1, 1, 1, 0), nr = 2,
+                              byrow = TRUE)))
> colnames(y) <- c("Sunspots", "Level", "Periodic")
> plot(y, yax.flip = TRUE, oma.multi = c(2, 0, 1, 0))
```

3.2.5 DLM representation of ARIMA models

Any ARIMA model can be expressed as a DLM. More precisely, for any ARIMA process, it is possible to find a DLM whose measurement process (Y_t) has the same distribution as the given ARIMA. The state space with its dynamics is not uniquely determined: several representations have been proposed in the literature and are in use. Here we will present only one of them, which is probably the most widely used. For alternative representations the reader can consult Gourieroux and Monfort (1997).

Let us start with the stationary case. Consider the ARMA(p,q) process defined by (3.4), assuming for simplicity that μ is zero. The defining relation can be written as

$$Y_t = \sum_{j=1}^{r} \phi_j Y_{t-j} + \sum_{j=1}^{r-1} \psi_j \epsilon_{t-j} + \epsilon_t,$$

with $r = \max\{p, q+1\}$, $\phi_j = 0$ for $j > p$ and $\psi_j = 0$ for $j > q$. Define the matrices

$$\begin{aligned} F &= \begin{bmatrix} 1 \; 0 \ldots 0 \end{bmatrix}, \\ G &= \begin{bmatrix} \phi_1 & 1 & 0 & \ldots & 0 \\ \phi_2 & 0 & 1 & \ldots & 0 \\ \vdots & \vdots & & \ddots & \\ \phi_{r-1} & 0 & \ldots & 0 & 1 \\ \phi_r & 0 & \ldots & 0 & 0 \end{bmatrix}, \\ R &= \begin{bmatrix} 1 \; \psi_1 \ldots \psi_{r-2} \; \psi_{r-1} \end{bmatrix}'. \end{aligned} \tag{3.23}$$

If one introduces an r-dimensional state vector $\theta_t = (\theta_{1,t}, \ldots, \theta_{r,t})'$, then the given ARMA model has the following DLM representation:

$$\begin{cases} Y_t = F\theta_t, \\ \theta_{t+1} = G\theta_t + R\epsilon_t. \end{cases} \tag{3.24}$$

This is a DLM with $V = 0$ and $W = RR'\sigma^2$, where σ^2 is the variance of the error sequence (ϵ_t). To verify this equivalence, note that the observation equation gives $y_t = \theta_{1,t}$ and the state equation is

$$\begin{aligned} \theta_{1,t} &= \phi_1 \theta_{1,t-1} + \theta_{2,t-1} + \epsilon_t \\ \theta_{2,t} &= \phi_2 \theta_{1,t-1} + \theta_{3,t-1} + \psi_1 \epsilon_t \\ &\vdots \\ \theta_{r-1,t} &= \phi_{r-1} \theta_{1,t-1} + \theta_{r,t-1} + \psi_{r-2} \epsilon_t \\ \theta_{r,t} &= \phi_r \theta_{1,t-1} + \psi_{r-1} \epsilon_t. \end{aligned} \tag{3.25}$$

Substituting the expression of $\theta_{2,t-1}$, obtained from the second equation, in the first equation, we have

$$\theta_{1,t} = \phi_1\theta_{1,t-1} + \phi_2\theta_{1,t-2} + \theta_{3,t-2} + \psi_1\epsilon_{t-1} + \epsilon_t$$

and proceeding by successive substitutions we eventually get

$$\theta_{1,t} = \phi_1\theta_{1,t-1} + \cdots + \phi_r\theta_{1,t-r} + \psi_1\epsilon_{t-1} + \cdots + \psi_{r-1}\epsilon_{t-r-1} + \epsilon_t\,.$$

Recalling that $r = \max\{p, q+1\}$ and $y_t = \theta_{1,t}$ we see that this is the ARMA model (3.4).

The DLM representation (3.24) might appear quite artificial. To develop a better understanding, let us look at the simpler case of pure autoregressive models, for example the AR(2) model

$$Y_t = \phi_1 Y_{t-1} + \phi_2 Y_{t-2} + \epsilon_t, \qquad \epsilon_t \overset{iid}{\sim} \mathcal{N}(0, \sigma^2). \tag{3.26}$$

One might think of a simpler representation of the AR(2) as a DLM with observation equation having $F_t = [Y_{t-1}, Y_{t-2}]$ and $\theta_t = [\phi_{1,t}, \phi_{2,t}]'$ (thus, possibly including a temporal evolution of the AR parameters, otherwise letting $W = 0$). However, the matrix F_t of a DLM cannot depend on past values of the observations: in our case, the above choice of F_t would imply

$$Y_t | y_{t-1}, y_{t-2}, \theta_t \sim N(\phi_1 y_{t-1} + \phi_2 y_{t-2}, \sigma^2),$$

that is, Y_t would not be independent on the past values y_{t-1}, y_{t-2} given θ_t, thus violating assumption (A.2) in the definition of state space model (p.40). In order to consider such a model we would have to extend the definition to include *conditionally* Gaussian state space models (Lipster and Shiryayev; 1972). In order to stay within the boundaries of the standard definition used in this book, the first trick that we have used in the DLM representation (3.4) is thus to "shift" the AR(2) dependence from Y_t to the state vector: $Y_t = \theta_{1,t}$. However, the second basic assumption of state space models is that the state process is Markovian, thus we need a second trick for representing the second order dependence. We augmented the state vector by a second component $\theta_{2,t}$, and we chose G and W such that

$$\begin{bmatrix}\theta_{1,t}\\ \theta_{2,t}\end{bmatrix} = \begin{bmatrix}\phi_1 & 1\\ \phi_2 & 0\end{bmatrix}\begin{bmatrix}\theta_{1,t-1}\\ \theta_{2,t-1}\end{bmatrix} + \begin{bmatrix}\epsilon_t\\ 0\end{bmatrix},$$

giving

$$\theta_{1,t} = \phi_1\theta_{1,t-1} + \phi_2\theta_{1,t-2} + \epsilon_t.$$

In this way, we obtain a DLM representation of AR(p) models in the framework of state-space models as defined by assumptions (A.1) and (A.2) in Section 2.3. A further step is needed for expressing the MA(q) component. For example, for the ARMA(1,1) model

$$Y_t = \phi_1 Y_{t-1} + \epsilon_t + \psi_1\epsilon_{t-1}, \qquad \epsilon_t \sim \mathcal{N}(0, \sigma^2), \tag{3.27}$$

$r = q + 1 = 2$ and the matrices of the corresponding DLM are

$$\begin{aligned} F &= \begin{bmatrix} 1 & 0 \end{bmatrix}, & V &= 0, \\ G &= \begin{bmatrix} \phi_1 & 1 \\ 0 & 0 \end{bmatrix}, & W &= \begin{bmatrix} 1 & \psi_1 \\ \psi_1 & \psi_1^2 \end{bmatrix} \sigma^2. \end{aligned} \tag{3.28}$$

Representing an ARMA model as a DLM is useful mainly for two reasons. The first is that an ARMA component in a DLM can explain residual autocorrelation not accounted for by other structural components such as trend and seasonal. The second reason is technical, and consists in the fact that the evaluation of the likelihood function of an ARMA model can be performed efficiently by applying the general recursion used to compute the likelihood of a DLM.

The DLM representation of an ARIMA(p, d, q) model, with $d > 0$, can be derived as an extension of the stationary case. In fact, if one considers $Y_t^* = \Delta^d Y_t$, then Y_t^* follows a stationary ARMA model, for which the DLM representation given above applies. In order to model the original series (Y_t) we need to be able to recover it from the Y_t^* and possibly other components of the state vector. For example, if $d = 1$, $Y_t^* = Y_t - Y_{t-1}$ and therefore $Y_t = Y_t^* + Y_{t-1}$. Suppose that Y_t^* satisfies the AR(2) model (3.26). Then a DLM representation for Y_t is given by the system

$$\begin{cases} Y_t = \begin{bmatrix} 1 & 1 & 0 \end{bmatrix} \theta_{t-1}, \\ \theta_t = \begin{bmatrix} 1 & 1 & 0 \\ 0 & \phi_1 & 0 \\ 0 & \phi_2 & 1 \end{bmatrix} \theta_{t-1} + w_t, \qquad w_t \sim \mathcal{N}(0, W), \end{cases} \tag{3.29}$$

with

$$\theta_t = \begin{bmatrix} Y_{t-1} \\ Y_t^* \\ \phi_2 Y_{t-1}^* \end{bmatrix} \tag{3.30}$$

and $W = \operatorname{diag}(0, \sigma^2, 0)$. For a general d, set $Y_t^* = \Delta^d Y_t$. It can be shown that the following relation holds:

$$\Delta^{d-j} Y_t = Y_t^* + \sum_{i=1}^{j} \Delta^{d-i} Y_{t-1}, \qquad j = 1, \dots, d. \tag{3.31}$$

Define the state vector as follows:

$$\theta_t = \begin{bmatrix} Y_{t-1} \\ \Delta Y_{t-1} \\ \vdots \\ \Delta^{d-1} Y_{t-1} \\ Y_t^* \\ \phi_2 Y_{t-1}^* + \cdots + \phi_r Y_{t-r+1}^* + \psi_1 \epsilon_t = \cdots + \psi_{r-1} \epsilon_{t-r+2} \\ \phi_3 Y_{t-1}^* + \cdots + \phi_r Y_{t-r+2}^* + \psi_2 \epsilon_t = \cdots + \psi_{r-1} \epsilon_{t-r+3} \\ \vdots \\ \phi_r Y_{t-1}^* + \psi_{r-1} \epsilon_t \end{bmatrix}. \tag{3.32}$$

Note that the definition of the last components of θ_t follows from equations (3.25). The system and observation matrices, together with the system variance, are defined by

$$F = \begin{bmatrix} 1\,1 \ldots 1\,0 \ldots 0 \end{bmatrix},$$

$$G = \begin{bmatrix} 1 & 1 & \ldots & 1 & 0 & \ldots\ldots & 0 \\ 0 & 1 & \ldots & 1 & 0 & \ldots\ldots & 0 \\ \ldots & \ldots & \ldots & \ldots & \ldots & \ldots & \ldots \\ \ldots & \ldots & 1 & 1 & 0 & \ldots\ldots & 0 \\ 0 & \ldots & 0 & \phi_1 & 1 & 0 \ldots & 0 \\ \ldots & \ldots & \ldots & \phi_2 & 0 & 1 \ldots & 0 \\ \vdots & & & \vdots & \vdots & \ddots & \\ \ldots & \ldots & \ldots & \phi_{r-1} & 0 & \ldots 0 & 1 \\ 0 & \ldots & 0 & \phi_r & 0 & \ldots 0 & 0 \end{bmatrix}, \tag{3.33}$$

$$R = \begin{bmatrix} 0 \ldots 0\; 1\; \psi_1 \ldots \psi_{r-2}\; \psi_{r-1} \end{bmatrix}',$$

$$W = RR'\sigma^2.$$

With the above definition the ARIMA model for (Y_t) has the DLM representation

$$\begin{cases} Y_t = F\theta_t, \\ \theta_t = G\theta_{t-1} + w_t, \qquad w_t \sim \mathcal{N}(0, W). \end{cases} \tag{3.34}$$

Since in DLM modeling a nonstationary behavior of the observations is usually accounted for directly, through the use of a polynomial trend or a seasonal component for example, the inclusion of nonstationary ARIMA components is less common than that of stationary ARMA components that, as we already mentioned, are typically used to capture correlated noise in the data.

3.2.6 Example: estimating the output gap

Measuring the so-called *output gap* is an important issue in monetary politics. The output gap is the difference between the observed Gross Domestic Product of a country (GDP, or output) and the *potential output* of the economy.

It is this discrepancy that is considered relevant in determining inflationary tendencies. Thus, it is crucial to be able to separate the two components and, since they are unobservable, treating them as latent states in a DLM results particularly attractive.

Variants of the following model for the deseasonalized output have been considered in the econometric literature; see, e.g., Kuttner (1994):

$$\begin{aligned} Y_t &= Y_t^{(p)} + Y_t^{(g)}, & & \\ Y_t^{(p)} &= Y_{t-1}^{(p)} + \delta_t + \epsilon_t, & \epsilon_t &\sim \mathcal{N}(0, \sigma_\epsilon^2), \\ \delta_t &= \delta_{t-1} + z_t, & z_t &\sim \mathcal{N}(0, \sigma_z^2), \\ Y_t^{(g)} &= \phi_1 Y_{t-1}^{(g)} + \phi_2 Y_{t-2}^{(g)} + u_t, & u_t &\sim \mathcal{N}(0, \sigma_u^2), \end{aligned} \tag{3.35}$$

where Y_t is the logarithm of the output, $Y_t^{(p)}$ represents the log-potential output and $Y_t^{(g)}$ is the log of the output gap. The above model can be seen as a DLM, obtained by adding, in the sense discussed in Section 3.2, a stochastic trend component for the potential output and a stationary AR(2) residual component. More specifically,

- $Y_t^{(p)}$ follows a linear growth model, observed without error. The state vector is $\theta_t^{(p)} = (Y_t^{(p)}, \delta_t)'$, with innovation vector $w_t^{(p)} = (\epsilon_t, z_t)'$. The observation matrix and observation variance are

$$F_{(p)} = \begin{bmatrix} 1 & 0 \end{bmatrix}, \qquad V^{(p)} = [0],$$

while the system evolution matrix and innovation variance are

$$G^{(p)} = \begin{bmatrix} 1 & 1 \\ 0 & 1 \end{bmatrix}, \qquad W^{(p)} = \operatorname{diag}(\sigma_\epsilon^2, \sigma_z^2).$$

The error terms ϵ_t and z_t are interpreted as shocks to the output level and to the output growth rate, respectively. Note that the case $\sigma_\epsilon^2 = \sigma_z^2 = 0$ corresponds to the global trend model $y_t^{(p)} = \mu_0 + t\delta_0$, while $\sigma_\epsilon^2 = 0$ gives an integrated random walk model.
- The output gap $Y_t^{(g)}$ is described by an AR(2) model, which can be written as a DLM with $\theta_t^{(g)} = (Y_t^{(g)}, \theta_{t,2}^{(g)})'$ and

$$\begin{aligned} F^{(g)} &= \begin{bmatrix} 1 & 0 \end{bmatrix}, & V^{(g)} &= [0], \\ G^{(g)} &= \begin{bmatrix} \phi_1 & 1 \\ \phi_2 & 0 \end{bmatrix}, & W^{(g)} &= \operatorname{diag}(\sigma_u^2, 0). \end{aligned}$$

The order of the AR process allows the residuals, i.e., the departures from the trend, to have a dumped cyclic autocorrelation function, which is often observed in economic time series. The DLM representation of model (3.35) for Y_t is obtained by adding the two components, as described in Section 3.2. The matrices of the resulting DLM are:

$$
\begin{aligned}
F &= \begin{bmatrix} 1 & 0 & 1 & 0 \end{bmatrix}, \\
G &= \begin{bmatrix} 1 & 1 & 0 & 0 \\ 0 & 1 & 0 & 0 \\ 0 & 0 & \phi_1 & 1 \\ 0 & 0 & \phi_2 & 0 \end{bmatrix}, \\
V &= [0] \\
W &= \operatorname{diag}(\sigma_\epsilon^2, \sigma_z^2, \sigma_u^2, 0).
\end{aligned}
\tag{3.36}
$$

The resulting DLM has five unknown parameters: σ_ϵ^2 and σ_z^2 for the trend component, and ϕ_1, ϕ_2 and σ_u^2 for the AR(2) component. Here, we estimate the unknown parameters by maximum likelihood (see Chapter 4) and then apply the Kalman filter and smoother using their MLE. In Section 4.6.1 we will illustrate a Bayesian approach to make inference both on the unknown parameters of the model and on the unobservable states at the same time.

For a practical application, we consider quarterly deseasonalized GDP of the US economy (source: Bureau of Economic Analysis), measured in billions of chained 2000 US dollars, from 1950.Q1 to 2004.Q4. A logarithmic transform is applied, and the resulting log GDP is plotted in Figure 3.10.

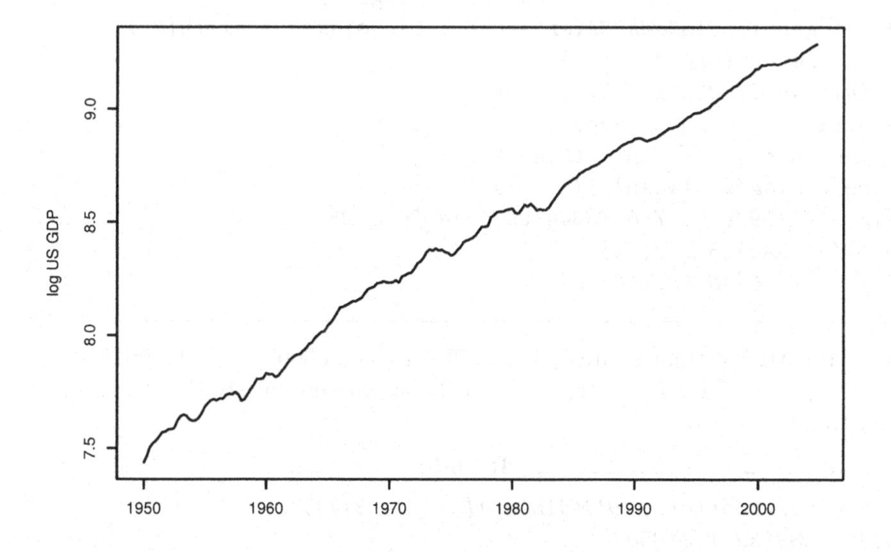

Fig. 3.10. Time plot of log US Gross Domestic Product

R code

```
> gdp <- read.table("Datasets/gdp5004.dat")
> gdp <- ts(gdp, frequency = 4, start = 1950)
```

```
> Lgdp <- log(gdp)
> plot(Lgdp, xlab = "", ylab = "log US GDP")
```

We first compute the MLE of the unknown parameters $\psi_1 = \log(\sigma_\epsilon^2), \psi_2 = \log(\sigma_z^2), \psi_3 = \log(\sigma_u^2), \psi_4 = \phi_1, \psi_5 = \phi_5$. For the moment, we do not impose stationarity restrictions on the AR(2) parameters ϕ_1, ϕ_2; we will return to this later. We choose convenient values for the initial mean $m_0^{(p)}$ of the states of the trend component, with a fairly large variance $C_0^{(p)}$, and use the default initial values in the R function *dlmModArma* for the AR(2) component. The MLE fitting function, as well as `dlmFilter`, do not allow singular observation variances, so we take a very small value for V, which can be considered zero for all practical purposes.

R code

```
> level0 <- Lgdp[1]
> slope0 <- mean(diff(Lgdp))
> buildGap <- function(u) {
+     trend <- dlmModPoly(dV = 1e-7, dW = exp(u[1 : 2]),
+                         m0 = c(level0, slope0),
+                         C0 = 2 * diag(2))
+     gap <- dlmModARMA(ar = u[4 : 5], sigma2 = exp(u[3]))
+     return(trend + gap)}
> init <- c(-3, -1, -3, .4, .4)
> outMLE <- dlmMLE(Lgdp, init, buildGap)
> dlmGap <- buildGap(outMLE$par)
> sqrt(diag(W(dlmGap))[1 : 3])
[1] 5.781792e-03 7.637864e-05 6.145361e-03
> GG(dlmGap)[3 : 4, 3]
[1]  1.4806246 -0.5468102
```

Thus the MLE estimates are $\hat{\sigma}_\epsilon = 0.00578$, $\hat{\sigma}_z = 0.00008$, $\hat{\sigma}_u = 0.00615$, $\hat{\phi}_1 = 1.481, \phi_2 = -0.547$. The estimated AR(2) parameters satisfy the stationarity constraints.

R code

```
> Mod(polyroot(c(1, -GG(dlmGap)[3 : 4, 3])))
[1] 1.289162 1.418587
> plot(ARMAacf(ar = GG(dlmGap)[3 : 4, 3], lag.max = 20),
+      ylim = c(0, 1), ylab = "acf", type = "h")
```

A note of caution is in order. There are clearly identifiability issues in separating the stochastic trend and the AR(2) residuals, and MLE is not stable; as a minimal check, one should repeat the fitting procedure starting from different sets of initial values.

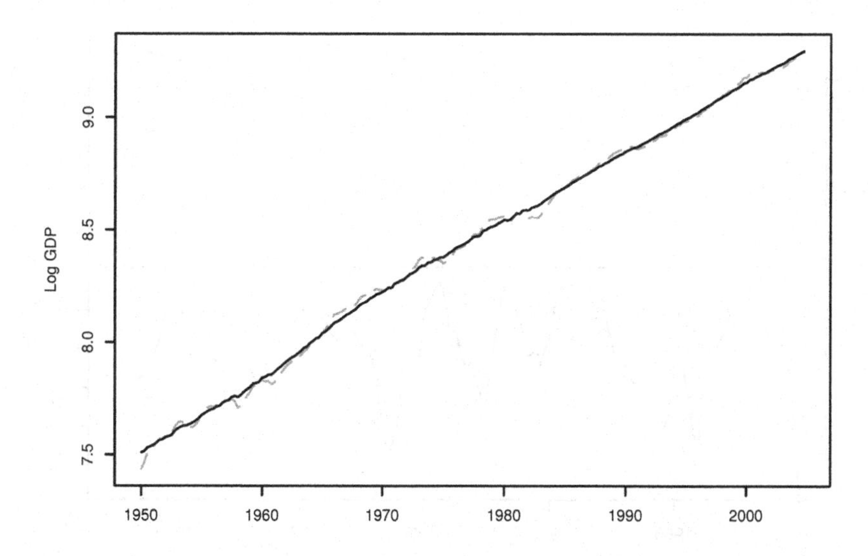

Fig. 3.11. Observed log GDP (grey) and smoothing estimate of the potential output

Plugging in the MLE, smoothing estimates of the model are finally computed. The observed output and the smoothing estimate of the potential output are plotted in Figure 3.11. Figure 3.12 shows the smoothing estimates of the potential output, together with the output gap (note the different scale).

R code

```
> gdpSmooth <- dlmSmooth(Lgdp, dlmGap)
> plot(cbind(Lgdp, dropFirst(gdpSmooth$s[, 1])),
+        xlab = "", ylab = "Log GDP", lty = c("longdash", "solid"),
+        col = c("darkgrey", "black"), plot.type = "single")
> plot(dropFirst(gdpSmooth$s[, 1:3]), ann = F, yax.flip = TRUE)
```

Finally, let us see how we could introduce stationarity constraints in the MLE, using the R function *ARtransPars*.

R code

```
> buildgapr <- function(u)
+ {
+      trend <- dlmModPoly(dV = 0.000001,
+                          dW = c(exp(u[1]), exp(u[2])),
+                          m0 = c(Lgdp[1], mean(diff(Lgdp))),
+                          C0 = 2 * diag(2))
+      gap <- dlmModARMA(ar = ARtransPars(u[4 : 5]),
+                        sigma2 = exp(u[3]))
+      return(trend + gap)
```

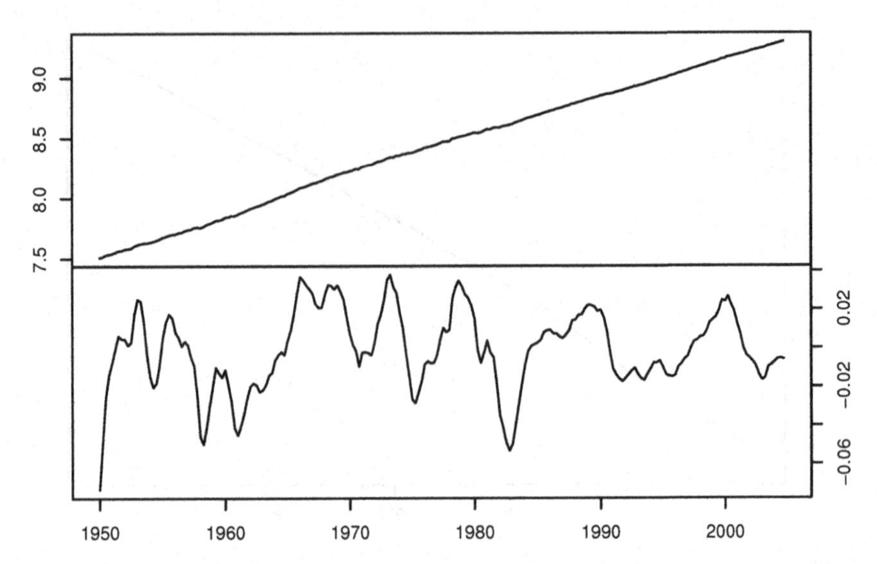

Fig. 3.12. Smoothing estimates of the potential output (top) and output-gap (bottom)

```
10  + }
    > init <- c(-3, -1, -3, .4, .4)
12  > outMLEr <- dlmMLE(Lgdp,init,buildgapr)
    > outMLEr$value
14  [1] -896.5234
    > parMLEr <- c(exp(outMLEr$par[1 : 3])^.5,
16  +              ARtransPars(outMLEr$par[4 : 5]))
    > round(parMLEr, 4)
18  [1]  0.0051  0.0000  0.0069  1.4972 -0.4972
```

We obtained a slightly better value of the likelihood. Yet, the estimates of the AR(2) parameters are now too close to the nonstationary region. It is anyway a useful exercise to look at the smoothing estimates of the potential output and of the output gap in this case: nonstationarity of the AR(2) residuals gives a nonrealistic estimated pattern for the output gap (figures not shown).

R code

```
    > Mod(polyroot(c(1, -ARtransPars(outMLEr$par[4 : 5]))))
2   [1] 1.000001 2.011090
    > plot(ARMAacf(ar = ARtransPars(outMLEr$par[4 : 5]),
4   +              lag.max = 10),
    +      ylim = c(0, 1), ylab = "acf")
6   > modr <- buildgapr(outMLEr$par)
```

```
> outFr <- dlmFilter(Lgdp, modr)
> outSr <- dlmSmooth(outFr)
> ts.plot(cbind(Lgdp, outSr$s[-1, 1]), col = 1 : 2)
> plot.ts(outSr$s[-1, c(1, 3)])
```

We will return to this example in Chapter 4, showing how we can impose stationarity constraints through a prior distribution on the AR parameters, in a Bayesian approach.

3.2.7 Regression models

Regression components can be quite easily incorporated in a DLM. It is of interest in many applications to include the effect of explanatory variables on a time series (Y_t). For example, in a clinical trial, one may be interested in studying a response Y_t to different drug doses x_t over time. Note that the explanatory variable x_t is nonstochastic; the case of stochastic regression requires a joint model for the bivariate time series $(X_t, Y_t), t \geq 1$ and will be discussed later.

The standard linear regression model for Y on x is defined as

$$Y_t = \beta_1 + \beta_2 x_t + \epsilon_t, \quad \epsilon_t \overset{iid}{\sim} \mathcal{N}(0, \sigma^2).$$

However, the assumption of i.i.d. errors is often not realistic if the observations are taken over time. A possible solution is to introduce a temporal dependence for the residuals, e.g., describing $(\epsilon_t : t \geq 1)$ as an autoregressive process. Another choice, which is in fact quite interesting in many problems, is to think that the relationship between y and x evolves over time. That is, to consider a *dynamic* linear regression model of the form

$$Y_t = \beta_{1,t} + \beta_{2,t} x_t + \epsilon_t, \quad \epsilon_t \overset{iid}{\sim} \mathcal{N}(0, \sigma^2),$$

and model the temporal evolution of $(\beta_{1,t}, \beta_{2,t})$, e.g., $\beta_{j,t} = \beta_{j,t-1} + w_{j,t}$, $j = 1, 2$, $w_{1,t}$ and $w_{2,t}$ independent. This gives a DLM with $F_t = \begin{bmatrix} 1 \ x_t \end{bmatrix}$, $\theta_t = \begin{bmatrix} \beta_{1,t} \ \beta_{2,t} \end{bmatrix}'$, $V = \sigma^2$, completed by a state equation for θ_t.

More generally, a dynamic linear regression model is described by

$$\begin{aligned} Y_t &= x_t' \theta_t + v_t, & v_t &\sim \mathcal{N}(0, \sigma_t^2) \\ \theta_t &= G_t \theta_{t-1} + w_t, & w_t &\sim \mathcal{N}_p(0, W_t) \end{aligned}$$

where $x_t' = \begin{bmatrix} x_{1,t} \cdots x_{p,t} \end{bmatrix}$ are the values of the p explanatory variables at time t. Again, note that x_t is not stochastic; in other terms, this is a conditional model for $Y_t | x_t$. A popular default choice for the state equation is to take the evolution matrix G_t as the identity matrix and W diagonal, which corresponds to modeling the regression coefficients as independent random walks.

A regression DLM, with the above assumptions for G and W, can be created in R through the function `dlmModReg`. The static linear regression model corresponds to the case where $W_t = 0$ for any t, so that θ_t is constant over time, $\theta_t = \theta$, with prior density $\theta \sim \mathcal{N}(m_0, C_0)$. Thus, the function `dlmModReg` can also be used for Bayesian inference in linear regression; the filtering density gives the posterior density of $\theta|y_{1:t}$, and $m_t = \mathrm{E}(\theta|y_{1:t})$ is the Bayesian estimate under a quadratic loss function of the regression coefficients. This also suggests that DLM techniques may be used to sequentially update the estimates of the parameters of a (static) regression model as new observations become available.

Example — Capital asset pricing model

The capital asset pricing model (CAPM) is a popular asset pricing tool in financial econometrics; see for example Campbell et al. (1996). In its simplest, univariate version, the CAPM model assumes that the returns on an asset depend linearly on the overall market returns, thus allowing us to study the behavior, in terms of risk and expected returns, of individual assets compared to the market as a whole. Here we consider a dynamic version of the standard, static CAPM model; in Section 3.3.3, it will be extended to a multivariate CAPM for a small portfolio of m assets.

Let $r_t, r_t^{(M)}$ and $r_t^{(f)}$ be the returns at time t of the asset under study, of the market and of a risk-free asset, respectively. Define the *excess returns* of the asset as $y_t = r_t - r_t^{(f)}$ and the market's excess returns as $x_t = r_t^{(M)} - r_t^{(f)}$. A univariate CAPM assumes that

$$y_t = \alpha + \beta x_t + v_t, \quad v_t \overset{iid}{\sim} N(0, \sigma^2). \tag{3.37}$$

The parameter β measures the sensitivity of the asset excess returns to changes in the market. A value of β greater than one suggests that the asset tends to amplify changes in the market returns, and it is therefore considered an aggressive investment; while assets having β smaller than one are thought of as conservative investments.

The data for this example consists of monthly returns[3] from January 1978 to December 1987 of four stocks (Mobil, IBM, Weyer, and Citicorp), of the 30-day Treasury Bill as a proxy for the risk-free asset, and of the value-weighted average returns for all the stocks listed at the New York and American Stock Exchanges, representing the overall market returns. The data are plotted in Figure 3.13. Let us consider the data for the IBM stock here; a multivariate CAPM will be illustrated later.

Least squares estimates for the static CAPM are obtained in R by the function `lm`. We can also compute Bayesian estimates of the regression coef-

[3] The data, originally from Berndt (1991), are available at the time of writing from `http://shazam.econ.ubc.ca/intro/P.txt`.

ficients by the function *dlmModReg*, letting $\text{diag}(W) = (0, 0)$ (while assuming, for simplicity, a known measurement variance).

R code

```
> capm <- read.table("http://shazam.econ.ubc.ca/intro/P.txt",
+                    header = TRUE)
> capm.ts <- ts(capm, start = c(1978, 1), frequency = 12)
> colnames(capm)
[1] "MOBIL"  "IBM"    "WEYER"  "CITCRP" "MARKET" "RKFREE"
> plot(capm.ts)
> IBM <- capm.ts[, "IBM"] - capm.ts[, "RKFREE"]
> x <- capm.ts[, "MARKET"] - capm.ts[, "RKFREE"]
> outLM <- lm(IBM ~ x)
> outLM$coef
  (Intercept)             x
-0.0004895937  0.4568207721
> acf(outLM$res)
> qqnorm(outLM$res)
```

Despite the data being taken over time, the residuals do not show relevant temporal dependence (plots not shown). Bayesian estimates of the regression coefficients are obtained below. For simplicity, we let $V = \hat{\sigma}^2 = 0.00254$. Prior information on the values of α and β can be introduced through the prior, $\mathcal{N}(m_0, C_0)$; below, we assume vague prior information on α, while the prior guess on β is 1.5 (aggressive investment), with a fairly small variance.

R code

```
> mod <- dlmModReg(x, dV = 0.00254, m0 = c(0, 1.5),
+                  C0 = diag(c(1e+07, 1)))
> outF <- dlmFilter(IBM, mod)
> outF$m[1 + length(IBM), ]
[1] -0.0005232801  0.4615301204
```

The filtering estimates at the end of the observation period are the Bayes estimates, under quadratic loss, of the regression coefficients; here, the results are very close to the OLS estimates. As a matter of fact, it seems more natural to allow the CAPM coefficients α and β to vary over time; for example, a stock might evolve from "conservative" to "aggressive." A dynamic version of the classical CAPM is obtained as

$$y_t = \alpha_t + \beta_t x_t + v_t, \quad v_t \overset{iid}{\sim} N(0, \sigma^2),$$

with a state equation modeling the dynamics of α_t and β_t. As mentioned above, the function *dlmModReg* assumes that the regression coefficients follow independent random walks. Below, the observation and state evolution

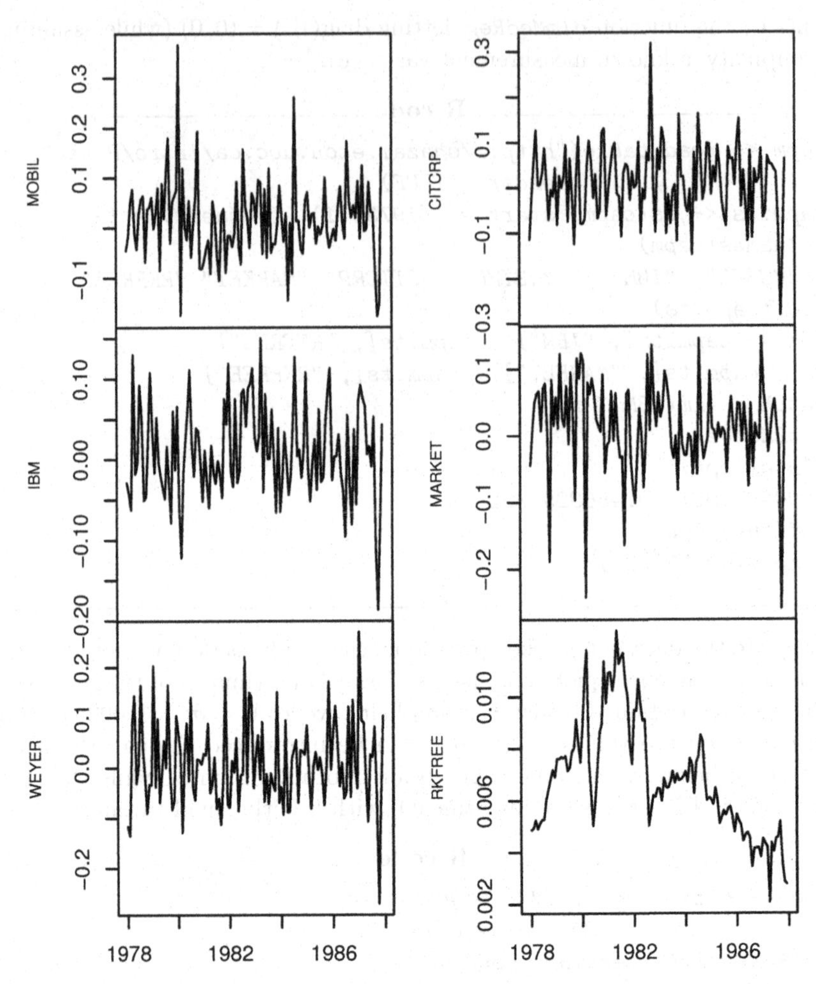

Fig. 3.13. Montly returns for Mobil, IBM, Weyer and Citicorp stocks; market index; 30-days Treasury bill (January 1978-December 1987)

variances are estimated by maximum likelihood. The smoothing estimates are shown in Figure 3.14. The results can be compared with those obtained in Section 3.3.3, where the model is estimated jointly for the four stocks.

R code

```
> buildCapm <- function(u) {
+     dlmModReg(x, dV = exp(u[1]), dW = exp(u[2 : 3]))
+ }
> outMLE <- dlmMLE(IBM, parm = rep(0, 3), buildCapm)
> exp(outMLE$par)
[1] 2.328397e-03 1.100181e-05 6.496259e-04
```

```
> outMLE$value
[1] -276.7014
> mod <- buildCapm(outMLE$par)
> outS <- dlmSmooth(IBM, mod)
> plot(dropFirst(outS$s))
```

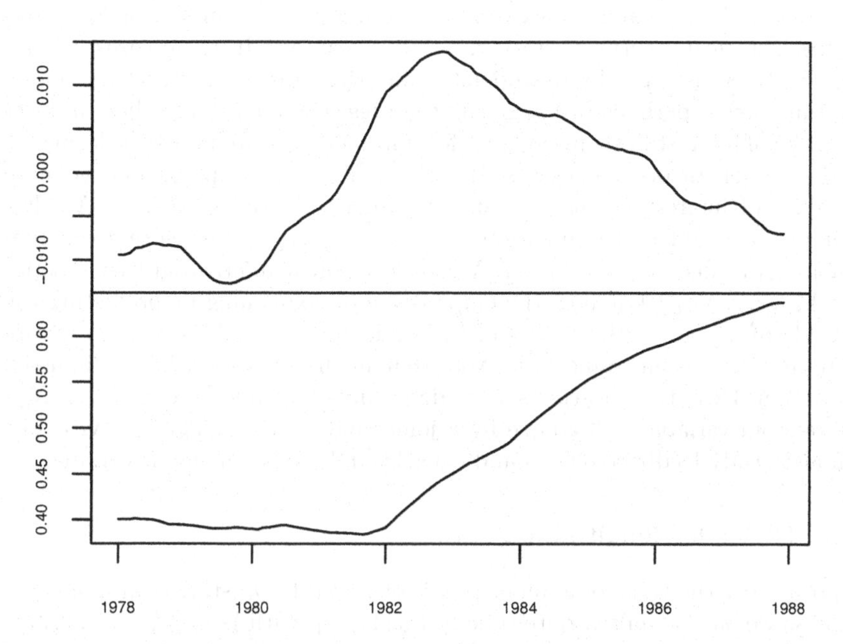

Fig. 3.14. Smoothing estimates of the coefficients of a dynamic CAPM model for the IBM stock returns

Further examples of dynamic regression with DLMs will be presented in the following sections, for more general, multivariate settings, such as longitudinal data, of the kind $(x_{i,t}, Y_{i,t}), t \geq 1, i = 1, \ldots, m$, or more simply $(x_t, Y_{i,t})$ (e.g., dose $x_{i,t}$ of a drug and response $Y_{i,t}$ at time t for patient i, $i = 1, \ldots, m$; or market returns x_t and excess returns $Y_{i,t}$ for asset i; see Section 3.3.3); or time series of cross-sectional data, where x and Y are observed for m statistical units, as above, but usually m is large; see Section 3.3.5.

3.3 Models for multivariate time series

Modeling multivariate time series is of course more interesting—and more challenging—than studying univariate models, and also in this case DLMs offer a very flexible framework for the analysis. In this section we just present a few examples of the extremely large variety of applications of DLMs to multivariate data, hoping that the reader can find tools and ideas for the analysis of her specific problems and models.

One can envision two basic types of data and problems in the analysis of multivariate time series. In many applications, one has data $Y_t = (Y_{1,t}, \ldots, Y_{m,t})'$ on one or more variables observed for different units; for example, Y_t could be the gross domestic product observed for m countries over time, or the income and the expenses for a group of m families, or $Y_{i,t}$ could be the historical returns of stock i, $i = 1, \ldots, m$, etc. In these cases, the focus of interest is typically in understanding the correlation structure among the time series, perhaps investigating the presence of clusters. These aspects might be of interest in themselves, or for improving the predictive performance of the model. In other contexts, the data are observations on one or more variables of interest Y and on some explanatory variables $X_1, \ldots, X_k$. For example, Y could be the inflation rate and $X_1, \ldots, X_k$ be relevant macroeconomic variables for a country. We have again a multivariate time series $(Y_t, X_{1,t}, \ldots, X_{k,t})$, but now the emphasis is on explaining or predicting the variable of interest Y_t by means of the explanatory variables $X_{j,t}$, so we are more in a regression framework. Note that in the regression DLM discussed in Section 3.2.7, the covariates were deterministic, while here $X_{1,t}, \ldots, X_{k,t}$ are random variables. Of course by a joint model for $(Y_t, X_{1,t}, \ldots, X_{k,t})$ one can also study feedback effects and causality relations among all variables.

3.3.1 DLMs for longitudinal data

Suppose that the value of a quantity y is observed for m statistical units over time, so we have a multivariate time series $(Y_t)_{t \geq 1}$, with $Y_t = (Y_{1,t}, \ldots, Y_{m,t})'$. In fact, the simplest approach would be to study the m series independently, specifying a univariate model for each of them. This approach might give fairly good forecasts, but it doesn't do what is sometimes called "borrowing strength;" that is, in predicting Y for individual i, say, we do not exploit the information provided by the similar time series $(Y_{j,t})$ for $j \neq i$. In order to use all the available information, we want to introduce dependence across the time series, that is, we want a joint model for the m-variate process $((Y_{1,t}, \ldots, Y_{m,t}) : t = 1, 2, \ldots)$.

One way of introducing dependence across the time series $(Y_{1,t}), \ldots, (Y_{m,t})$ is the following. Suppose that we can reasonably model each of the m time series using the same type of DLM, possibly with different variances but with the same time-invariant matrices G and F; that is

$$\begin{aligned} Y_{i,t} &= F\theta_t^{(i)} + v_{i,t}, \qquad v_{i,t} \sim \mathcal{N}(0, V_i) \\ \theta_t^{(i)} &= G\theta_{t-1}^{(i)} + w_t^{(i)}, \quad w_{i,t} \sim \mathcal{N}_p(0, W_i), \end{aligned} \tag{3.38}$$

$i = 1, \ldots, m$. This corresponds to the qualitative assumption that all series follow the same type of dynamics. It also implies that the components of the state vectors have similar interpretations across the different DLM, but they can assume different values for each time series $(Y_{i,t})$. Furthermore, as we shall discuss more extensively in Chapter 4, the variance matrices of the individual DLMs (3.38) might not be completely known, but be dependent on some unknown parameters, ψ_i say. For example, in a regression DLM of the kind discussed in Section 3.2.7, we would have

$$\begin{aligned}
Y_{i,t} &= \alpha_{i,t} + \beta_{i,t} x_t + v_{i,t}, & v_{i,t} &\overset{iid}{\sim} N(0, \sigma_i^2)\\
\alpha_{i,t} &= \alpha_{i,t-1} + w_{1t,i}, & w_{1t,i} &\overset{iid}{\sim} N(0, \sigma_{w_1,i}^2)\\
\beta_{i,t} &= \beta_{i,t-1} + w_{2t,i}, & w_{2t,i} &\overset{iid}{\sim} N(0, \sigma_{w_2,i}^2)
\end{aligned}$$

so that the model for $Y_{i,t}$ is characterized by its own state vector $\theta_t^{(i)} = (\alpha_{i,t}, \beta_{i,t})$ and by its parameters $\psi_i = (\sigma_i^2, \sigma_{w_1,i}^2, \sigma_{w_2,i}^2)$. In this framework we can assume that the time series $(Y_{1,t}), \ldots, (Y_{m,t})$ are conditionally independent given the state processes $(\theta_t^{(1)}), \ldots, (\theta_t^{(m)})$ and the parameters $(\psi_1, \ldots, \psi_m)$, with $(Y_{i,t})$ depending only on its state $(\theta_t^{(i)})$ and parameters ψ_i; in particular

$$(Y_{1,t}, \ldots, Y_{m,t}) | \theta_t^{(1)}, \ldots, \theta_t^{(m)}, \psi_1, \ldots, \psi_m \sim \prod_{i=1}^{m} \mathcal{N}(y_{i,t}; F\theta_t^{(i)}, V_i(\psi_i)).$$

Note that this framework is similar to the one studied in Section 1.3. A dependence among $Y_{1,t}, \ldots, Y_{m,t}$ can be introduced through the joint probability law of $(\theta_t^{(1)}, \ldots, \theta_t^{(m)})$ and/or of the parameters $(\psi_1, \ldots, \psi_m)$. If the states and the parameters are independent across individuals, then the time series $(Y_{1,t}), \ldots, (Y_{m,t})$ are independent, and there is no borrowing strength; on the other hand, a dependence across the individual states and parameters implies dependence across the m time series. In the next sections we will provide some examples, where, for simplicity, the parameters, i.e., the matrices F, G, V_i, W_i, are known; their estimate (MLE and Bayesian) will be discussed in Chapter 4. A recent reference for mixtures of DLMs is Frühwirth-Schnatter and Kaufmann (2008). Also related are the articles by Caron et al. (2008) and Lau and So (2008).

3.3.2 Seemingly unrelated time series equations

Seemingly unrelated time series equations (SUTSE) are a class of models which specify the dependence structure among the state vectors $\theta_t^{(1)}, \ldots, \theta_t^{(m)}$ as follows. As we said, the model (3.38) corresponds to the qualitative assumption that all series follow the same type of dynamics, and that the components of the state vectors have similar interpretations across the different DLMs. For example, each series might be modeled using a linear growth model, so that for each of them the state vector has a level and a slope component and, although not strictly required, it is commonly assumed for simplicity that the variance matrix of the system errors is diagonal. This means that the evolution of level and slope is governed by independent random inputs. Clearly, the individual DLMs can be combined to give a DLM for the multivariate observations. A simple way of doing so is to assume that the evolution of the levels of the series is driven by correlated inputs, and the same for the slopes. In other words, at any fixed time, the components of the system error corresponding to the levels of the different series may be correlated, and the components of the system error corresponding to the different slopes may be correlated as well. To keep the model simple, we retain the assumption that levels and slopes evolve in an uncorrelated way. This suggests describing the joint evolution of the state vectors by grouping together all the levels and then all the slopes in an overall state vector $\theta_t = (\mu_{1,t}, \ldots, \mu_{m,t}, \beta_{1,t}, \ldots, \beta_{m,t})'$. The system error of the dynamics of this common state vector will then be characterized by a block-diagonal variance matrix having a first $m \times m$ block accounting for the correlation among levels and a second $m \times m$ block accounting for the correlation among slopes. To be specific, suppose one has $m = 2$ series. Then $\theta_t = (\mu_{1,t}, \mu_{2,t}, \beta_{1,t}, \beta_{2,t})'$ and the system equation is

$$\begin{bmatrix} \mu_{1,t} \\ \mu_{2,t} \\ \beta_{1,t} \\ \beta_{2,t} \end{bmatrix} = \begin{bmatrix} 1 & 0 & 1 & 0 \\ 0 & 1 & 0 & 1 \\ 0 & 0 & 1 & 0 \\ 0 & 0 & 0 & 1 \end{bmatrix} \begin{bmatrix} \mu_{1,t-1} \\ \mu_{2,t-1} \\ \beta_{1,t-1} \\ \beta_{2,t-1} \end{bmatrix} + \begin{bmatrix} w_{1,t} \\ w_{2,t} \\ w_{3,t} \\ w_{4,t} \end{bmatrix}, \tag{3.39a}$$

where $(w_{1,t}, w_{2,t}, w_{3,t}, w_{4,t})' \sim \mathcal{N}(0, W)$ and

$$W = \left[\begin{array}{c|c} W_\mu & \begin{matrix} 0 & 0 \\ 0 & 0 \end{matrix} \\ \hline \begin{matrix} 0 & 0 \\ 0 & 0 \end{matrix} & W_\beta \end{array}\right]. \tag{3.39b}$$

The observation equation for the bivariate time series $((Y_{1,t}, Y_{2,t})' : t = 1, 2, \ldots)$ is

$$\begin{bmatrix} Y_{1,t} \\ Y_{2,t} \end{bmatrix} = \begin{bmatrix} 1 & 0 & 0 & 0 \\ 0 & 1 & 0 & 0 \end{bmatrix} \theta_t + \begin{bmatrix} v_{1,t} \\ v_{2,t} \end{bmatrix}, \tag{3.39c}$$

with $(v_{1,t}, v_{2,t})' \sim \mathcal{N}(0, V)$. In order to introduce a further correlation between the series, the observation error variance V can be taken nondiagonal.

The previous example can be extended to the general case of m univariate time series. Let Y_t denote the multivariate observation at time t, and suppose that the ith component of Y_t follows a time invariant DLM as in (3.38), where $\theta_t^{(i)} = (\theta_{1,t}^{(i)}, \ldots, \theta_{p,t}^{(i)})'$ for $i = 1, \ldots, m$. Then a SUTSE model for (Y_t) has the form[4]

[4] Given two matrices A and B, of dimensions $m \times n$ and $p \times q$ respectively, the Kronecker product $A \otimes B$ is the $mp \times nq$ matrix defined as

$$\begin{bmatrix} a_{1,1}B & \cdots & a_{1,n}B \\ \vdots & \vdots & \vdots \\ a_{m,1}B & \cdots & a_{m,n}B \end{bmatrix}.$$

$$\begin{cases} Y_t = (F \otimes I_m)\,\theta_t + v_t, & v_t \sim \mathcal{N}(0, V), \\ \theta_t = (G \otimes I_m)\,\theta_{t-1} + w_t, & w_t \sim \mathcal{N}(0, W), \end{cases} \tag{3.40}$$

with $\theta_t = (\theta_{1,t}^{(1)}, \theta_{1,t}^{(2)}, \ldots, \theta_{p,t}^{(m-1)}, \theta_{p,t}^{(m)})'$. When the $w_t^{(i)}$ have diagonal variances, it is common to assume for W a block-diagonal structure with p blocks of size m. An immediate implication of the structure of the model is that forecasts made at time t of $\theta_{t+k}^{(i)}$ or $Y_{i,t+k}$ are based on the distribution of $\theta_t^{(i)}$ given all the observations $y_{1:t}$.

Example — Annual Denmark and Spain investments

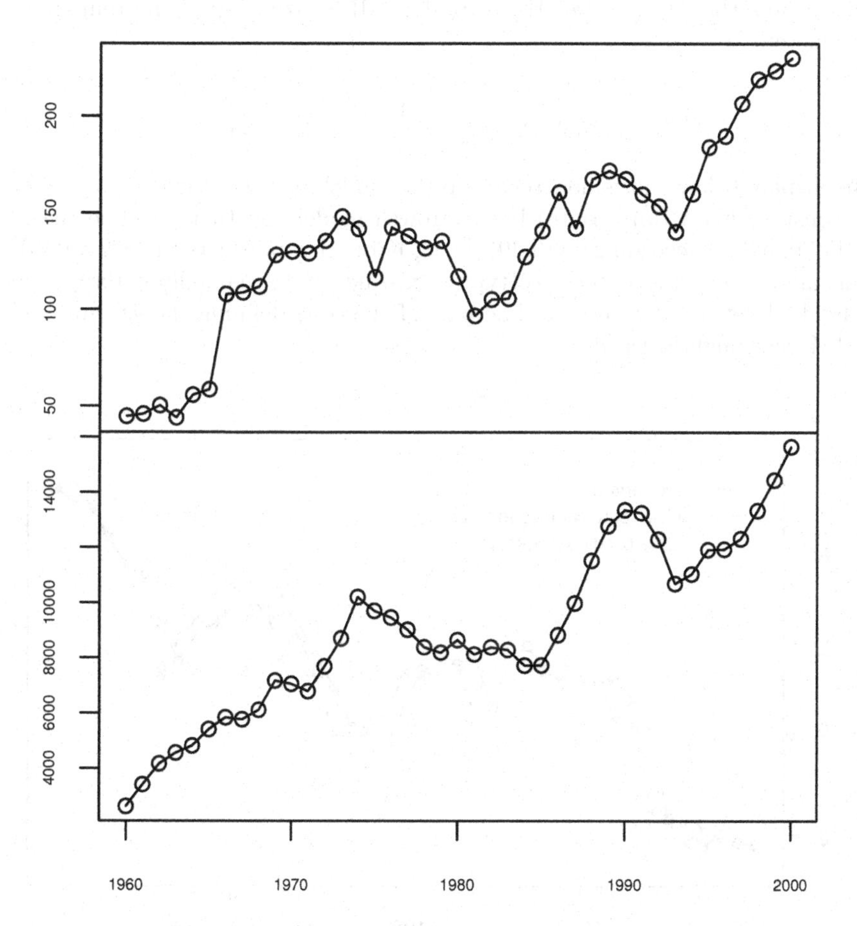

Fig. 3.15. Annual Denmark and Spain investments

Figure 3.15 shows the annual investment in Denmark and Spain from 1960 to 2000[5]. From visual inspection it appears that the two series display the same

[5] Source: `http://www.fgn.unisg.ch/eumacro/macrodata`.

type of qualitative behavior, that can be modeled by a linear growth DLM. This is the model we used on page 97 for the investments in Spain series alone. To set up a multivariate model for the two series one can combine the two linear growth models in a comprehensive SUTSE model. This turns out to be exactly of the form described by (3.39). There are six variances and three covariances in the model, for a total of nine parameters that need to be specified—or estimated from the data, as we will see in the next chapter. It is convenient to simplify slightly the model in order to reduce the overall number of parameters. So, for this example, we are going to assume that the two individual linear growth models are in fact integrated random walks. This means that, in (3.39b), $W_\mu = 0$. The MLE estimates of the remaining parameters are

$$W_\beta = \begin{bmatrix} 49 & 155 \\ 155 & 437266 \end{bmatrix}, \qquad V = \begin{bmatrix} 72 & 1018 \\ 1018 & 14353 \end{bmatrix}.$$

The display below shows how to set up the model in R. In doing this we start by constructing a (univariate) linear growth model and then redefine the F and G matrices according to (3.40), using Kronecker products (lines 2 and 3). This approach is less subject to typing mistakes than manually entering the individual entries of F and G. The part of the code defining the variances V and W is straightforward.

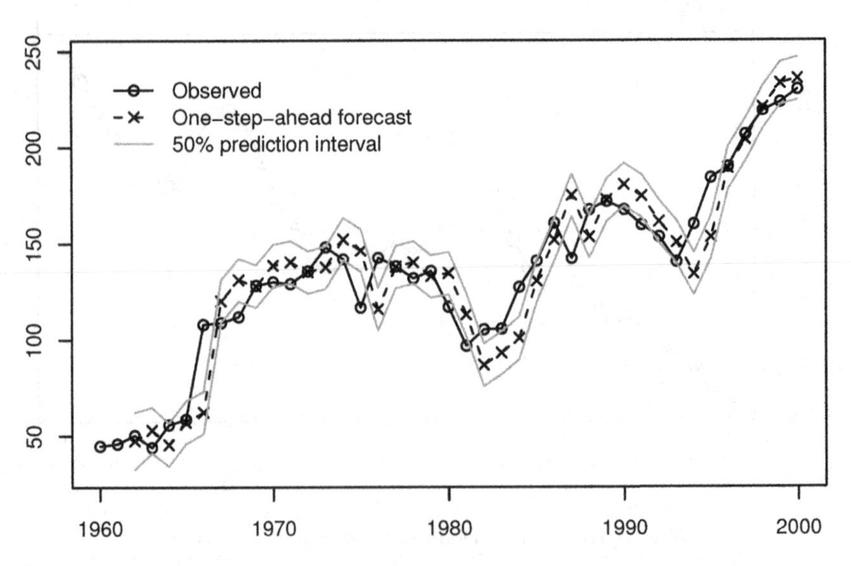

Fig. 3.16. Denmark investments and one step-ahead forecasts

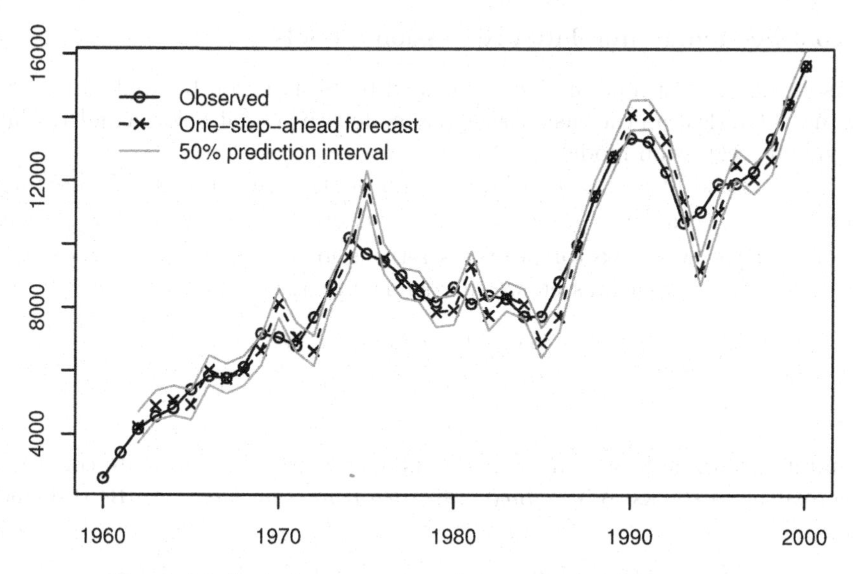

Fig. 3.17. Spain investments and one step-ahead forecasts

R code

```
> mod <- dlmModPoly(2)
> mod$FF <- mod$FF %x% diag(2)
> mod$GG <- mod$GG %x% diag(2)
> W1 <- matrix(0, 2, 2)
> W2 <- diag(c(49, 437266))
> W2[1, 2] <- W2[2, 1] <- 155
> mod$W <- bdiag(W1, W2)
> V <- diag(c(72, 14353))
> V[1, 2] <- V[2, 1] <- 1018
> mod$V <- V
> mod$m0 <- rep(0, 4)
> mod$C0 <- diag(4) * 1e7
> investFilt <- dlmFilter(invest, mod)
> sdev <- residuals(investFilt)$sd
> lwr <- investFilt$f + qnorm(0.25) * sdev
> upr <- investFilt$f - qnorm(0.25) * sdev
```

The code also illustrates how to compute probability intervals for the one-step-ahead forecasts, shown in Figures 3.16 and 3.17. Note that conditionally on $y_{1:t-1}$, Y_t and e_t have the same variance, see Section 2.8. This justifies the use of the innovation variances in lieu of the one-step-ahead observation forecast variances on line 14.

3.3.3 Seemingly unrelated regression models

As an example of how the idea expressed by SUTSE can be applied to more general DLMs than the basic structural model, we present below a multivariate dynamic regression model.

Let us consider a multivariate version of the dynamic CAPM illustrated on pages 122-125. Again, let $x_t = r_t^{(M)} - r_t^{(f)}$ be the excess market returns, but we have now a vector of excess returns for m assets, $y_{i,t} = r_{i,t} - r_t^{(f)}$, $i = 1, \ldots, m$. For each asset, we can define a regression DLM

$$\begin{aligned} y_{i,t} &= \alpha_{i,t} + \beta_{i,t} x_t + v_{i,t} \\ \alpha_{i,t} &= \alpha_{i,t-1} + w_{1i,t} \\ \beta_{i,t} &= \beta_{i,t-1} + w_{2i,t}, \end{aligned}$$

and it is sensible to assume that the intercepts and the slopes are correlated across the m stocks. A *seemingly unrelated regression model* (SUR) is defined as

$$\begin{aligned} y_t &= (F_t \otimes I_m)\theta_t + v_t, & v_t &\overset{iid}{\sim} \mathcal{N}(0, V), \\ \theta_t &= (G \otimes I_m)\theta_{t-1} + w_t, & w_t &\overset{iid}{\sim} \mathcal{N}(0, W), \end{aligned}$$

with

$$y_t = \begin{bmatrix} y_{1,t} \\ \vdots \\ y_{m,t} \end{bmatrix}, \quad \theta_t = \begin{bmatrix} \alpha_{1,t} \\ \vdots \\ \alpha_{m,t} \\ \beta_{1,t} \\ \vdots \\ \beta_{m,t} \end{bmatrix}, \quad v_t = \begin{bmatrix} v_{1,t} \\ \vdots \\ v_{m,t} \end{bmatrix}, \quad w_t = \begin{bmatrix} w_{1,t} \\ \vdots \\ w_{2m,t} \end{bmatrix},$$

$F_t = \begin{bmatrix} 1 & x_t \end{bmatrix}$, $G = I_2$, and $W = \text{blockdiag}(W_\alpha, W_\beta)$.

The data we analyze are the monthly returns for the Mobil, IBM, Weyer, and Citicorp stocks, 1978.1-1987.12, described on page 122. We assume for simplicity that the $\alpha_{i,t}$ are time-invariant, which amounts to assuming that $W_\alpha = 0$. The correlation between the different excess returns is explained in terms of the nondiagonal variance matrices V and W_β, estimated from the data:

$$V = \begin{bmatrix} 41.06 & 0.01571 & -0.9504 & -2.328 \\ 0.01571 & 24.23 & 5.783 & 3.376 \\ -0.9504 & 5.783 & 39.2 & 8.145 \\ -2.328 & 3.376 & 8.145 & 39.29 \end{bmatrix},$$

$$W_\beta = \begin{bmatrix} 8.153 \cdot 10^{-7} & -3.172 \cdot 10^{-5} & -4.267 \cdot 10^{-5} & -6.649 \cdot 10^{-5} \\ -3.172 \cdot 10^{-5} & 0.001377 & 0.001852 & 0.002884 \\ -4.267 \cdot 10^{-5} & 0.001852 & 0.002498 & 0.003884 \\ -6.649 \cdot 10^{-5} & 0.002884 & 0.003884 & 0.006057 \end{bmatrix}.$$

Smoothing estimates of the $\beta_{i,t}$'s, shown in Figure 3.18, can be obtained using the code below.

R code

```
> tmp <- ts(read.table("http://shazam.econ.ubc.ca/intro/P.txt",
+                      header = TRUE),
+           start = c(1978, 1), frequency = 12) * 100
> y <- tmp[, 1 : 4] - tmp[, "RKFREE"]
> colnames(y) <- colnames(tmp)[1 : 4]
> market <- tmp[, "MARKET"] - tmp[, "RKFREE"]
> rm("tmp")
> m <- NCOL(y)
> ### Set up the model
> CAPM <- dlmModReg(market)
> CAPM$FF <- CAPM$FF %x% diag(m)
> CAPM$GG <- CAPM$GG %x% diag(m)
> CAPM$JFF <- CAPM$JFF %x% diag(m)
> CAPM$W <- CAPM$W %x% matrix(0, m, m)
> CAPM$W[-(1 : m), -(1 : m)] <-
+     c(8.153e-07,  -3.172e-05, -4.267e-05, -6.649e-05,
+       -3.172e-05,  0.001377,   0.001852,   0.002884,
+       -4.267e-05,  0.001852,   0.002498,   0.003884,
+       -6.649e-05,  0.002884,   0.003884,   0.006057)
> CAPM$V <- CAPM$V %x% matrix(0, m, m)
> CAPM$V[] <- c(41.06,     0.01571, -0.9504, -2.328,
+               0.01571, 24.23,     5.783,   3.376,
+              -0.9504,   5.783,   39.2,     8.145,
+              -2.328,    3.376,    8.145,  39.29)
> CAPM$m0 <- rep(0, 2 * m)
> CAPM$C0 <- diag(1e7, nr = 2 * m)
> ### Smooth
> CAPMsmooth <- dlmSmooth(y, CAPM)
> ### Plot
> plot(dropFirst(CAPMsmooth$s[, m + 1 : m]),
+      lty = c("13", "6413", "431313", "B4"),
+      plot.type = "s", xlab = "", ylab = "Beta")
> abline(h = 1, col = "darkgrey")
> legend("bottomright", legend = colnames(y), bty = "n",
+        lty = c("13", "6413", "431313", "B4"), inset = 0.05)
```

Apparently, while Mobil's beta remained essentially constant during the period under consideration, starting around 1980 the remaining three stocks became less and less conservative, with Weyer and Citicorp reaching the status of aggressive investments around 1984. Note in Figure 3.18 how the estimated betas for the different stocks move in a clearly correlated fashion (with the

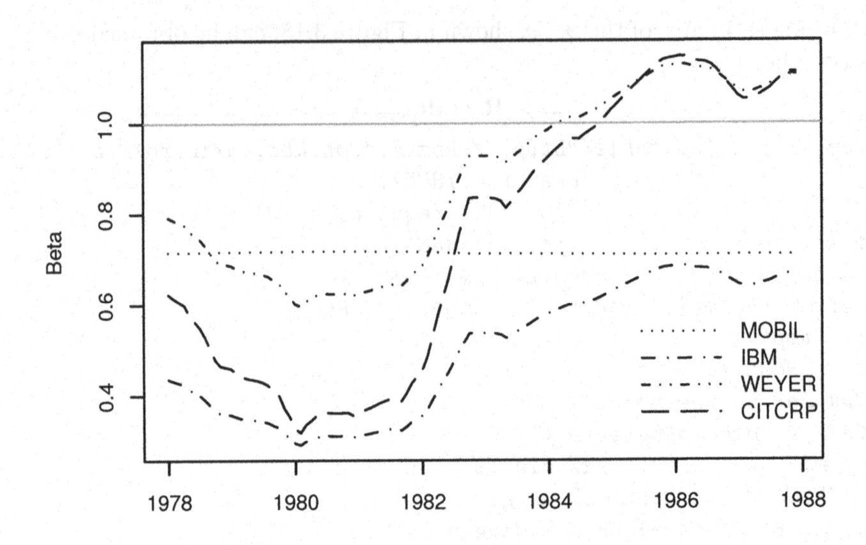

Fig. 3.18. Estimated betas for four stocks

exception of Mobil, that does not move at all)—a consequence of the positive covariances specified in the matrix W_β.

3.3.4 Hierarchical DLMs

Another general class of models for panel data and longitudinal studies is given by *dynamic hierarchical models* (Gamerman and Migon (1993) and references therein), which extend to dynamic systems the hierarchical linear models introduced by Lindley and Smith (1972).

A two-stage hierarchical DLM is specified as follows

$$\begin{aligned} Y_t &= F_{y,t}\theta_t + v_t, & v_t &\sim \mathcal{N}_m(0, V_{y,t}), \\ \theta_t &= F_{\theta,t}\lambda_t + \epsilon_t, & \epsilon_t &\sim \mathcal{N}_P(0, V_{\theta,t}), \\ \lambda_t &= G_t\lambda_{t-1} + w_t, & w_t &\sim \mathcal{N}_k(0, W_t), \end{aligned} \tag{3.41}$$

where the disturbance sequences $(v_t), (\epsilon_t), (w_t)$ are independent, and the matrices $F_{y,t}$ and $F_{\theta,t}$ are of full rank. Thus, in a two-stage DLM the state vector θ_t is itself modeled by a DLM. A key aspect is the progressive reduction in the dimension of the state parameters as the level becomes higher, that is $P > k$.

One application of hierarchical DLM is in modeling random effects in multivariate time series. Suppose that $Y_t = (Y_{1,t}, \ldots, Y_{m,t})'$ are observations of a variable Y for m units at time t, and $Y_{i,t}$ is modeled as

$$Y_{i,t} = F_{1,t}\theta_{i,t} + v_{i,t}, \quad v_{i,t} \sim \mathcal{N}(0, \sigma^2_{i,t}) \tag{3.42a}$$

$i = 1, \ldots, m$, with $v_{1,t}, \ldots, v_{m,t}$ independent, for any given t. In the previous section, we illustrated SUTSE models for introducing a dependence across the individual time series $(Y_{i,t})$; however, these models require us to specify or estimate blocks of $m \times m$ matrices in the covariance matrix W, which might become complicated if m is large. In many applications it is in fact sufficient to model a simpler dependence, in particular to allow individual *random effects*. In its simpler form, this means assuming that

$$\theta_{i,t} = \lambda_t + \epsilon_{i,t}, \qquad \epsilon_{i,t} \sim \mathcal{N}_p(0, \Sigma_t), \tag{3.42b}$$

$$\lambda_t = G\lambda_{t-1} + w_t, \qquad w_t \sim \mathcal{N}_p(0, W_t), \tag{3.42c}$$

with $\epsilon_{1,t}, \ldots, \epsilon_{m,t}$ independent. In other words, for any t the cross-sectional state vectors $\theta_{1,t}, \ldots, \theta_{m,t}$ are a random sample from a $\mathcal{N}_p(\lambda_t, \Sigma_t)$ (i.e., they are conditionally i.i.d. given λ_t, with common distribution $\mathcal{N}_p(\lambda_t, \Sigma_t)$). Thus, we assume the same observation equation for the individual time series $Y_{i,t}$, modeling however heterogeneity across individuals by allowing random effects in the state processes.

Equations (3.42) can be expressed as a hierarchical DLM of the form (3.41), letting $\theta_t = (\theta'_{1,t}, \ldots, \theta'_{m,t})'$, $v_t = (v_{1,t}, \ldots, v_{m,t})'$, $\epsilon_t = (\epsilon'_{1,t}, \ldots, \epsilon'_{m,t})'$, $V_{y,t} = \mathrm{diag}(\sigma^2_{1,t}, \ldots, \sigma^2_{m,t})$, $V_{\theta,t} = \mathrm{diag}(\Sigma_t, \ldots, \Sigma_t)$, $F_{y,t}$ block-diagonal with m blocks given by $F_{1,t}$, and

$$F_{\theta,t} = \left[I_p \mid \cdots \mid I_p\right]'.$$

A dimensionality reduction is obtained by passing from the mp-dimensional vectors θ_t to their p-dimensional common mean λ_t.

An example are dynamic regression models with random effects. Consider

$$Y_{i,t} = x'_{i,t}\theta_{i,t} + v_{i,t}.$$

Here, $Y_{i,t}$ are individual response variables, explained by the same regressors $X_1, \ldots, X_p$ with known value $x_{i,t} = (x_{1,it}, \ldots, x_{p,it})'$ for unit i at time t. Again, random effects in the regression coefficients can be modeled by assuming that, for fixed t, the coefficients for the same regressor are exchangeable, more precisely

$$\theta_{1,t}, \ldots, \theta_{m,t} | \lambda_t \overset{iid}{\sim} \mathcal{N}_p(\lambda_t, \Sigma_t).$$

A dynamics is then specified for (λ_t), e.g., $\lambda_t = \lambda_{t-1} + w_t$, with $w_t \sim \mathcal{N}_p(0, W_t)$.

One way of obtaining filtering and smoothing estimates for hierarchical DLMs is to substitute the expression for θ_t in the observation equation (3.41), obtaining

$$Y_t = F_{y,t}F_{\theta,t}\lambda_t + v^*_t, \qquad v^*_t \sim \mathcal{N}_m(0, F_{y,t}V_{\theta,t}F'_{y,t} + V_{y,t})$$

$$\lambda_t = G_t\lambda_{t-1} + w_t, \qquad w_t \sim \mathcal{N}(0, W_t).$$

This has the form of a DLM and can be estimated by the usual procedures. In particular, for model (3.42), the observation equation reduces to

$$Y_t = F_{y,t}\, F_{\theta,t}\, \lambda_t + v_t^*,$$
$$v_t^* \sim \mathcal{N}(0, \mathrm{diag}(F_{1,t}\Sigma_t F_{1,t}' + \sigma_y^2, \ldots, F_{1,t}V_{\epsilon,t}F_{1,t}' + \sigma_y^2)).$$

Recursive formulae for filtering and prediction for hierarchical DLM are further discussed in Gamerman and Migon (1993). Landim and Gamerman (2000) present extensions to multivariate time series.

3.3.5 Dynamic regression

In many applications one wants to study the dependence of a variable Y on one or more explanatory variables x, over time. Suppose that, for each time t, we observe Y for different values of x, so that we have a time series of cross sectional data of the kind $((Y_{i,t}, x_i), i = 1, \ldots, m, t \geq 1)$ (the values $x_1, \ldots, x_m$ are deterministic, and, for simplicity, we suppose that they are constant over time). For example, in financial applications, $Y_{i,t}$ might be the yield at time t of a zero coupon bond that gives one euro at time-to-maturity x_i. Data of this kind are plotted in Figure 3.19. Here the data are monthly yields from January 1985 to December 2000, with times-to-maturity of 3, 6, 9, 12, 15, 18, 21, 24, 30, 36, 48, 60, 72, 84, 96, 108, and 120 months [6].

Several issues arise for data of this nature. From the cross-sectional observations available at time t, we can estimate the regression function of Y_t on x, that is $m_t(x) = E(Y_t|x)$, for example with the aim of interpolating y at a new value x_0. On the other hand, it is clearly of interest to study the temporal evolution of the m time series $(Y_{i,t} : t \geq 1)$. One might want to predict $Y_{i,t+1}$ based on the data $(Y_{i,1}, \ldots, Y_{i,t})$, by a univariate time series model; or specify a multivariate model for the m-variate time series $(Y_{1,t}, \ldots, Y_{m,t}), t \geq 1$. However, this approach does not fully exploit all the available information, since it ignores the dependence of the Y_i's on the covariate x. In fact, estimating the dynamics of the regression curve is often the main objective of the analysis.

To consider both aspects of the problem, that is the cross-sectional and the temporal nature of the data, the proposal we illustrate in this section is a (semiparametric) dynamic regression model, written in the form of a DLM.

Suppose for simplicity that x is univariate. A flexible cross-sectional regression model is obtained by considering a basis functions expansion of the regression function at time t, of the form

$$m_t(x) = E(Y_t|x) = \sum_{j=1}^{k} \beta_{j,t}\, h_j(x) \tag{3.43}$$

where $h_j(x)$ are known basis functions (e.g., powers of x: $m_t(x) = \sum_{j=1}^{\infty} \beta_{j,t} x^j$, or trigonometric functions, or splines) and $\beta_t = (\beta_{1,t}, \ldots, \beta_{k,t})'$ is the vector

[6] Source: `http://www.ssc.upenn.edu/~fdiebold/papers/paper49/FBFITTED.txt`

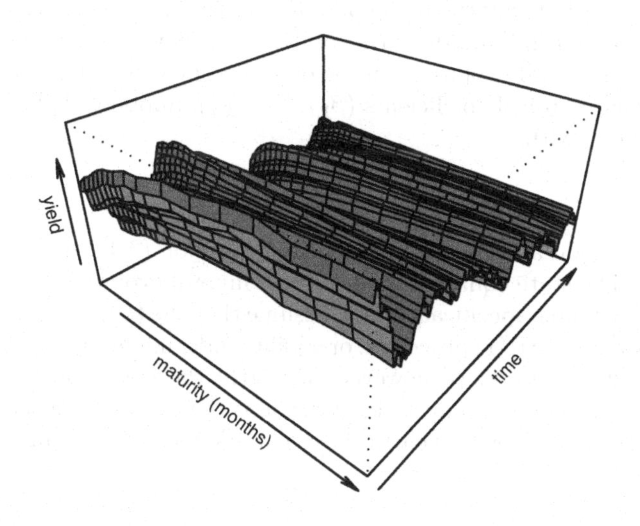

Fig. 3.19. Monthly yields, 1985.01-2000.12, for times-to-maturity 3, 6, 9, 12, 15, 18, 21, 24, 30, 36, 48, 60, 72, 84, 96, 108, and 120 months

of the expansion coefficients. The idea is that, for k large enough (in principle, $k \to \infty$), (3.43) can approximate any interesting shape of the regression function $m_t(x)$; for example, any continuous function on a closed interval can be approximated by polynomials. Models of the kind (3.43) are nevertheless simple since they are linear in the parameters $\beta_{j,t}$; at a given time t, we have a regression model

$$Y_{i,t} = \sum_{j=1}^{k} \beta_{j,t}\, h_j(x_i) + \epsilon_{i,t}, \quad i = 1, \ldots, m,$$

with $\epsilon_{i,t} \overset{iid}{\sim} \mathcal{N}(0, \sigma^2)$; in matrix notation,

$$Y_t = F\beta_t + \epsilon_t, \qquad \epsilon_t \sim \mathcal{N}(0, \sigma^2 I_m), \tag{3.44}$$

where

$$Y_t = \begin{bmatrix} Y_{1,t} \\ \vdots \\ Y_{m,t} \end{bmatrix}, \quad F = \begin{bmatrix} h_1(x_1) & \ldots & h_k(x_1) \\ \vdots & & \vdots \\ h_1(x_m) & \ldots & h_k(x_m) \end{bmatrix} \quad \beta_t = \begin{bmatrix} \beta_{1,t} \\ \vdots \\ \beta_{k,t} \end{bmatrix} \quad \epsilon_t = \begin{bmatrix} \epsilon_{1,t} \\ \vdots \\ \epsilon_{k,t} \end{bmatrix}.$$

Thus, β_t can be easily estimated, e.g., by least squares.

However, the regression curve evolves over time. Clearly, day-by-day cross-sectional estimates do not give a complete picture of the problem. We might have information on the dynamics of the curve that should be included in the analysis. Modeling the dynamics of the curve $m_t(x)$ is not simple at first, since the curve is infinite-dimensional. However, having expressed $m_t(x)$ as in (3.43), its temporal evolution can be described by the dynamics of the finite-dimensional vector of coefficients $(\beta_{1,t}, \ldots, \beta_{k,t})$. Suppose that the series (β_t) is Markovian, with

$$\beta_t = G\beta_{t-1} + w_t, \; w_t \sim \mathcal{N}_k(0, W_t).$$

Then, we obtain a DLM with observation equation (3.44) and state equation as above. The state equation models the temporal evolution of the regression function. A simple specification is to assume that the $(\beta_{j,t})_{t \geq 1}$ are independent random walks or AR(1) processes; or a joint model might be used for the vector β_t. A word of caution is, however, of worth; the state equation introduces additional information, but it also constrains the dynamics of the curve. So, it is delicate: a poor specification of the dynamics may result in an unsatisfactory fit of the data.

3.3.6 Common factors

Sometimes it is conceptually useful to think of a number of observed series as driven by a small number of common factors. This is a common approach for example in economics, where one assumes that many observable series reflect the current state of the economy, which in turn can be expressed as a lower dimensional unobservable time series. For example, suppose that m observed series depend linearly on p $(p < m)$ correlated random walks. The model can be written as

$$\begin{aligned} Y_t &= A\mu_t + v_t, & v_t &\sim \mathcal{N}(0, V), \\ \mu_t &= \mu_{t-1} + w_t, & w_t &\sim \mathcal{N}(0, W), \end{aligned} \tag{3.45}$$

where A is a fixed $m \times p$ matrix of factor loadings. The model can be seen as a dynamic generalization of factor analysis, where the common factors μ_t evolve with time. Note that (3.45) is nothing else than a DLM, with $\theta_t = \mu_t$ and $F_t = A$. One important difference with other DLMs that we have seen in this chapter is that here $p < m$, i.e., the state has a lower dimension than the observation. In addition, the system matrix A does not have any particular structure. As in standard factor analysis, in order to achieve identifiability of the unknown parameters, some constraints have to be imposed. In fact, if H is a $p \times p$ invertible matrix, defining $\tilde{\mu}_t = H\mu_t$ and $\tilde{A} = AH^{-1}$ and multiplying the second equation in (3.45) on the left by H, we obtain the equivalent model

$$\begin{aligned} Y_t &= \tilde{A}\tilde{\mu}_t + v_t & v_t &\sim \mathcal{N}(0, V), \\ \tilde{\mu}_t &= \tilde{\mu}_{t-1} + \tilde{w}_t & \tilde{w}_t &\sim \mathcal{N}(0, HWH'). \end{aligned}$$

Since A and W contain mp and $\frac{1}{2}p(p+1)$ parameters, respectively, but each combination of parameters belongs to a manifold of dimension p^2 (the number of elements of H) of equivalent models, the effective number of free parameters (not including those in V) is $mp - \frac{1}{2}p(p-1)$. One way to parametrize the model and achieve identifiability is to set W equal to the identity matrix, and to impose that $A_{i,j}$, the (i,j) element of A, be zero for $j > i$. The fact that A is $m \times p$, with $p < m$, implies that A can be written in a partitioned form as

$$A = \begin{bmatrix} T \\ B \end{bmatrix},$$

with T being a $p \times p$ lower triangular matrix, and B an $(m-p) \times p$ rectangular matrix. This clearly shows that with this parametrization there are only $\frac{1}{2}p(p+1) + p(m-p) = mp - \frac{1}{2}p(p-1)$ parameters, which is exactly the number of free parameters of the unrestricted model. An alternative parametrization that achieves identifiability is obtained by assuming that W is a diagonal matrix, that $A_{i,i} = 1$ and $A_{i,j} = 0$ for $j > i$.

The model expressed by (3.45) is related to the notion of cointegrated series, introduced by Granger (1981) (see also Engle and Granger; 1987). The components of a vector time series x_t are said to be cointegrated of order d, b, written $x_t \sim \mathrm{CI}(d,b)$, if (i) all the components of x_t are integrated of order d (i.e., $\Delta^d x_t^{(i)}$ is stationary for any i), and (ii) there exists a nonzero vector α such that $\alpha' x_t$ is integrated of order $d-b < d$. The components of Y_t in (3.45), as linear combinations of independent random walks (assuming for simplicity that the components of μ_0 are independent), are integrated of order 1. The columns of A are p vectors in $\mathbb{R}^m$, hence there are at least $m-p$ other linearly independent vectors in $\mathbb{R}^m$ that are orthogonal to the columns of A. For any such α we have $\alpha' A = 0$ and, therefore, $\alpha' Y_t = \alpha' v_t$, i.e., $\alpha' Y_t$ is stationary—in fact, white noise. This shows that $Y_t \sim \mathrm{CI}(1,1)$. In a model where the common factors are stochastic linear trends instead of random walks, one can see that the observable series are $\mathrm{CI}(2,2)$.

Other DLM components that are commonly used as common factors include seasonal components and cycles, especially in economic applications. Further details on dynamic factor models can be found in Harvey (1989). For more recent developments, see also Forni et al. (2000).

3.3.7 Multivariate ARMA models

ARMA models for multivariate, m-dimensional, observations are formally defined as in the univariate case, through the recursive relation

$$Y_t = \sum_{j=1}^{p} \Phi_j Y_{t-j} + \epsilon_t + \sum_{j=1}^{q} \Psi_j \epsilon_{t-j}, \tag{3.46}$$

where (ϵ_t) is an m-variate Gaussian white noise sequence with variance Σ and the Φ_j and Ψ_j are $m \times m$ matrices. Here, without loss of generality, we have taken the mean of the process to be zero. In order for (3.46) to define a stationary process, all the roots of the complex polynomial

$$\det(I - \Phi_1 z - \cdots - \Phi_p z^p) \tag{3.47}$$

must lie outside the unit disk. A DLM representation of a multivariate ARMA process can be formally obtained by a simple generalization of the representation given earlier (Section 3.2.5) for univariate ARMA processes. Namely, in the G matrix each ϕ_j needs to be replaced by a block containing the matrix Φ_j; similarly for the ψ_j in the matrix R, that have to be replaced by Ψ_j blocks. Finally, all the occurrences of a "one" in F, G, and R must be replaced by the identity matrix of order m, and all the occurrences of a "zero" with a block of zeroes of order $m \times m$. For example, let us consider the bivariate ARMA(2,1) process

$$Y_t = \Phi_1 Y_{t-1} + \Phi_2 Y_{t-2} + \epsilon_t + \Psi_1 \epsilon_{t-1}, \qquad \epsilon_t \sim \mathcal{N}(0, \Sigma), \tag{3.48}$$

with

$$\Psi_1 = \begin{bmatrix} \Psi_{11} & \Psi_{12} \\ \Psi_{21} & \Psi_{22} \end{bmatrix}, \qquad \Phi_i = \begin{bmatrix} \Phi_{11,i} & \Phi_{12,i} \\ \Phi_{21,i} & \Phi_{22,i} \end{bmatrix}, \quad i = 1, 2. \tag{3.49}$$

Then the system and observation matrices needed to define the DLM representation of (3.48) are the following:

$$F = \begin{bmatrix} 1 & 0 & 0 & 0 \\ 0 & 1 & 0 & 0 \end{bmatrix},$$
$$G = \begin{bmatrix} \Phi_{11,1} & \Phi_{12,1} & 1 & 0 \\ \Phi_{21,1} & \Phi_{22,1} & 0 & 1 \\ \Phi_{11,2} & \Phi_{12,2} & 0 & 0 \\ \Phi_{21,2} & \Phi_{22,2} & 0 & 0 \end{bmatrix}, \tag{3.50}$$
$$R = \begin{bmatrix} 1 & 0 \\ 0 & 1 \\ \Psi_{11} & \Psi_{12} \\ \Psi_{21} & \Psi_{22} \end{bmatrix}, \qquad W = R \Sigma R'.$$

In R, the function ***dlmModARMA*** can be used to create a DLM representation of an ARMA model also in the multivariate case. The contributed package dse1 also provides tools for the analysis of multivariate ARMA models.

For a detailed treatment of multivariate ARMA models the reader can consult Reinsel (1997) and Lütkepohl (2005).

In many applications, in particular in econometrics, ARMA models with no moving average part are often considered, as they are more easily interpretable. Multivariate AR models are commonly called *vector autoregressive* (VAR) models. In econometrics, VAR models are widely used both for forecasting and for detecting relationships between macroeconomic variables or groups of variables. Suppose we partition Y_t into a group of k variables and a group of $m - k$ variables, X_t and Z_t say, so that $Y_t' = (X_t', Z_t')$. It is often of interest to study causal relationships between (X_t) and (Z_t). There are several ways of rigorously defining this problem. One possibility is to appeal to the notion of *Granger causality*, which is based on assessing the improvement in the forecasting of one variable produced by the inclusion of the other in the information set upon which the forecast is based. More precisely, we say that (X_t) is Granger-causal for (Z_t) if for all t

$$\mathrm{Var}\big(\mathrm{E}(Z_{t+h}|z_{1:t})\big) > \mathrm{Var}\big(\mathrm{E}(Z_{t+h}|z_{1:t}, x_{1:t})\big)$$

for some $h \geq 1$.

Instantaneous causality is another notion of causal relationship between two groups of variables that is often used in macroeconometric analysis. We say that there is instantaneous causality between (X_t) and (Z_t) if for all t

$$\mathrm{Var}\big(\mathrm{E}(Z_{t+1}|y_{1:t})\big) > \mathrm{Var}\big(\mathrm{E}(Z_{t+1}|y_{1:t}, x_{t+1})\big).$$

Yet another way of studying relationships among variables is the approach based on impulse response analysis. Loosely speaking, and without entering in the details, in this type of analysis one tries to understand how one variable responds to a shock in another variable in a system that includes other variables as well.

It turns out that also the class of VAR models is too large and the interpretation of a particular model may not be easy. In particular, impulse responses are generally not unique. To overcome this issue, structural restrictions can be imposed on VAR models, leading to the so-called *structural* VARs; see, e.g., Amisano and Giannini (1997). VARs and structural VARs provide also an appropriate setting to apply the notion of cointegration (see p. 139), which has received a lot of attention from econometricians in the last twenty years.

We will not treat VAR models or cointegration in this book; the interested reader can consult Lütkepohl (2005), Canova (2007), or Pfaff (2008a), the latter showing also implementations in R. The contributed package vars (Pfaff; 2008b) can be used for the analysis in R of VAR models.

To conclude this section, let us point out that macroeconomic analysis is an area where expert opinion, in the form of informative prior distributions for unknown model parameters, is often available and can be fruitfully incorporated into a model to improve forecasts and inference in general. However, specifying in a meaningful way a prior for the many parameters of a VAR model may not be an easy task, and several simplified ways of doing so have been suggested, starting with the so-called "Minnesota prior" (Litterman; 1986; Doan et al.; 1984). There is now a vast literature on Bayesian VAR's and we refer the reader to Canova (2007) and references therein. In R, the contributed package MSBVAR (Brandt; 2008) can be used for the analysis of Bayesian VAR models.

Problems

3.1. Simulate patterns of the random walk-plus-noise model for different values of V and W.

3.2. Show that, for the local level model, $\lim_{t\to\infty} C_t = KV$, where K is defined by (3.8).

3.3. Let $(Y_t, t = 1, 2, \ldots)$ be described by a random walk plus noise model. Show that the first differences $Z_t = Y_t - Y_{t-1}$ are stationary and have the same autocorrelation function of a MA(1) model.

3.4. Simulate patterns of the linear growth model for different values of V and W_1, W_2.

3.5. Let $(Y_t, t = 1, 2, \ldots)$ be described by a linear growth model. Show that the second differences $Z_t = Y_t - 2Y_{t-1} + Y_{t-2}$ are stationary and have the same autocorrelation function of a MA(2) model.

3.6. Show that the forecast function for the polynomial model of order n is polynomial of order $n - 1$.

3.7. Verify that the second differences of (Y_t) for a linear growth model can be written in terms of the innovations as in (3.12).

3.8. Consider the data studied in Section 3.2.6. Estimate a global trend model, $y_t = \mu_0 + \delta t + \epsilon_t$. Look at the one-step-ahead forecast residuals, and make your comments. Then, compare with the integrated random walk and the linear growth model for these data.

3.9. Show that a Fourier form seasonal DLM has a periodic forecast function, even when W is a general variance matrix.

3.10. Prove that (3.31) holds.

3.11. Simulate paths of a stationary bivariate VAR(1) process. Compare with paths of a nonstationary VAR(1) process.

4

Models with unknown parameters

In the previous chapters we presented some basic DLMs for time series analysis, assuming that the system matrices F_t, G_t, V_t, and W_t were known. This was done to more easily study their behavior and general properties. In fact, in time series applications the matrices in the DLM are very rarely completely known. In this chapter we let the model matrices depend on a vector of unknown parameters, ψ say. The unknown parameters are usually constant over time, but we also give some examples where they may have a temporal evolution. Anyway, the dynamics of ψ_t will be such as to maintain the linear, Gaussian structure of DLMs.

In a classical framework one typically starts by estimating ψ, usually by maximum likelihood. If the researcher is only interested in the unknown parameters, the analysis terminates here; if, on the other hand, he is interested in smoothing or forecasting the values of the observed series or those of the state vectors, the customary way to proceed is to use the estimated value of ψ as if it were a known constant, and apply the relevant techniques of Chapter 2 for forecasting or smoothing.

From a Bayesian standpoint, unknown parameters are instead random quantities, as we discussed in Chapter 1: therefore, in the context of DLMs, the posterior distribution of interest is the joint conditional distribution of the state vectors—or of future measurements—and the unknown parameter ψ, given the observations. As we shall see, Bayesian inference, even if simple in principle, involves computations that are usually not analytically manageable; however, Markov chain Monte Carlo and modern sequential Monte Carlo methods can be quite efficient in providing an approximation of the posterior distributions of interest.

In Section 4.1 we discuss maximum likelihood estimation of an unknown parameter occurring in the specification of a DLM, while the rest of the chapter is devoted to Bayesian inference.

G. Petris et al., *Dynamic Linear Models with R*, Use R, DOI: 10.1007/b135794_4,

4.1 Maximum likelihood estimation

Suppose that we have n random vectors, $Y_1, \ldots, Y_n$, whose distribution depends on an unknown parameter ψ. We will denote the joint density of the observations for a particular value of the parameter, by $p(y_1, \ldots, y_n; \psi)$. The likelihood function is defined to be, up to a constant factor, the probability density of the observed data read as a function of ψ, i.e., denoting the likelihood by L, we can write $L(\psi) = \text{const.} \cdot p(y_1, \ldots, y_n; \psi)$. For a DLM it is convenient to write the joint density of the observations in the form

$$p(y_1, \ldots, y_n; \psi) = \prod_{t=1}^{n} p(y_t | y_{1:t-1}; \psi), \tag{4.1}$$

where $p(y_t | y_{1:t-1}; \psi)$ is the conditional density of y_t given the data up to time $t-1$, assuming that ψ is the value of the unknown parameter. We know from Chapter 2 that the terms occurring in the RHS of (4.1) are Gaussian densities with mean f_t and variance Q_t. Therefore we can write the loglikelihood as

$$\ell(\psi) = -\frac{1}{2} \sum_{t=1}^{n} \log |Q_t| - \frac{1}{2} \sum_{t=1}^{n} (y_t - f_t)' Q_t^{-1} (y_t - f_t), \tag{4.2}$$

where the f_t and the Q_t depend implicitly on ψ. The expression (4.2) can be numerically maximized to obtain the MLE of ψ:

$$\hat{\psi} = \underset{\psi}{\operatorname{argmax}}\, \ell(\psi). \tag{4.3}$$

Denote by H the Hessian matrix of $-\ell(\psi)$, evaluated at $\psi = \hat{\psi}$. The matrix H^{-1} provides an estimate of the variance of the MLE, $\operatorname{Var}(\hat{\psi})$. Conditions for consistency as well as asymptotic normality of the MLE can be found in Caines (1988) and Hannan and Deistler (1988). See also Shumway and Stoffer (2000) for an introduction. For most of the commonly used DLM, however, the usual consistency and asymptotic normality properties of MLE hold.

A word of caution about numerical optimization is in order. The likelihood function for a DLM may present many local maxima. This implies that starting the optimization routine from different starting points may lead to different maxima. It is therefore a good idea to start the optimizer several times from different starting values and compare the corresponding maxima. A rather flat likelihood is another problem that one may face when looking for a MLE. In this case the optimizer, starting from different initial values, may end up at very different points corresponding to almost the same value of the likelihood. The estimated variance of the MLE will typically be very large. This is a signal that the model is not well identifiable. The solution is usually to simplify the model, eliminating some of the parameters, especially when one is interested in making inference and interpreting the parameters themselves. On the other hand, if smoothing or forecasting is the focus, then sometimes even a model that is poorly identified in terms of its parameters may produce good results.

R provides an extremely powerful optimizer with the function `optim`, which is used inside the function `dlmMLE` in package dlm. In the optimization world it is customary to minimize functions, and `optim` is no exception: by default it seeks a minimum. Statisticians too, when looking for an MLE, tend to think in terms of minimizing the negative loglikelihood. In line with this point of view, the function `dlmLL` returns the *negative* loglikelihood of a specified DLM for a given data set. In terms of the parameter ψ occurring in the definition of the DLM of interest, one can think of minimizing the compound function obtained in two steps by building a DLM first, and then evaluating its negative loglikelihood, as a function of the matrices defining it. A suggestive graphical representation is the following:

$$\psi \quad \overset{\text{build}}{\Longrightarrow} \quad \text{DLM} \quad \overset{\text{loglik.}}{\Longrightarrow} \quad -\ell(\psi).$$

That is exactly what `dlmMLE` does: it takes a user-defined function `build` that creates a DLM, defines a new function by composing it with `dlmLL`, and passes the result to `optim` for the actual minimization. Consider, for example, the annual precipitation data for Lake Superior (see page 91). By plotting the data, it seems that a polynomial model of order one can provide an adequate description of the phenomenon. The code below shows how to find the MLE of V and W.

R code

```
> y <- ts(read.table("Datasets/lakeSuperior.dat",
+                    skip = 3))[, 2], start = c(1900, 1)
> build <- function(parm) {
+   dlmModPoly(order = 1, dV = exp(parm[1]), dW = exp(parm[2]))
+ }
> fit <- dlmMLE(y, rep(0, 2), build)
> fit$convergence
[1] 0
> unlist(build(fit$par)[c("V", "W")])
        V         W
9.4654447 0.1211534
```

We have parametrized the two unknown variances in terms of their log, so as to avoid problems in case the optimizer went on to examine negative values of the parameters. The value returned by `dlmMLE` is the list returned by the call to `optim`. In particular, the component `convergenge` needs always to be checked: a nonzero value signals that convergence to a minimum has not been achieved. `dlmMLE` has a `...` argument that can be used to provide additional named arguments to `optim`. For example, a call to `optim` including the argument `hessian=TRUE` forces `optim` to return a numerically evaluated

Hessian at the minimum. This can be used to estimate standard errors of the components of the MLE, or more generally its estimated variance matrix, as detailed above. In the previous example we parametrized the model in terms of $\psi = (\log(V), \log(W))$, so that standard errors estimated from the Hessian refer to the MLE of these parameters. In order to get standard errors for the MLE of V and W, one can apply the delta method. Let us recall the general multivariate form of the delta method. Suppose that ψ is h-dimensional, and $g : \mathbb{R}^h \to \mathbb{R}^k$ is a function that has continuous first derivatives. Write $g(\psi) = (g_1(\psi), \ldots, g_k(\psi))$ for any $\psi = (\psi_1, \ldots, \psi_h) \in \mathbb{R}^h$, and define the derivative of g to be the k by h matrix

$$Dg = \begin{bmatrix} \frac{\partial g_1}{\partial \psi_1} & \cdots & \frac{\partial g_1}{\partial \psi_h} \\ \cdots & \cdots & \cdots \\ \frac{\partial g_k}{\partial \psi_1} & \cdots & \frac{\partial g_k}{\partial \psi_h} \end{bmatrix}, \tag{4.4}$$

that is, the ith row of Dg is the gradient of g_i. If $\widehat{\Sigma}$ is the estimated variance matrix of the MLE $\hat{\psi}$, then the MLE of $g(\psi)$ is $g(\hat{\psi})$, and its estimated variance is $Dg(\hat{\psi})\widehat{\Sigma}Dg(\hat{\psi})'$. In the example, $g(\psi) = (\exp(\psi_1), \exp(\psi_2))$, so that

$$Dg(\psi) = \begin{bmatrix} \exp(\psi_1) & 0 \\ 0 & \exp(\psi_2) \end{bmatrix}. \tag{4.5}$$

We can use the Hessian of the negative loglikelihood at the minimum and the delta method to compute in R standard errors of the estimated variances, as the code below shows.

R code

```
> fit <- dlmMLE(y, rep(0, 2), build, hessian = TRUE)
> avarLog <- solve(fit$hessian)
> avar <- diag(exp(fit$par)) %*% avarLog %*%
+   diag(exp(fit$par)) # Delta method
> sqrt(diag(avar)) # estimated standard errors
[1] 1.5059107 0.1032439
```

As an alternative to using the delta method, one can numerically compute the Hessian of the loglikelihood, expressed as a function of the new parameters $g(\psi)$, at $g(\hat{\psi})$. The recommended package `nlme` provides the function `fdHess`, which we put to use in the following piece of code.

R code

```
> avar1 <- solve(fdHess(exp(fit$par), function(x)
+                      dlmLL(y, build(log(x))))$Hessian)
> sqrt(diag(avar1))
[1] 1.5059616 0.1032148 # estimated standard errors
```

In this example one could parametrize the model in terms of V and W, and then use the Hessian returned by *dlmMLE* to compute the estimated standard errors directly. In this case, however, one needs to be careful about the natural restriction of the parameter space, and provide a lower bound for the two variances. Note that the default optimization method, *L-BFGS-B*, is the only method that accepts restrictions on the parameter space, expressed as bounds on the components of the parameter. In the following code, the lower bound 10^{-6} for V reflects the fact that the functions in `dlm` require the matrix V to be nonsingular. On the scale of the data, however, 10^{-6} can be considered zero for all practical purposes.

R code

```
> build <- function(parm) {
+    dlmModPoly(order = 1, dV = parm[1], dW = parm[2])
+ }
> fit <- dlmMLE(y, rep(0.23, 2), build, lower = c(1e-6, 0),
+                  hessian = T)
> fit$convergence
[1] 0
> unlist(build(fit$par)[c("V", "W")])
        V         W
9.4654065 0.1211562
> avar <- solve(fit$hessian)
> sqrt(diag(avar))
[1] 1.5059015 0.1032355
```

To conclude, let us mention the function *StructTS*, in base R. This function can be used to find MLE for the variances occurring in some particular univariate DLM. The argument *type* selects the model to use. The available models are the first order polynomial model (*type="level"*), the second order polynomial model (*type="trend"*), and a second order polynomial model plus a seasonal component (*type="BSM"*). Standard errors are not returned by *StructTS*, nor are they easy to compute from its output.

R code

```
> StructTS(y, "level")

Call:
StructTS(x = y, type = "level")

Variances:
  level  epsilon
 0.1212   9.4654
```

4.2 Bayesian inference

The common practice of plugging the MLE $\hat{\psi}$ in the filtering and smoothing recursions suffers from the difficulties in taking properly into account the uncertainty about ψ. The Bayesian approach offers a more consistent formulation of the problem. The unknown parameters ψ are regarded as a *random* vector. The general hypotheses of state space models for the processes (Y_t) and (θ_t) (assumptions (A.1) and (A.2) on page 40) are assumed to hold *conditionally* on the parameters ψ. Prior knowledge about ψ is expressed through a probability law $\pi(\psi)$. Thus, for any $n \geq 1$, we assume that

$$(\theta_0, \theta_1, \ldots, \theta_n, Y_1, \ldots, Y_n, \psi) \sim \pi(\theta_0|\psi)\pi(\psi) \prod_{t=1}^{n} \pi(y_t|\theta_t, \psi)\pi(\theta_t|\theta_{t-1}, \psi) \tag{4.6}$$

(compare with (2.3)).

Given the data $y_{1:t}$, inference on the unknown state θ_s at time s and on the parameters is solved by computing their joint posterior distribution

$$\pi(\theta_s, \psi|y_{1:t}) = \pi(\theta_s|\psi, y_{1:t})\pi(\psi|y_{1:t}), \tag{4.7}$$

where, as usual, one might be interested in $s = t$, in filtering problems; $s > t$, for state prediction; or $s < t$, for smoothing. The marginal conditional density of θ_s is obtained from (4.7); for example, the filtering density is given by

$$\pi(\theta_t|y_{1:t}) = \int \pi(\theta_t|\psi, y_{1:t})\pi(\psi|y_{1:t})d\psi.$$

Here, the recursion formulae for filtering, given in Chapter 2, can be used to compute the conditional density $\pi(\theta_t|\psi, y_{1:t})$; however, this is now averaged with respect to the posterior distribution of ψ, given the data.

Often, one is interested in reconstructing all the unknown state history up to time t; inference on $\theta_{0:t}$ and ψ, given the data $y_{1:t}$, is expressed through their joint posterior density

$$\pi(\theta_{0:t}, \psi|y_{1:t}) = \pi(\theta_{0:t}|\psi, y_{1:t})\pi(\psi|y_{1:t}). \tag{4.8}$$

In principle, the posterior density (4.8) is obtained from the Bayes rule. In some simple models and using conjugate priors, it can be computed in closed form; examples are given in the following section. More often, computations are analytically intractable. However, MCMC methods and sequential Monte Carlo algorithms provide quite efficient tools for approximating the posterior distributions of interest, and this is one reason for the enormous impulse enjoyed by Bayesian inference for state space models in recent years.

Posterior distribution. MCMC and, in particular, Gibbs sampling algorithms are widely used to approximate the joint posterior $\pi(\theta_{0:t}, \psi|y_{1:t})$. Gibbs sampling from π requires us to iteratively simulate from the full conditional distributions $\pi(\theta_{0:t}|\psi, y_{1:t})$ and $\pi(\psi|\theta_{0:t}, y_{1:t})$. Efficient algorithms for sampling from the full conditional $\pi(\theta_{0:t}|\psi, y_{1:t})$ have been developed, and will be presented in Section 4.4.1. Furthermore, exploiting the conditional independence assumptions of DLMs, the full conditional density $\pi(\psi|\theta_{0:t}, y_{1:t})$ is usually easier to compute than $\pi(\psi|y_{1:t})$. Clearly, this full conditional is problem-specific, but we will provide several examples in the next sections.

We can thus implement Gibbs sampling algorithms to approximate π. Samples from $\pi(\theta_{0:t}, \psi|y_{1:t})$ can also be used for approximating the filtering density $\pi(\theta_t|y_{1:t})$ and the marginal smoothing densities $\pi(\theta_s|y_{1:t}), s < t$. As we shall see, they also allow us to simulate samples from the predictive distribution of the states and observables, $\pi(\theta_{t+1}, y_{t+1}|y_{1:t})$. Thus, this approach solves at the same time the filtering, smoothing, and forecasting problems for a DLM with unknown parameters.

Filtering and on-line forecasting. The shortcoming of the MCMC procedures described above is that they are not designed for recursive or on-line inference. When a new observation y_{t+1} is available, the distribution of interest becomes $\pi(\theta_{0:t+1}, \psi|y_{1:t+1})$ and one has to run a new MCMC all over again to sample from it. This is computationally inefficient, especially in applications that require an *on-line* type of analysis, in which new data arrive rather frequently. As discussed in Chapter 2, one of the attractive properties of DLM is the recursive nature of the filter formulae, which allows us to update the inference efficiently as new data become available. In the case of no unknown parameters in the DLM, one could compute $\pi(\theta_{t+1}|y_{1:t+1})$ from $\pi(\theta_t|y_{1:t})$ by the estimation-error correction formulae given by the Kalman filter, without doing all the computations again. Analogously, when there are unknown parameters ψ, one would like to exploit the samples generated from $\pi(\theta_{0:t}, \psi|y_{1:t})$ in simulating from $\pi(\theta_{0:t+1}, \psi|y_{1:t+1})$, without running the MCMC all over again. Modern sequential Monte Carlo techniques, in particular the family of algorithms that go under the name of *particle filters*, can be used to this aim and allow efficient on-line analysis and simulation-based sequential updating of the posterior distribution of states and unknown parameters. A description of these techniques is postponed until Chapter 5.

4.3 Conjugate Bayesian inference

In some simple cases, Bayesian inference can be carried out in closed form using conjugate priors. We illustrate an example here.

Even in simple structural models such as those presented in Chapter 3, where the system matrices F_t and G_t are known, very rarely are the covariance matrices V_t and W_t completely known. Thus, a basic problem is estimating V_t and W_t. Here we consider a simple case where V_t and W_t are known up to a common scale factor; that is, $V_t = \sigma^2\tilde{V}_t$ and $W_t = \sigma^2\tilde{W}_t$, with σ^2 unknown.

This specification of the covariance matrices has been discussed in Section 1.5 of Chapter 1 for the static linear regression model. A classical example is $V_t = \sigma^2 I_m$; an interesting way of specifying $\tilde{W}_t$ is discussed later, using discount factors.

4.3.1 Unknown covariance matrices: conjugate inference

Let $((Y_t, \theta_t) : t = 1, 2, \ldots)$ be described by a DLM with

$$V_t = \sigma^2 \tilde{V}_t, \qquad W_t = \sigma^2 \tilde{W}_t, \qquad C_0 = \sigma^2 \tilde{C}_0. \tag{4.9}$$

Here all the matrices $\tilde{V}_t$, $\tilde{W}_t$, as well as $\tilde{C}_0$ and all the F_t and G_t, are assumed to be known. The scale parameter σ^2, on the other hand, is unknown. As usual in Bayesian inference, it is convenient to work with its inverse $\phi = 1/\sigma^2$. The uncertainty, therefore, is all in the state vectors and in the parameter ϕ. The DLM provides the conditional probability law of (Y_t, θ_t) given ϕ; in particular the model assumes, for any $t \geq 1$,

$$Y_t | \theta_t, \phi \sim \mathcal{N}_m(F_t \theta_t, \phi^{-1} \tilde{V}_t)$$

$$\theta_t | \theta_{t-1}, \phi \sim \mathcal{N}_p(G_t \theta_{t-1}, \phi^{-1} \tilde{W}_t).$$

As the prior for (ϕ, θ_0), a convenient choice is a conjugate Normal-Gamma prior (see Appendix A), that is

$$\phi \sim \mathcal{G}(\alpha_0, \beta_0), \quad \theta_0 | \phi \sim \mathcal{N}(m_0, \phi^{-1} \tilde{C}_0),$$

denoted as $(\theta_0, \phi) \sim \mathcal{NG}(m_0, \tilde{C}_0, \alpha_0, \beta_0)$. Then we have the following recursive formulae for filtering. Note the analogy with the recursive formulae valid for a DLM with no unknown parameters.

Proposition 4.1. *For the DLM described above, if*

$$(\theta_{t-1}, \phi) | y_{1:t-1} \sim \mathcal{NG}(m_{t-1}, \tilde{C}_{t-1}, \alpha_{t-1}, \beta_{t-1})$$

where $t \geq 1$, then:

(i) The one-step-ahead predictive density of $(\theta_t, \phi) | y_{1:t-1}$ is Normal-Gamma with parameters $(a_t, \tilde{R}_t, \alpha_{t-1}, \beta_{t-1})$, where

$$a_t = G_t m_{t-1}, \qquad \tilde{R}_t = G_t \tilde{C}_{t-1} G_t' + \tilde{W}_t; \tag{4.10}$$

(ii) The one-step-ahead predictive density of $Y_t | y_{1:t-1}$ is Student-t, with parameters $(f_t, \tilde{Q}_t \beta_{t-1}/\alpha_{t-1}, 2\alpha_{t-1})$, where

$$f_t = F_t a_t, \qquad \tilde{Q}_t = F_t \tilde{R}_t F_t' + \tilde{V}_t; \tag{4.11}$$

(iii) The filtering density of $(\theta_t, \phi | y_{1:t})$ *is Normal-Gamma, with parameters*

$$\begin{aligned} m_t &= a_t + \tilde{R}_t F_t \tilde{Q}^{-1}(y_t - f_t) \\ \tilde{C}_t &= \tilde{R}_t - \tilde{R}_t F_t' \tilde{Q}_t^{-1} \tilde{R}_t', \\ \alpha_t &= \alpha_{t-1} + \frac{m}{2}, \\ \beta_t &= \beta_{t-1} + \frac{1}{2}(y_t - f_t)' \tilde{Q}_t^{-1}(y_t - f_t). \end{aligned} \tag{4.12}$$

Proof. (i) Suppose that $\theta_{t-1}, \phi | y_{1:t-1} \sim \mathcal{NG}(m_{t-1}, \tilde{C}_{t-1}, \alpha_{t-1}, \beta_{t-1})$ (this is true for $t = 1$). In particular, this implies that $\phi | y_{1:t-1} \sim \mathcal{G}(\alpha_{t-1}, \beta_{t-1})$.
By (i) of Proposition 2.2, we have that

$$\theta_t | \phi, y_{1:t-1} \sim \mathcal{N}_p(a_t, \phi^{-1} \tilde{R}_t)$$

with a_t and $\tilde{R}_t$ given by (4.10). Therefore $(\theta_t, \phi) | y_{1:t-1} \sim \mathcal{NG}(a_t, \tilde{R}_t, \alpha_t, \beta_t)$.
(ii) It also follows, from part (ii) of Proposition 2.2, that

$$Y_t | \phi, y_{1:t-1} \sim \mathcal{N}_m(f_t, \phi^{-1} \tilde{Q}_t)$$

where f_t and $\tilde{Q}_t$ are given by (4.11). Thus we obtain that $(Y_t, \phi) | y_{1:t-1} \sim \mathcal{NG}(f_t, \tilde{Q}_t, \alpha_{t-1}, \beta_{t-1})$; the corresponding marginal density of $Y_t | y_{1:t-1}$ is Student-t, as given in (ii).
(iii) For a new observation Y_t, the likelihood is

$$Y_t | \theta_t, \phi \sim \mathcal{N}_m(F_t \theta_t, \phi^{-1} \tilde{V}_t).$$

The theory of linear regression with a Normal-Gamma prior discussed in Chapter 1 (page 21; use (1.11) and (1.12)) applies, and leads to the conclusion that (θ_t, ϕ) given $y_{1:t}$ has again a $\mathcal{NG}(m_t, \tilde{C}_t, \alpha_t, \beta_t)$ distribution, with parameters defined by (4.12). □

By the properties of the Student-t distribution we have that the one-step-ahead forecast of Y_t given $y_{1:t-1}$ is $\mathrm{E}(Y_t | y_{1:t-1}) = f_t$, with covariance matrix $\mathrm{Var}(Y_t | y_{1:t-1}) = \tilde{Q}_t \beta_{t-1} / (\alpha_{t-1} - 1)$.

From (iii), the marginal filtering density of $\sigma^2 = \phi^{-1}$ given $y_{1:t}$ is Inverse-Gamma, with parameters (α_t, β_t), so that, for $\alpha_t > 2$,

$$\mathrm{E}(\sigma^2 | y_{1:t}) = \frac{\beta_t}{\alpha_t - 1}, \quad \mathrm{Var}(\sigma^2 | y_{1:t}) = \frac{\beta_t^2}{(\alpha_t - 1)^2 (\alpha_t - 2)}.$$

The marginal filtering density of the states is Student-t:

$$\theta_t | y_{1:t} \sim \mathcal{T}(m_t, \tilde{C}_t \beta_t / \alpha_t, 2\alpha_t),$$

with

$$E(\theta_t|y_{1:t}) = m_t, \quad \text{Var}(\theta_t|y_{1:t}) = \frac{\beta_t}{\alpha_t - 1}\tilde{C}_t. \tag{4.13}$$

Were σ^2 known, the Kalman filter would give the same point estimate, $E(\theta_t|y_{1:t}) = m_t$, with covariance matrix $\text{Var}(\theta_t|y_{1:t}) = \sigma^2\tilde{C}_t$. Instead, the unknown value of σ^2 is replaced in (4.13) by its conditional expectation $\beta_t/(\alpha_t - 1)$. The larger uncertainty is reflected in a filtering density with thicker tails than the Gaussian density (see Exercise 4.2).

As far as smoothing is concerned, note that

$$(\theta_T, \phi|y_{1:T}) \sim \mathcal{NG}(s_T, \tilde{S}_T, \alpha_T, \beta_T), \tag{4.14}$$

with $s_T = m_T$ and $\tilde{S}_T = \tilde{C}_T$, and write

$$\pi(\theta_t, \phi|y_{1:T}) = \pi(\theta_t|\phi, y_{1:T})\pi(\phi|y_{1:T}), \qquad t = 0, \ldots, T. \tag{4.15}$$

Conditional on ϕ, the Normal theory of Chapter 1 applies, showing that (θ_t, ϕ), conditional on $y_{1:T}$ has a Normal-Gamma distribution. The parameters can be computed using recursive formulae that are the analog of those developed for the Normal case. Namely, for $t = T - 1, \ldots, 0$, let

$$\begin{aligned} s_t &= m_t + \tilde{C}_t G'_{t+1} \tilde{R}^{-1}_{t+1}(s_{t+1} - a_{t+1}), \\ \tilde{S}_t &= \tilde{C}_t - \tilde{C}_t G'_{t+1} \tilde{R}^{-1}_{t+1}(\tilde{R}_{t+1} - \tilde{S}_{t+1})\tilde{R}^{-1}_{t+1} G_{t+1} \tilde{C}_t. \end{aligned}$$

Then

$$\theta_t, \phi|y_{1:T} \sim \mathcal{NG}(s_t, \tilde{S}_t, \alpha_T, \beta_T). \tag{4.16}$$

4.3.2 Specification of W_t by discount factors

The Bayesian conjugate analysis discussed in the previous section applies if the matrices $\tilde{V}_t$, $\tilde{W}_t$ and $\tilde{C}_0$ are known, which is a restrictive assumption. However, an interesting case is when $\tilde{V}_t = I_m$ and $\tilde{W}_t$ is specified by a technique that is known as *discount factor*. We give here an informal explanation of discount-factors; for an in-depth discussion, see West and Harrison (1997, Section 6.3).

As often underlined (see, e.g., Chapter 2, pages 57-58), the structure and magnitude of the state covariance matrices W_t has a crucial role in determining the role of past observations in state estimation and forecasting. Think of W_t as diagonal, for simplicity. Large values in the diagonal of W_t imply high uncertainty in the state evolution, so that a lot of sample information is lost in passing from θ_{t-1} to θ_t: the past observations $y_{1:t-1}$ give information about θ_{t-1}, which, however, becomes of little relevance in forecasting θ_t; in fact, the current observation y_t is what mainly determines the estimate of $\theta_t|y_{1:t}$. In the Kalman filter recursions, the uncertainty about θ_{t-1} given $y_{1:t-1}$ is summarized in the conditional covariance matrix $\text{Var}(\theta_{t-1}|y_{1:t-1}) = C_{t-1}$. Moving from θ_{t-1} to θ_t via the state equation $\theta_t = G_t\theta_{t-1} + w_t$, the uncertainty increases and we have $\text{Var}(\theta_t|y_{1:t-1}) = R_t = G_t'C_{t-1}G_t + W_t$. Thus, if $W_t = 0$, i.e., there is no error in the state equation, we have $R_t = \text{Var}(G_t\theta_{t-1}|y_{1:t-1}) = P_t$, say. Otherwise, P_t is increased in $R_t = P_t + W_t$. In this sense, W_t expresses the loss of information in passing from θ_{t-1} to θ_t due to the stochastic error component in the state evolution, the loss depending on the magnitude of W_t with respect to P_t. Therefore, one can think of expressing W_t as a proportion of P_t:

$$W_t = \frac{1-\delta}{\delta} P_t$$

where $\delta \in (0, 1]$. It follows that $R_t = 1/\delta \, P_t$, with $1/\delta > 1$. The parameter δ is called *discount factor* since it "discounts" the matrix P_t that one would have with a deterministic state evolution into the matrix R_t. If $\delta = 1$, $W_t = 0$ and we have no loss of information from θ_{t-1} to θ_t: $\text{Var}(\theta_t|y_{1:t}) = \text{Var}(G_t\theta_{t-1}|y_{1:t-1}) = P_t$. For $\delta < 1$, for example $\delta = 0.8$, we have $\text{Var}(\theta_t|y_{1:t}) = (1/0.8)\ \text{Var}(G_t\theta_{t-1}|y_{1:t-1}) = 1.25 P_t$, showing bigger uncertainty. In practice, the value of the discount factor is usually fixed between 0.9 and 0.99, or it is chosen by model selection diagnostics, e.g., looking at the predictive performance of the model for different values of δ.

The discount factor specification may be used with conjugate priors, that is one can let

$$\tilde{W}_t = \frac{1-\delta}{\delta} G_t'\tilde{C}_{t-1}G_t \tag{4.17}$$

in (4.9). Given $\tilde{C}_0$ and $\tilde{V}_t$ (e.g., $\tilde{V}_t = I_m$), the value of $\tilde{W}_t$ can be recursively computed for every t. The evolution covariance matrix is not time-invariant; however, its dynamics is completely determined once $\tilde{C}_0$, $\tilde{V}_t$ and the system matrices F_t and G_t are given. Further refinements consider different discount factors δ_i for the different components of the state vector.

Example. As a simple illustration, let us consider again the data on the annual precipitations at Lake Superior, which we modeled as a random walk plus noise (see page 145). In Section 4.1, the unknown variances $V_t = \sigma^2$ and $W_t = W$ were estimated by maximum likelihood. Here, we consider Bayesian conjugate inference with $\tilde{W}_t$ specified by the discount factor.

Note that the quantities a_t, f_t, m_t, as well as $\tilde{C}_t$ and $\tilde{R}_t$, can be computed by the Kalman filter for a DLM with known covariance matrices $\tilde{C}_0$ and $\tilde{V}$ (as in (4.9), with $\sigma^2 = 1$) and $\tilde{W}_t$. In fact, the evolution variance $\tilde{W}_t$ is known for $t = 1$, while it is assigned sequentially according to (4.17) for $t > 1$. That is, for each $t = 2, 3, \ldots$, one has to compute $\tilde{W}_t$ from the results obtained at $t-1$, and then use the Kalman filter recursions (4.10)-(4.12). These steps can be easily implemented in R, with a slight modification of the function `dlmFilter`. For the user's convenience, we provide them in the function `dlmFilterDF`, available from the book website. The arguments of `dlmFilterDF` are the data y, the model `mod` (where F_t, G_t and V_t are assumed to be time-invariant) and the discount factor `DF`; the function returns, as for `dlmFilter`, the values m_t,

a_t, f_t and the SVDs of $\tilde{C}_t$ and $\tilde{R}_t$ for any t. In addition, it returns the SVDs of the matrices W_t obtained using the discount factor.

The parameters of the filtering distribution for the unknown precision ϕ can be computed from (4.12), which gives

$$\begin{aligned} \alpha_t &= \alpha_0 + \tfrac{t}{2} \\ \beta_t &= \beta_0 + \tfrac{1}{2}\textstyle\sum_{i=1}^t (y_i - f_i)^2 \tilde{Q}_i^{-1} = \beta_0 + \tfrac{1}{2}\sum_{i=1}^t \tilde{e}_i^2, \end{aligned}$$

where the standardized innovations $\tilde{e}_t$ and the standard deviation $\tilde{Q}_t^{1/2}$ can be obtained with a call to the ***residuals*** function. Finally, one can compute

$$Q_t = \mathrm{Var}(Y_t|y_{1:t-1}) = \tilde{Q}_t \frac{\beta_{t-1}}{\alpha_{t-1} - 1}$$

$$C_t = \mathrm{Var}(\theta_t|y_{1:t}) = \tilde{C}_t \frac{\beta_t}{\alpha_t - 1}.$$

As for smoothing, we can compute s_t and $\tilde{S}_t$ with a call to the ***dlmSmooth*** function. Note that we need, as inputs, the matrices W_t obtained as output of ***dlmFilterDF***. Finally, we obtain

$$S_t = \mathrm{Var}(\theta_t|y_{1:T}) = \tilde{S}_t \frac{\beta_T}{\alpha_T - 1}.$$

For illustration with the Lake Superior data, we choose the prior hyperparameters $m_0 = 0$, $\tilde{C}_0 = 10^7$, $\alpha_0 = 2$, $\beta_0 = 20$, so that $\mathrm{E}(\sigma^2) = \beta_0/(\alpha_0 - 1) = 20$, and a discount factor $\delta = 0.9$. Figure 4.1 shows the filtering and smoothing estimates of the level, m_t and s_t, respectively, with 90% probability intervals (see Problem 4.2), as well as the one-step-ahead point forecasts f_t, with 90% probability intervals.

R code

```
> y <- ts(read.table("Datasets/lakeSuperior.dat",
+                    skip = 3)[, 2], start = c(1900, 1))
> mod <- dlmModPoly(1, dV = 1)
> modFilt <- dlmFilterDF(y, mod, DF = 0.9)
> beta0 <- 20; alpha0 <- 2
> ## Filtering estimates
> out <- residuals(modFilt)
> beta <- beta0 + cumsum(out$res^2) / 2
> alpha <- alpha0 + (1 : length(y)) / 2
> Ctilde <- unlist(dlmSvd2var(modFilt$U.C, modFilt$D.C))[-1]
> prob <- 0.95
> tt <- qt(prob, df = 2 * alpha)
> lower <-  dropFirst(modFilt$m) - tt * sqrt(Ctilde * beta /
+                                                 alpha)
> upper <-  dropFirst(modFilt$m) + tt * sqrt(Ctilde * beta /
```

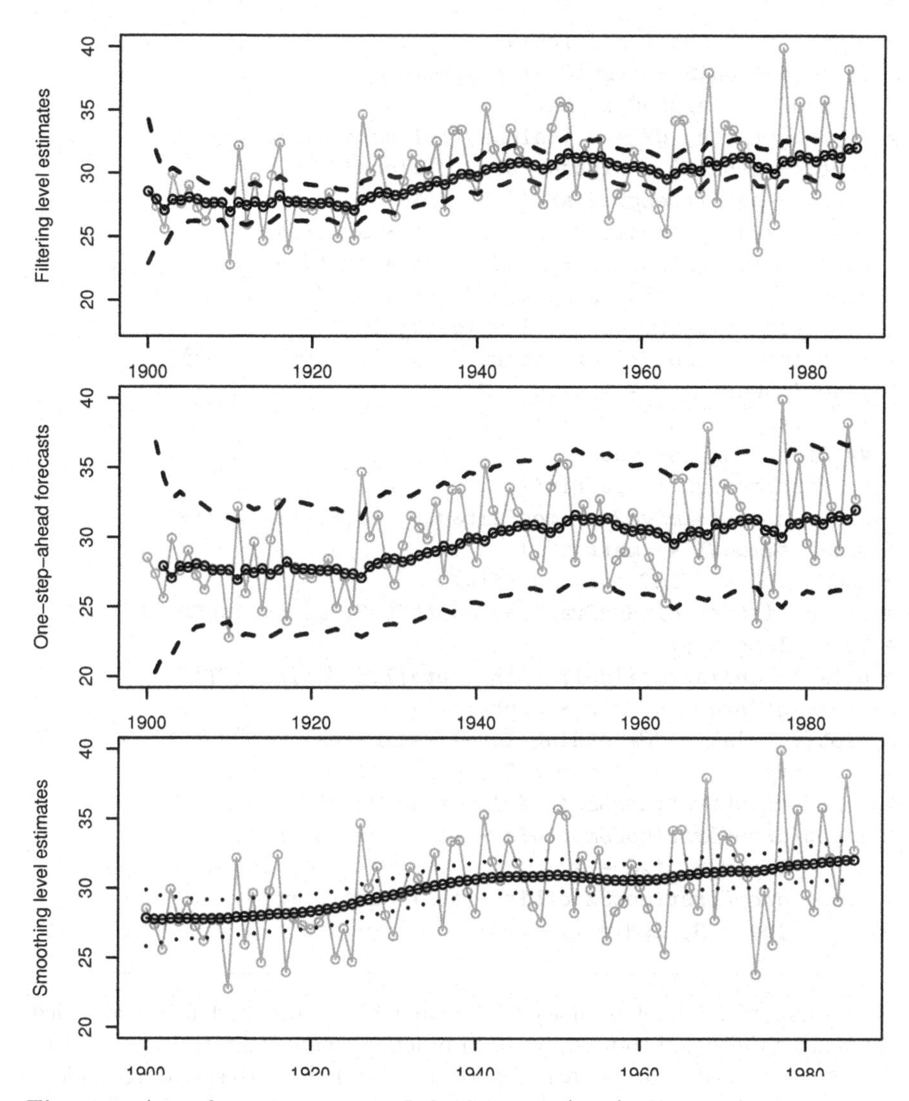

Fig. 4.1. Annual precipitation at Lake Superior (gray), filtering and smoothing level estimates, and one-step-ahead forecasts

```
16 +                                                    alpha)
   > plot(y, ylab = "Filtering level estimates", type = "o",
18 +   ylim = c(18, 40), col = "darkgray")
   > lines(dropFirst(modFilt$m), type = "o")
20 > lines(lower, lty = 2, lwd = 2)
   > lines(upper, lty = 2, lwd = 2)
22 > ## One-step-ahead forecasts
```

```
   > sigma2 <- c(beta0 / (alpha0-1), beta / (alpha-1))
24 > Qt <- out$sd^2 * sigma2[-length(sigma2)]
   > alpha0T = c(alpha0,alpha)
26 > tt <- qt(prob, df = 2 * alpha0T[-length(alpha0T)])
   > parf <- c(beta0 / alpha0, beta / alpha)
28 > parf <- parf[-length(parf)] * out$sd^2
   > lower <- dropFirst(modFilt$f) - tt * sqrt(parf)
30 > upper <- dropFirst(modFilt$f) + tt * sqrt(parf)
   > plot(y, ylab = "One-step-ahead forecasts", type = "o",
32 +      ylim = c(20, 40), col = "darkgray")
   > lines(window(modFilt$f, start = 1902),  type = "o")
34 > lines(lower, lty = 2, lwd = 2)
   > lines(upper, lty = 2, lwd = 2)
36 > ## Smoothing estimates
   > modFilt$mod$JW <- matrix(1)
38 > X <- unlist(dlmSvd2var(modFilt$U.W, modFilt$D.W))[-1]
   > modFilt$mod$X <- matrix(X)
40 > modSmooth <- dlmSmooth(modFilt)
   > Stildelist <- dlmSvd2var(modSmooth$U.S, modSmooth$D.S)
42 > TT <- length(y)
   > pars <- unlist(Stildelist) * (beta[TT] / alpha[TT])
44 > tt <- qt(prob, df = 2 * alpha[TT])
   > plot(y, ylab = "Smoothing level estimates",
46 +      type = "o", ylim = c(20, 40), col = "darkgray")
   > lines(dropFirst(modSmooth$s),  type = "o")
48 > lines(dropFirst(modSmooth$s - tt * sqrt(pars)),
   +       lty = 3, lwd = 2)
50 > lines(dropFirst(modSmooth$s + tt * sqrt(pars)),
   +       lty = 3, lwd = 2)
```

An important point is choosing the value of the discount factor δ, which determines the signal-to-noise ratio. In practice, one usually tries several values of δ, comparing the predictive performance of the corresponding models. Figure 4.2 shows the one-step-ahead point forecasts for the time window 1960-85, for different choices of δ ($\delta = 1$, which corresponds to the static model with $W_t = 0$ and $\theta_t = \theta_0$; $\delta = 0.9; \delta = 0.8$ and $\delta = 0.3$, a quite small value). The degree of adaptation to new data increases as δ becomes smaller. For $\delta = 0.3$ the forecasts $f_{t+1} = \mathrm{E}(Y_{t+1}|y_{1:t})$ are mainly determined by the current observation y_t. For δ closer to one, smoother forecasts are obtained.

Finally, Figure 4.3 shows the sequence of point estimates $\mathrm{E}(\sigma^2|y_{1:t}) = \beta_t/(\alpha_t - 1)$ of the observation variance, for $t = 1, 2, \ldots, 87$; the final estimates at $t = 87$ are reported in the table below.

discount factor	1.0	0.9	0.8	0.7
$\mathrm{E}(\sigma^2|y_{1:87})$	12.0010	9.6397	8.9396	8.3601

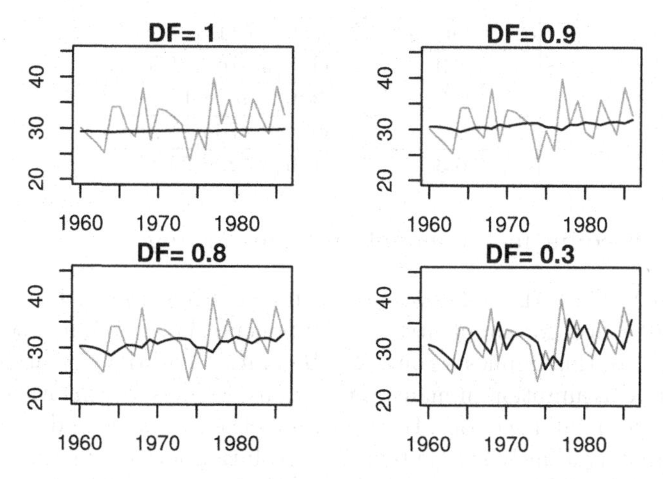

Fig. 4.2. Annual precipitation at Lake Superior (gray) and one-step-ahead forecasts, for different values of the discount factor (DF)

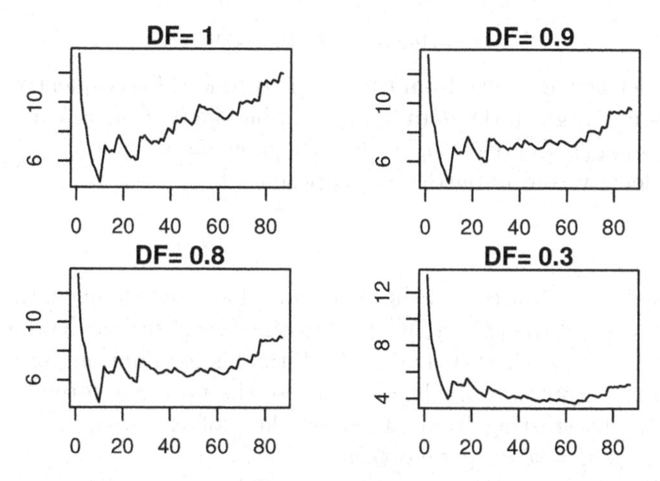

Fig. 4.3. Posterior expectation $E(\sigma^2|y_{1:t})$, for different values of the discount factor (DF)

Smaller values of δ imply larger evolution variance; correspondingly, a smaller observation variance is expected. Note that, for $\delta = 0.9$, the Bayesian estimate $E(\sigma^2|y_{1:87}) = 9.6397$ is close to the MLE $\hat{\sigma}^2 = 9.4654$, obtained on page 145.

The table below shows some measures of forecast accuracy for the four different choices of the discount factor, which suggest a better predictive performance for $\delta = 0.9$.

discount factor	MAPE	MAD	MSE
1.0	0.0977	3.0168	21.5395
0.9	0.0946	2.8568	19.9237
0.8	0.0954	2.8706	20.2896
0.3	0.1136	3.4229	25.1182

4.3.3 A discount factor model for time-varying V_t

In the DLM (4.9), the unknown precision factor ϕ is assumed to be constant over time. Since, for simplicity, the components $\tilde{V}_t$ are often taken as time-invariant too, this implies a constant observation covariance matrix, which is a restrictive assumption in many applications. Models for time varying V_t will be discussed in later sections. Here, we apply the technique of discount factors for introducing a fairly simple temporal evolution for the precision ϕ in (4.9) (see West and Harrison; 1997, Section 10.8).

Consider the DLM described in Section 4.3.1, and suppose again that at time $t-1$

$$\phi_{t-1}|y_{1:t-1} \sim \mathcal{G}(\alpha_{t-1}, \beta_{t-1}).$$

However, let now ϕ evolve from time $t-1$ to time t. Consequently, the uncertainty about ϕ_t, given the data $y_{1:t-1}$, will be bigger, that is $\mathrm{Var}(\phi_t|y_{1:t-1}) > \mathrm{Var}(\phi_{t-1}|y_{1:t-1})$. Let us *suppose* for the moment that $\phi_t|y_{1:t-1}$ has still a Gamma density, and in particular suppose that

$$\phi_t|y_{1:t-1} \sim \mathcal{G}(\delta^*\alpha_{t-1}, \delta^*\beta_{t-1}) \,, \tag{4.18}$$

where $0 < \delta^* < 1$. Notice that the expected value is not changed: $\mathrm{E}(\phi_t|y_{1:t-1}) = \mathrm{E}(\phi_{t-1}|y_{1:t-1}) = \alpha_{t-1}/\beta_{t-1}$, while the variance is bigger: $\mathrm{Var}(\phi_t|y_{1:t-1}) = 1/\delta^* \ \mathrm{Var}(\phi_{t-1}|y_{1:t-1})$, with $1/\delta^* > 1$. With this assumption, once a new observation y_t becomes available, one can use the updating formulae of Proposition 4.1, but starting from (4.18) in place of the $\mathcal{G}(\alpha_{t-1}, \beta_{t-1})$. Letting $\alpha_t^* = \delta^*\alpha_{t-1}, \beta_t^* = \delta^*\beta_{t-1}$, we obtain

$$Y_t|y_{1:t-1} \sim \mathcal{T}(f_t, \tilde{Q}_t\frac{\beta_t^*}{\alpha_t^*}, 2\alpha_t^*)$$

and

$$\theta_t|y_{1:t} \sim \mathcal{T}(m_t, \tilde{C}_t\frac{\beta_t}{\alpha_t}, 2\alpha_t)$$

where

$$\alpha_t = \alpha_t^* + \frac{m}{2}, \qquad \beta_t = \beta_t^* + \frac{1}{2}(y_t - f_t)'\tilde{Q}_t^{-1}(y_t - f_t). \tag{4.19}$$

For $m = 1$, the above expressions give

$$\begin{aligned} \alpha_t &= (\delta^*)^t\alpha_0 + \tfrac{1}{2}\textstyle\sum_{i=1}^t(\delta^*)^{i-1} \\ \beta_t &= (\delta^*)^t\beta_0 + \tfrac{1}{2}\textstyle\sum_{i=1}^t(\delta^*)^{t-i}\tilde{e}_i^2 \end{aligned}$$

where $\tilde{e}_t^2 = (y_t - f_t)^2/\tilde{Q}_t$.

It remains to motivate the assumption (4.18), and in fact we have not specified yet the dynamics that leads from ϕ_{t-1} to ϕ_t. It can be proved that assumption (4.18) is equivalent to the following multiplicative model for the dynamics of ϕ_t

$$\phi_t = \frac{\gamma_t}{\delta^*}\phi_{t-1} \, ,$$

where γ_t is a random variable independent on ϕ_{t-1}, having a beta distribution with parameters $(\delta^*\alpha_{t-1}, (1-\delta^*)\alpha_{t-1})$, so that $\mathrm{E}(\gamma_t) = \delta^*$. Therefore, ϕ_t is equivalent to ϕ_{t-1} multiplied by a random impulse with expected value 1 ($\mathrm{E}(\gamma_t/\delta^*) = 1$).

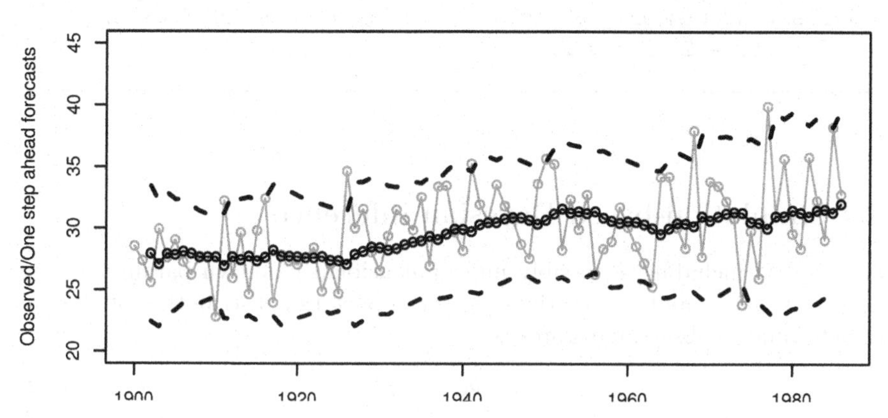

Fig. 4.4. Lake Superior data (gray) and one-step-ahead forecasts, with 90% probability intervals. Both V_t and W_t are specified through discount factors

For illustration with the Lake Superior data, let again $m_0 = E(\theta_0|\phi_0) = 0$, $\tilde{C}_0 = 10^7$, $\alpha_0 = 2$, $\beta_0 = 20$ and use discount factors $\delta = 0.9$, for $\tilde{W}_t$, and $\delta^* = 0.9$, so that $\mathrm{Var}(\phi_t|y_{1:t-1}) = (1/0.9)\mathrm{Var}(\phi_{t-1}|y_{1:t-1})$. Figure 4.4 shows the one-step-ahead point forecasts f_t, with 90% probability intervals.

R code

```
> y <- ts(read.table("Datasets/lakeSuperior.dat",
+                    skip = 3)[, 2], start = c(1900, 1))
> beta0 <- 20; alpha0 <- 2 ; TT <- length(y)
> mod <- dlmModPoly(1, dV = 1)
> modFilt <- dlmFilterDF(y, mod, DF = 0.9)
> out <- residuals(modFilt)
> DFstar <- 0.9
> delta <- DFstar^0 : TT
> alpha <- delta[-1] * alpha0 + cumsum(delta[-TT]) / 2
> res <- as.vector(out$res)
```

```
> beta <- delta[-1] * beta0
> for (i in 1 : TT)
+     beta[i] <- beta[i] + 0.5 * sum(delta[i:1] * res[1:i]^2)
> alphaStar <- DFstar * c(alpha0, alpha)
> betaStar <- DFstar * c(beta0, beta)
> tt <- qt(0.95, df = 2 * alphaStar[1 : TT])
> param <- sqrt(out$sd) * (betaStar / alphaStar)[1 : TT]
> plot(y, ylab = "Observed/One step ahead forecasts",
+        type = "o", col = "darkgray", ylim =c(20, 45))
> lines(window(modFilt$f, start = 1902), type = "o")
> lines(window(modFilt$f, start = 1902) - tt * sqrt(param),
+         lty = 2, lwd = 2)
> lines(window(modFilt$f, start = 1902) + tt * sqrt(param),
+         lty = 2, lwd = 2)
```

4.4 Simulation-based Bayesian inference

For a DLM including a possibly multidimensional unknown parameter ψ in its specification, and observations $y_{1:T}$, the posterior distribution of the parameter and unobservable states is

$$\pi(\psi, \theta_{0:T}|y_{1:T}). \tag{4.20}$$

As mentioned in Section 4.2, in general it is impossible to compute this distribution in closed form. Therefore, in order to come up with posterior summaries one has to resort to numerical methods, almost invariably stochastic, Monte Carlo methods. The customary MCMC approach to analyze the posterior distribution (4.20) is to generate a (dependent) sample from it and evaluate posterior summaries from the simulated sample. The inclusion of the states in the posterior distribution usually simplifies the design of an efficient sampler, even when one is only interested in the posterior distribution of the unknown parameter, $\pi(\psi|y_{1:T})$. In fact, drawing a random variable/vector from $\pi(\psi|\theta_{0:T}, y_{1:T})$ is almost invariably much easier than drawing it from $\pi(\psi|y_{1:T})$; in addition, efficient algorithms to generate the states conditionally on the data and the unknown parameter are available, see Section 4.4.1. This suggests that a sample from (4.20) can be obtained from a Gibbs sampler alternating draws from $\pi(\psi|\theta_{0:T}, y_{1:T})$ and $\pi(\theta_{0:T}|\psi, y_{1:T})$. The simulated sample from the posterior can in turn be used as input to generate a sample from the predictive distribution of states and observables, $\pi(\theta_{T+1:T+k}, y_{T+1:T+k}|y_{1;T})$. In fact,

$$\pi(\theta_{T+1:T+k}, y_{T+1:T+k}, \psi, \theta_T|y_{1;T}) =$$
$$\pi(\theta_{T+1:T+k}, y_{T+1:T+k}|\psi, \theta_T) \cdot \pi(\psi, \theta_T|y_{1;T}).$$

Therefore, for every pair (ψ, θ_T) drawn from $\pi(\psi, \theta_T|y_{1:T})$, one can generate the "future" $\theta_{T+1:T+k}, y_{T+1:T+k}$ from $\pi(\theta_{T+1:T+k}, y_{T+1:T+k}|\psi, \theta_T)$ (see Section 2.8) to obtain a sample from the predictive distribution.

The approach sketched above completely solves the filtering, smoothing, and forecasting problems for a DLM with unknown parameters. However, if one needs to update the posterior distribution after one or more new observations become available, then one has to run the Gibbs sampler all over again, and this can be extremely inefficient. As already mentioned, on-line analysis and simulation-based sequential updating of the posterior distribution of states and unknown parameters are best dealt with employing sequential Monte Carlo techniques.

4.4.1 Drawing the states given $y_{1:T}$: forward filtering backward sampling

In a Gibbs sampling from $\pi(\theta_{0:T}, \psi|y_{1:T})$, one needs to simulate from the full conditional densities $\pi(\psi|\theta_{0:T}, y_{1:T})$ and $\pi(\theta_{0:T}|\psi, y_{1:T})$. While the first density is problem specific, the general expression of the latter density and efficient algorithms for sampling from it are available.

The smoothing recursions provide an algorithm for computing the mean and variance of the distribution of θ_t conditional on $y_{1:T}$ and ψ $(t = 0, 1, \ldots, T)$. Since all the involved distributions are Normal, this completely determines the *marginal* posterior distribution of θ_t given $y_{0:T}$ and ψ. If one is interested in the joint posterior distribution of $\theta_{0:T}$ given $(y_{1:T}, \psi)$, then also the posterior covariances between θ_t and θ_s have to be computed. General formulae to recursively evaluate these covariances are available, see Durbin and Koopman (2001). However, when $\pi(\theta_{0:T}|\psi, y_{1:T})$ has the role of full conditional in a Gibbs sampling from $\pi(\theta_{0:T}, \psi|y_{1:T})$, the main question becomes: how can one generate a draw from the distribution of $\theta_{0:T}$ given $(y_{1:T}, \psi)$? We will use a method due to Carter and Kohn (1994), Früwirth-Schnatter (1994), and Shephard (1994), which is now widely known as the forward filtering backward sampling (FFBS) algorithm. By reading the description that follows, the reader will realize that FFBS is essentially a simulation version of the smoothing recursions.

We can write the joint distribution of $\theta_{0:T}$ given $y_{1:T}$ as

$$\pi(\theta_{0:T}|y_{1:T}) = \prod_{t=0}^{T} \pi(\theta_t|\theta_{t+1:T}, y_{1:T}), \tag{4.21}$$

where the last factor in the product is simply $\pi(\theta_T|y_{1:T})$, i.e., the filtering distribution of θ_T, which is $\mathcal{N}(m_T, C_T)$. Formula (4.21) suggests that in order to obtain a draw from the distribution on the left-hand side, one can start by drawing θ_T from a $\mathcal{N}(m_T, C_T)$ and then, for $t = T-1, T-2, \ldots, 0$, recursively draw θ_t from $\pi(\theta_t|\theta_{t+1:T}, y_{1:T})$. We have seen in the proof of Proposition 2.4

that $\pi(\theta_t|\theta_{t+1:T}, y_{1:T}) = \pi(\theta_t|\theta_{t+1}, y_{1:t})$, and we showed that this distribution is $\mathcal{N}(h_t, H_t)$, with

$$h_t = m_t + C_t G'_{t+1} R_{t+1}^{-1}(\theta_{t+1} - a_{t+1}),$$
$$H_t = C_t - C_t G'_{t+1} R_{t+1}^{-1} G_{t+1} C_t.$$

Therefore, having already $(\theta_{t+1}, \ldots, \theta_T)$, the next step consists in drawing θ_t from a $\mathcal{N}(h_t, H_t)$. Note that h_t explicitly depends on the value of θ_{t+1} already generated. The FFBS algorithm is summarized in Algorithm 4.1

```
1. Run Kalman filter.
2. Draw θ_T ~ N(m_T, C_T).
3. For t = T − 1,...,0, draw θ_t ~ N(h_t, H_t).
```

Algorithm 4.1: Forward filtering backward sampling

FFBS is commonly used as a building block of a Gibbs sampler, as we will illustrate in many examples in the remaining of the chapter, in particular in Section 4.5. However, it can be of interest also in DLMs containing no unknown parameters. In this case, the marginal smoothing distribution of each θ_t is usually enough to evaluate posterior probabilities of interest. However, the posterior distribution of a nonlinear function of the states may be difficult or impossible to derive, even when all the parameters of the model are known. In this case FFBS provides an easy way to generate an independent sample from the posterior of the nonlinear function of interest. Note that in this type of application the "forward filtering" part of the algorithm only needs to be performed once.

4.4.2 General strategies for MCMC

For a completely specified DLM, i.e., one not containing unknown parameters, draws from the posterior distribution of the states, and possibly from the forecast distribution of states and observations, can be obtained using the FFBS algorithm described in the previous subsection. In the more realistic situation of a DLM containing an unknown parameter vector ψ, with prior distribution $\pi(\psi)$, in the observation, system, or variance matrices, one typically uses MCMC to obtain posterior summaries of the distributions of interest. Almost all Markov chain samplers for posterior analysis of a DLM fall in one of the following categories: Gibbs samplers, which include the states as latent variables; marginal samplers; and hybrid samplers, which combine aspects of both. Note that, depending on the context, the analyst may be interested in making inference about the unobservable states, the unknown parameter, or both. Of the three types of samplers, two (Gibbs and hybrid samplers) generate draws from the joint posterior of the states and the parameter, while the other (marginal samplers) only generates draws from the posterior of the parameter. Keep in mind, however, that once a sample from the posterior distribution of the parameter is available, a sample from the joint posterior of states and parameter can be easily obtained in view of the decomposition

$$\pi(\theta_{0:T}, \psi | y_{1:T}) = \pi(\theta_{0:T} | \psi, y_{1:T}) \cdot \pi(\psi | y_{1:T}).$$

More specifically, for each $\psi^{(i)}$ in the sample ($i = 1, \ldots, N$), it is enough to draw $\theta_{0:T}^{(i)}$ from $\pi(\theta_{0:T} | \psi = \psi^{(i)}, y_{1:T})$ using FFBS, and $\{(\theta_{0:T}^{(i)}, \psi^{(i)}) : i = 1, \ldots, N\}$ will be the required sample from the joint posterior distribution.

The Gibbs sampling approach, consisting in drawing in turn the states from their conditional distribution given the parameter and observations, and the parameter from its conditional distribution given the states and observations, is summarized in Algorithm 4.2. Package dlm provides the function

```
0. Initialize: set ψ = ψ^(0).
1. For i = 1,...,N:
   a) Draw θ_{0:T}^(i) from π(θ_{0:T}|y_{1:T}, ψ = ψ^(i-1)) using FFBS.
   b) Draw ψ^(i) from π(ψ|y_{1:T}, θ_{0:T} = θ_{0:T}^(i)).
```

Algorithm 4.2: Forward Filtering Backward Sampling in a Gibbs sampler

`dlmBSample` which, in conjunction with `dlmFilter`, can be used to perform step (a). Step (b), on the other hand, depends heavily on the model under consideration, including the prior distribution on ψ. In fact, when ψ is an r-dimensional vector, it is often simpler to perform a Gibbs step for each component of ψ instead of drawing ψ at once. The intermediate approach of drawing blocks of components of ψ together is another option. In any case, when a full conditional distribution is difficult to sample from, a Metropolis–Hastings step can replace the corresponding Gibbs step. A generic sampler for nonstandard distributions is ARMS (Section 1.6), available in package dlm in the function `arms`.

The second approach, marginal sampling, is conceptually straightforward, consisting of drawing a sample from $\pi(\psi | y_{0:T})$. The actual implementation of the sampler depends on the model under consideration. Typically, if ψ is multivariate, one can use a Gibbs sampler, drawing each component, or block of components, from its full conditional distributions, possibly using a Metropolis–Hastings step when the relevant full conditional is not a standard distribution. Again, ARMS can be used in the latter case.

A hybrid sampler can be used when the parameter can be decomposed in two components, that is, when ψ can be written as (ψ_1, ψ_2), where each component may be univariate or multivariate. Algorithm 4.3 gives an algo-

```
0. Initialize: set ψ2 = ψ2^(0).
1. For i = 1,...,N:
   a) Draw ψ1^(i) from π(ψ1|y0:T, ψ2 = ψ2^(i-1)).
   b) Draw θ0:T^(i) from π(θ0:T|y1:T, ψ1 = ψ1^(i), ψ2 = ψ2^(i-1)) using FFBS.
   c) Draw ψ2^(i) from π(ψ2|y1:T, θ0:T = θ0:T^(i), ψ1 = ψ1^(i)).
```

Algorithm 4.3: Forward Filtering Backward Sampling in a hybrid sampler

rithmic description of a generic hybrid sampler. As for the previous schemes, Metropolis–Hastings steps, and ARMS in particular, can be substituted for direct sampling in steps (a) and (c). Step (b) can always be performed using `dlmFilter` followed by `dlmBSample`. For the theoretically inclined reader, let us point out a subtle difference between this sampler and a Gibbs sampler. In a Gibbs sampler, each step consists of applying a Markov transition kernel whose invariant distribution is the target distribution, so that the latter is also invariant for the composition of all the kernels. In a hybrid sampler, on the other hand, the target distribution is not invariant for the Markov kernel corresponding to step (a), so the previous argument does not apply directly. However, it is not difficult to show that the composition of step (a) and (b) does preserve the target distribution and so, when combined with step (c), which is a standard Gibbs step, it produces a Markov kernel having the correct invariant distribution. An example of Bayesian inference with hybrid sampling will be given in Section 4.6.1.

The output produced by a Markov chain sampler must always be checked to assess convergence to the stationary distribution and mixing of the chain. Given that the chain has practically reached the stationary distribution, mixing can be assessed by looking at autocorrelation functions of parameters or functions of interest. Ideally, one would like to have as low a correlation as possible between draws. Correlation may be reduced by *thinning* the simulated chain, i.e., discarding a fixed number of iterations between every saved value. Although this method is very easy to implement, the improvements are usually only marginal, unless the number of discarded simulations is substantial, which significantly increases the time required to run the entire sampler. As far as assessing convergence, a fairly extensive body of literature exists on diagnostic tools for MCMC. In R the package BOA provides a suite of functions implementing many of these diagnostics. In most cases, a visual inspection of the output, after discarding a first part of the simulated chain as *burn in*, can reveal obvious departures from stationarity. For an organic treatment of MCMC diagnostics, we refer to Robert and Casella (2004) and references therein.

4.4.3 Illustration: Gibbs sampling for a local level model

Before moving on to more general models, we give in the present section a simple illustration of how a Gibbs sampler can be implemented in practice. The example that we are going to examine is the local level model with unknown observation and evolution variances. A convenient, yet flexible, family of priors for each variance is the inverse-gamma family. More specifically, let $\psi_1 = V^{-1}$ and $\psi_2 = W^{-1}$, and assume that ψ_1 and ψ_2 are a priori independent, with

$$\psi_i \sim \mathcal{G}(a_i, b_i), \qquad i = 1, 2.$$

The parameters of the prior, a_i, b_i $(i = 1, 2)$ can be specified so as to match the prior opinion of the analyst about the unknown precisions, expressed in terms of their means and variances. Most people find it easier to specify prior moments (mean and variance) of an unknown variance than it is to specify those of a precision. In this case, just recall that a $\mathcal{G}(a, b)$ prior for ψ_i is the same as an Inverse Gamma prior with the same parameters for ψ_i^{-1}.

A Gibbs sampler for this model will iterate the following three steps: draw $\theta_{0:T}$, draw ψ_1, and draw ψ_2. The random quantities generated in each of the three steps should be drawn from their *full conditional* distribution, i.e., the distribution of that quantity given all the other random variables in the model, including the observations. Note that the states $\theta_{0:T}$ must be included among the conditioning variables when drawing ψ_1 and ψ_2. To generate $\theta_{0:T}$ we can use the FFBS algorithm, setting ψ_1 and ψ_2 to their most recently simulated value. This is something that can be done in a straightforward way using the function `dlmBSample`. On the other hand, a simple calculation is needed to determine the full conditional distribution of ψ_1. The standard approach is based on the fact that the full conditional distribution is proportional to the joint distribution of all the random variables considered. For ψ_1, this is

$$\begin{aligned}
\pi(\psi_1|\psi_2, \theta_{0:T}, y_{1:T}) &\propto \pi(\psi_1, \psi_2, \theta_{0:T}, y_{1:T}) \\
&\propto \pi(y_{1:T}|\theta_{0:T}, \psi_1, \psi_2)\pi(\theta_{0:T}|\psi_1, \psi_2)\pi(\psi_1, \psi_2) \\
&\propto \prod_{t=1}^{T} \pi(y_t|\theta_t, \psi_1) \prod_{t=1}^{T} \pi(\theta_t|\theta_{t-1}, \psi_2)\pi(\psi_1)\pi(\psi_2) \\
&\propto \pi(\psi_1)\psi_1^{T/2} \exp\left(-\frac{\psi_1}{2}\sum_{t=1}^{T}(y_t - \theta_t)^2\right) \\
&\propto \psi_1^{a_1+T/2-1} \exp\left(-\psi_1 \cdot \left[b_1 + \frac{1}{2}\sum_{t=1}^{T}(y_t - \theta_t)^2\right]\right).
\end{aligned}$$

From the previous equations we deduce that ψ_1 and ψ_2 are conditionally independent, given $\theta_{0:T}$ and $y_{1:T}$, and

$$\psi_1|\theta_{0:T}, y_{1:T} \sim \mathcal{G}\left(a_1 + \frac{T}{2}, b_1 + \frac{1}{2}\sum_{t=1}^{T}(y_t - \theta_t)^2\right).$$

A similar argument shows that

$$\psi_2|\theta_{0:T}, y_{1:T} \sim \mathcal{G}\left(a_2 + \frac{T}{2}, b_2 + \frac{1}{2}\sum_{t=1}^{T}(\theta_t - \theta_{t-1})^2\right).$$

The following code implements the Gibbs sampler discussed above for the Nile River data. The actual sampler consists of the loop starting on line 18; what comes before is just book keeping, space allocation for the output, and the definition of starting values and variables that do not change within the main loop.

R code

```
   > a1 <- 2
 2 > b1 <- 0.0001
   > a2 <- 2
 4 > b2 <- 0.0001
   > ## starting values
 6 > psi1 <- 1
   > psi2 <- 1
 8 > mod_level <- dlmModPoly(1, dV = 1 / psi1, dW = 1 / psi2)
   >
10 > mc <- 1500
   > psi1_save <- numeric(mc)
12 > psi2_save <- numeric(mc)
   > n <- length(Nile)
14 > sh1 <- a1 + n / 2
   > sh2 <- a2 + n / 2
16 > set.seed(10)
   >
18 > for (it in 1 : mc)
   + {
20 +     ## draw the states: FFBS
   +     filt <- dlmFilter(Nile, mod_level)
22 +     level <- dlmBSample(filt)
   +     ## draw observation precision psi1
24 +     rate <- b1 + crossprod(Nile - level[-1]) / 2
   +     psi1 <- rgamma(1, shape = sh1, rate = rate)
26 +     ## draw system precision psi2
   +     rate <- b2 + crossprod(level[-1] - level[-n]) / 2
28 +     psi2 <- rgamma(1, shape = sh2, rate = rate)
   +     ## update and save
30 +     V(mod_level) <- 1 / psi1
```

```
+       W(mod_level) <- 1 / psi2
+       psi1_save[it] <- psi1
+       psi2_save[it] <- psi2
+ }
```

A visual analysis of trace plots (not shown) shows that the chain reached convergence after the first two or three hundred iterations. In the code we did not save the simulated states, assuming the focus of the analysis was on the unknown variances. Note, however, that a sample from the posterior distribution of the states does not require us to run the Gibbs sampler all over again, as one can just run the FFBS algorithm once for each simulated value of (ψ_1, ψ_2), thanks to the equality

$$\pi(\theta_{0:T}, \psi_1, \psi_2 | y_{1:T}) = \pi(\theta_{0:T} | \psi_1, \psi_2 y_{1:T}) \pi(\psi_1, \psi_2 | y_{1:T}).$$

4.5 Unknown variances

In many of the models analyzed in Chapter 3, the system and observation matrices G_t and F_t are set to specific values as part of the model specification. This is the case for polynomial and seasonal factor models, for example. The only possibly unknown parameters are therefore part of the variance matrices W_t and V_t. In Section 4.3.1, we discussed Bayesian conjugate inference for the simple case of V_t and W_t known up to a scale factor; in more general cases, analytical computations become more complex and MCMC approximations are used, as we illustrate in this section.

4.5.1 Constant unknown variances: *d* Inverse Gamma prior

This is the simplest and most commonly used model for unknown variances. Let us assume that the observations are univariate ($m = 1$). If the observation variance and the evolution covariance matrices are unknown, the simplest assumption is to consider them as time-invariant, with W diagonal. More specifically, and working as usual with the precisions, a *d-inverse-gamma* prior assumes that

$$V_t = \phi_y^{-1}, \qquad W_t = \mathrm{diag}(\phi_{\theta,1}^{-1}, \ldots, \phi_{\theta,p}^{-1})$$

and $\phi_y, \phi_{\theta,1}, \ldots, \phi_{\theta,p}$ have independent Gamma distributions. This implies that the prior on the vector of the variances $(\phi_y^{-1}, \phi_{\theta,1}^{-1}, \ldots, \phi_{\theta,p}^{-1})$ is the product of $d = (p+1)$ Inverse-Gamma densities. To fix the prior hyperparamters, it is usually convenient to express a prior guess on the unknown precisions, $\mathrm{E}(\phi_y) = a_y$, and $\mathrm{E}(\phi_{\theta,i}) = a_{\theta,i}$, with prior uncertainty summarized by the prior variances $\mathrm{Var}(\phi_y) = b_y$, $\mathrm{Var}(\phi_{\theta,i}) = b_{\theta,i}$, $i = 1 \ldots, p$, so that the Gamma priors can be parametrized as

$$\phi_y \sim \mathcal{G}(\alpha_y, \beta_y), \qquad \phi_{\theta,i} \sim \mathcal{G}(\alpha_{\theta,i}, \beta_{\theta,i}) \qquad i = 1, \ldots, p,$$

with

$$\alpha_y = \frac{a_y^2}{b_y}, \quad \beta_y = \frac{a_y}{b_y}, \qquad \alpha_{\theta,i} = \frac{a_{\theta,i}^2}{b_{\theta,i}}, \quad \beta_{\theta,i} = \frac{a_{\theta,i}}{b_{\theta,i}}, \qquad i = 1, \ldots, p.$$

As particular cases, this framework includes a Bayesian treatment of nth order polynomial models as well as the structural time series models of Harvey and coauthors, discussed in Chapter 3.

Given the observations $y_{1:T}$, the joint posterior of the states $\theta_{0:T}$ and the unknown parameters $\psi = (\phi_y, \phi_{\theta,1}, \ldots, \phi_{\theta,p})$ is proportional to the joint density

$$\begin{aligned}
\pi(y_{1:T}, \theta_{0:T}, \psi) &= \pi(y_{1:T}|\theta_{0:T}, \psi) \cdot \pi(\theta_{0:T}|\psi) \cdot \pi(\psi) \\
&= \prod_{t=1}^{T} \pi(y_t|\theta_t, \phi_y) \cdot \prod_{t=1}^{T} \pi(\theta_t|\theta_{t-1}, \phi_{\theta,1}, \ldots, \phi_{\theta,p}) \cdot \pi(\theta_0) \cdot \pi(\phi_y) \cdot \prod_{i=1}^{p} \pi(\phi_{\theta,i}).
\end{aligned}$$

Note that the second product in the factorization can also be written as a product over $i = 1, \ldots, p$, due to the diagonal form of W. This alternative factorization is useful when deriving the full conditional distribution of the $\phi_{\theta,i}$'s. A Gibbs sampler for the d-Inverse-Gamma model draws from the full conditional distribution of the states and from the full conditional distributions of $\phi_y, \phi_{\theta,1}, \ldots, \phi_{\theta,p}$ in turn. Sampling the states can be done using the FFBS algorithm of Section 4.4.1. Let us derive the full conditional distribution[1] of ϕ_y:

$$\begin{aligned}
\pi(\phi_y|\ldots) &\propto \prod_{t=1}^{T} \pi(y_t|\theta_t, \phi_y) \cdot \pi(\phi_y) \\
&\propto \phi_y^{\frac{T}{2}+\alpha_y-1} \exp\left\{-\phi_y \cdot \left[\frac{1}{2}\sum_{t=1}^{T}(y_t - F_t\theta_t)^2 + \beta_y\right]\right\}.
\end{aligned}$$

The full conditional of ϕ_y is therefore again a Gamma distribution,

$$\phi_y|\ldots \sim \mathcal{G}\left(\alpha_y + \frac{T}{2}, \beta_y + \frac{1}{2}SS_y\right),$$

with $SS_y = \sum_{t=1}^{T}(y_t - F_t\theta_t)^2$. Similarly, it is easy to show that the full conditionals of the $\phi_{\theta,i}$'s are as follows:

$$\phi_{\theta,i}|\ldots \sim \mathcal{G}\left(\alpha_{\theta,i} + \frac{T}{2}, \beta_{\theta,i} + \frac{1}{2}SS_{\theta,i}\right), \qquad i = 1, \ldots, p,$$

[1] The dots on the right-hand side of the conditioning vertical bar in $\pi(\phi_y|\ldots)$ stay for every other random variable in the model except ϕ_y, including the states $\theta_{0:T}$. This standard convention will be used throughout.

with $SS_{\theta,i} = \sum_{t=1}^{T}(\theta_{t,i} - (G_t\theta_{t-1})_i)^2$.

Example. Let us consider again the data on Spain investment (Section 3.2.1). We are going to fit a 2nd-order polynomial model—local linear growth—to the data. The priors for the precisions of the observation and evolution errors are (independent) gamma distributions with means a_y, a_μ, a_β and variances b_y, b_μ, b_β. We decide for the values $a_y = 1$, $a_\mu = a_\beta = 10$, with a common variance equal to 1000, to express a large uncertainty in the prior estimate of the precisions. The function `dlmGibbsDIG` can be called to generate a sample from the posterior distribution of the parameters and the states. The means and variances of the gamma priors are passed to the function via the arguments `a.y`, `b.y` (prior mean and variance of observation precision), `a.theta`, `b.theta` (prior mean(s) and variance(s) of evolution precision(s)). Alternatively, the prior distribution can be specified in terms of the usual shape and rate parameters of the gamma distribution. The arguments to pass in this case are `shape.y`, `rate.y`, `shape.theta`, and `rate.theta`. The number of samples from the posterior to generate is determined by the argument `n.sample`, while the logical argument `save.states` is used to determine whether to include the generated unobservable states in the output. In addition, a thinning parameter can be specified via the integer argument `thin`. This gives the number of Gibbs iterations to discard for every saved one. Finally, the data and the model are passed via the arguments `y` and `mod`, respectively. The following display show how `dlmGibbsDIG` works in practice.

R code

```
> invSpain <- ts(read.table("Datasets/invest2.dat",
+                            colClasses = "numeric")[,2]/1000,
+                 start = 1960)
> set.seed(5672)
> MCMC <- 12000
> gibbsOut <- dlmGibbsDIG(invSpain, mod = dlmModPoly(2),
+                         a.y = 1, b.y = 1000,
+                         a.theta = 10, b.theta = 1000,
+                         n.sample = MCMC,
+                         thin = 1, save.states = FALSE)
```

Setting `thin = 1` means that the function actually generates a sample of size 24,000 but only keeps in the output every other value. In addition, the states are not returned (`save.states = FALSE`). Considering the first 2000 saved iterations as burn in, one can proceed to graphically assess the convergence and mixing properties of the sampler. Figure 4.5 displays a few diagnostic plots obtained from the MCMC output for the variances V, W_{11}, and W_{22}. The first row shows the traces of the sampler, i.e., the simulated values, the second the running ergodic means of the parameters (starting

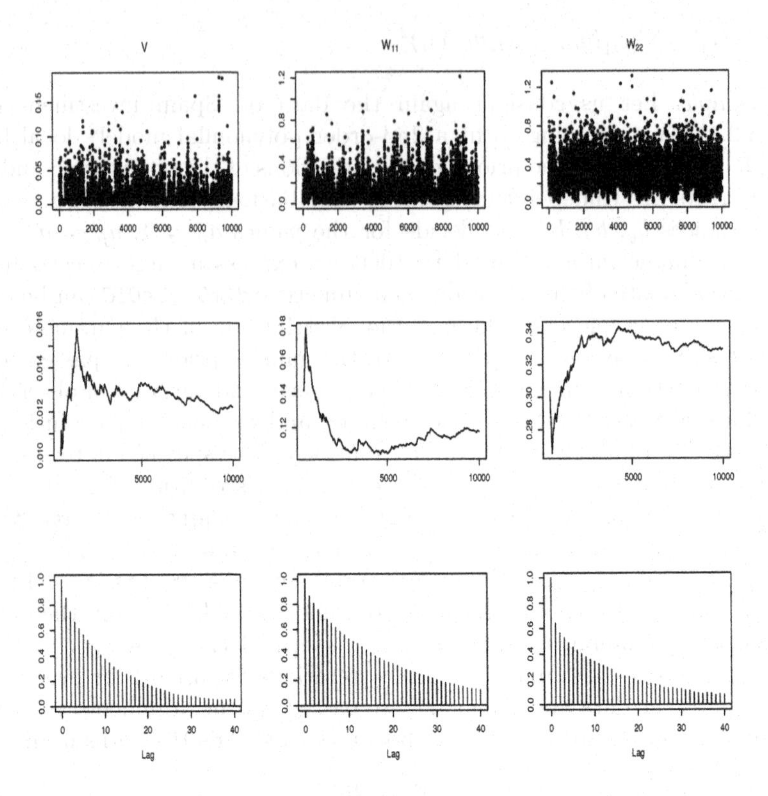

Fig. 4.5. Diagnostic plots for *d*-inverse-Gamma model applied to Spain investments

at iteration 500), and the last the estimated autocovariance functions. We obtained the running ergodic means using the function `ergMean`. For example, the plot in the first column and second row was created using the following commands.

R code

```
> use <- MCMC - burn
> from <- 0.05 * use
> plot(ergMean(gibbsOut$dV[-(1 : burn)], from), type = "l",
+      xaxt = "n", xlab = "", ylab = "")
> at <- pretty(c(0, use), n = 3)
> at <- at[at >= from]
> axis(1, at = at - from, labels = format(at))
```

From a visual assessment of the MCMC output it seems fair to deduce that convergence has been achieved and, while the acf's of the simulated variances do not decay very fast, the ergodic means are nonetheless pretty stable in the last part of the plots. One can therefore go ahead and use the MCMC output to estimate the posterior means of the unknown variances. The function `mcmcMean` computes the (column) means of a matrix of simulated values, together with an estimate of the Monte Carlo standard deviation, obtained using Sokal's method (Section 1.6).

R code

```
> mcmcMean(cbind(gibbsOut$dV[-(1 : burn), ],
+               gibbsOut$dW[-(1 : burn), ]))
      V.1         V.2         V.3
   0.012197    0.117391    0.329588
  (0.000743)  (0.007682)  (0.007833)
```

Bivariate plots of the simulated parameters may provide additional insight. Consider the plots in Figure 4.6. The joint realizations seem to suggest that

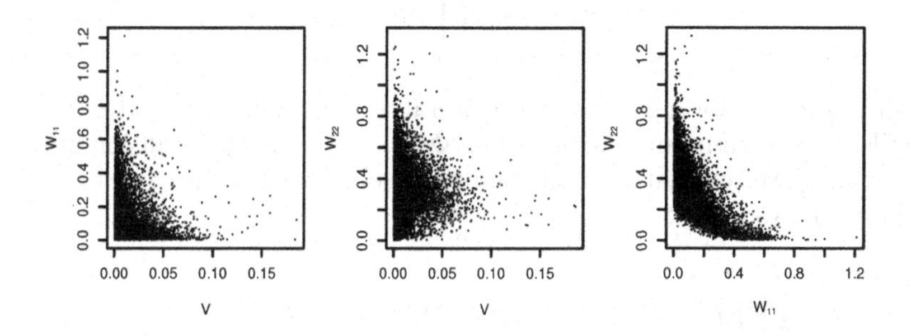

Fig. 4.6. Bivariate plots for d-inverse-Gamma model applied to Spain investments

one, or maybe two, of the model variances may be zero. From the third plot we can see that when W_{11} is close to zero then W_{22} is clearly positive, and vice versa: W_{11} and W_{22} cannot both be zero. The second plot suggests that this is also true for V and W_{22}. From the first plot it seems that V and W_{11} may be zero, possibly at the same time. In summary, by looking at the bivariate plots, three reduced models come up as alternative to the full model worth exploring: the submodel obtained by setting $V = 0$, the submodel $W_{11} = 0$, the submodel $V = W_{11} = 0$, and the submodel $W_{22} = 0$. We do not pursue the issue of Bayesian model selection here; a recent reference is Früwirth-Schnatter and Wagner (2008), who suggest a different prior on the unknown variances, more suitable for model comparison.

4.5.2 Multivariate extensions

Multivariate extensions of the d Inverse-Gamma model can be devised, using independent inverse-Wishart priors. Suppose that Y_t is m-variate, $m \geq 1$, and

W is block-diagonal with elements $(W_1, \ldots, W_h)$, W_i having dimension $p_i \times p_i$. Examples of DLMs with a block-diagonal state covariance matrix W include additive compositions of structural models or SUTSE models, presented in Chapter 3. At least in principle, the case of a general matrix W is obtained by letting $h = 1$.

Let us parametrize the model in the precision matrices $\Phi_0 = V^{-1}$ and $\Phi = W^{-1}$, the latter being block-diagonal with elements $\Phi_i = W_i^{-1}$, $i = 1, \ldots, h$. We assume that $\Phi_0, \Phi_1, \ldots, \Phi_h$ have independent Wishart priors, $\Phi_i \sim \mathcal{W}(\nu_i, S_i)$, $i = 0, \ldots, h$, where S_i is a symmetric positive definite matrix of dimensions $p_i \times p_i$, with $p_0 = m$. Then the posterior density $\pi(\theta_{0:T}, \Phi_0, \ldots, \Phi_h | y_{1:T})$ is proportional to

$$\prod_{t=1}^{T} \mathcal{N}(y_t; F_t\theta_t, \Phi_0^{-1})\, \mathcal{N}(\theta_t; G\theta_{t-1}, \Phi^{-1}) \mathcal{N}(\theta_0; m_0, C_0) \\ \cdot \mathcal{W}(\Phi_0; \nu_0, S_0) \prod_{i=1}^{h} \mathcal{W}(\Phi_i; \nu_i, S_i). \tag{4.22}$$

A Gibbs sampler for π is obtained by iteratively sampling the states $\theta_{0:T}$ (via the FFBS algorithm) and the precisions $\Phi_0, \ldots, \Phi_h$ from their full conditionals. From (4.22) we see that the full conditional density of Φ_i, for $i = 1, \ldots, h$, is proportional to

$$\prod_{t=1}^{T}\prod_{j=1}^{h} |\Phi_j|^{1/2} \exp\{-\frac{1}{2}(\theta_t - G_t\theta_{t-1})'\Phi(\theta_t - G_t\theta_{t-1})\} \cdot \\ |\Phi_i|^{\nu_i - (p_i+1)/2} \exp\{-\mathrm{tr}(S_i\Phi_i)\}$$

$$\propto |\Phi_i|^{T/2+\nu_i-(p_i+1)/2} \exp\{-\mathrm{tr}(\frac{1}{2}\sum_{t=1}^{T}(\theta_t - G_t\theta_{t-1})(\theta_t - G_t\theta_{t-1})'\Phi) - \mathrm{tr}(S_i\Phi_i)\}$$

(see Section 1.5). Let

$$SS_t = (\theta_t - G_t\theta_{t-1})(\theta_t - G_t\theta_{t-1})'$$

and partition it in accordance with Φ:

$$SS_t = \begin{bmatrix} SS_{11,t} & \cdots & SS_{1h,t} \\ \vdots & \ddots & \vdots \\ SS_{h1,t} & \cdots & SS_{hh,t} \end{bmatrix}. \tag{4.23}$$

Then $\mathrm{tr}(SS_t\Phi) = \sum_{j=1}^{h} \mathrm{tr}(SS_{jj,t}\Phi_j)$, so that the full conditional of Φ_i results to be proportional to

$$|\Phi_i|^{T/2+\nu_i-(p_i+1)/2} \exp\left\{-\mathrm{tr}\left(\left(\frac{1}{2}SS_{i\cdot} + S_i\right)\Phi_i\right)\right\},$$

with $SS_{i\cdot} = \sum_{t=1}^{T} SS_{ii,t}$. That is, for $i = 1, \ldots, h$, the full conditional of Φ_i is Wishart, with parameters $(\nu_i + T/2, 1/2SS_{i\cdot} + S_i)$. In particular, for a DLM obtained by combining component models as in Section 3.2, we have

$$\theta_t = \begin{bmatrix} \theta_{1,t} \\ \vdots \\ \theta_{h,t} \end{bmatrix} \qquad G_t = \begin{bmatrix} G_{1,t} & \cdots & 0 \\ 0 & \ddots & 0 \\ 0 & \cdots & G_{h,t} \end{bmatrix},$$

with $\theta_{i,t}$ and $G_{i,t}$ referring to the ith component model. Then, the full conditional of Φ_i is $\mathcal{W}(\nu_i + T/2, S_i + 1/2SS_{i\cdot})$, and $S_{ii,t} = (\theta_{i,t} - G_{i,t}\theta_{i,t-1})(\theta_{i,t} - G_{i,t}\theta_{i,t-1})'$.

Similarly, one finds that the full conditional of Φ_0 is

$$\mathcal{W}(\nu_0 + T/2, S_0 + 1/2SS_y),$$

with $SS_y = \sum_{t=1}^{T}(y_t - F_t\theta_t)(y_t - F_t\theta_t)'$.

Example: SUTSE models.

As an illustration, let us consider again the data on Spain and Denmark investments. In Section 3.3.2, we used a SUTSE system where each series was described through a linear growth model; there, we just plugged in the MLE estimates of the unknown variances, while here we illustrate Bayesian inference. The prior distributions for the precisions $\Phi_0 = V^{-1}$, $\Phi_1 = W_\mu^{-1}$ and $\phi_2 = W_\beta^{-1}$ are independent Wishart, namely $\Phi_j \sim \mathcal{W}(\nu_j, S_j)$, $j = 0, 1, 2$.

It is convenient to express the Wishart hyperparameters as $\nu_0 = (\delta_0 + m - 1)/2 = (\delta_0 + 1)/2$, $\nu_j = (\delta_j + p_j - 1)/2 = (\delta_j + 1)/2$, $j = 1, 2$ and $S_0 = V_0/2$, $S_1 = W_{\mu,0}/2$, $S_2 = W_{\beta,0}/2$, so that, if $\delta_j > 2$, $j = 0, 1, 2$,

$$\mathrm{E}(V) = \frac{1}{\delta_0 - 2} V_0, \quad \mathrm{E}(W_\mu) = \frac{1}{\delta_1 - 2} W_{\mu,0} \quad \mathrm{E}(W_\beta) = \frac{1}{\delta_2 - 2} W_{\beta,0}$$

(see Appendix A). Thus, the matrix V_0 can be fixed as $V_0 = (\delta_0 - 2)\mathrm{E}(V)$, and similarly for $W_{\mu,0}$ and $W_{\beta,0}$. The parameters δ_i give an idea of the prior uncertainty: note that, from the full conditional distributions, we have

$$\mathrm{E}(V | y_{1:T}, W_\mu, W_\beta, \theta_{0:t}) =$$

$$= \frac{\delta_0 - 2}{(\delta_0 + T) - 2} \mathrm{E}(V) + \frac{T}{(\delta_0 + T) - 2} \frac{\sum_{t=1}^{T}(y_t - F_t\theta_t)(y_t - F_t\theta_t)'}{T},$$

and analogous expressions hold for the conditional expectations of W_μ and W_β. So, values of δ_i close to 2 imply a smaller weight of the prior in the updating. Expressing honest prior information on a covariance matrix is difficult, and data-dependent priors are sometimes used; but for this exercise let's suppose that

$$V_0 = (\delta_0 - 2) \begin{bmatrix} 10^2 & 0 \\ 0 & 500^2 \end{bmatrix},$$

$$W_{\mu,0} = (\delta_1 - 2) \begin{bmatrix} 0.01^2 & 0 \\ 0 & 0.01^2 \end{bmatrix}, \quad W_{\beta,0} = (\delta_2 - 2) \begin{bmatrix} 5^2 & 0 \\ 0 & 100^2 \end{bmatrix},$$

with $\delta_0 = \delta_2 = 3$ and $\delta_1 = 100$. The choice of $W_{\mu,0}$ and δ_1 expresses the prior assumption that the two individual linear growth models are in fact (independent) integrated random walks; the other choices give a prior idea of the signal-to-noise ratio.

The Gibbs sampling from the joint posterior $\pi(\theta_{0:T}, \Phi_0, \Phi_1, \Phi_2 | y_{1:T})$ generates in turn the state vectors $\theta_{0:T}$ and the precision Φ_0, Φ_1, Φ_2. The full conditional of Φ_0 is

$$\mathcal{W}\left(\frac{\delta_0 + 1 + T}{2}, \frac{1}{2}(V_0 + SS_y)\right)$$

and the full conditionals of $\Phi_1 = W_\mu^{-1}$ and $\Phi_2 = W_\beta^{-1}$ are, respectively, $\mathcal{W}((\delta_1 + 1 + T)/2, (W_{\mu,0} + SS_{1\cdot})/2)$ and $\mathcal{W}((\delta_2 + 1 + T)/2, (W_{\beta,0} + SS_{2\cdot})/2)$. Note that the function *rwishart* in the package dlm takes as inputs the degrees of freedom and the scale matrix (if $\Phi \sim \mathcal{W}(\delta/2, V_0/2)$, the degrees of freedom are δ and the scale matrix is V_0^{-1}).

R code

```
> inv <- read.table("Datasets/invest2.dat",
+                   col.names = c("Denmark", "Spain"))
> y <- ts(inv, frequency = 1, start = 1960)
> # prior hyperparameters
> delta0 <- delta2 <- 3 ; delta1 <- 100
> V0 <- (delta0 - 2) *diag(c(10^2, 500^2))
> Wmu0 <- (delta1 - 2) * diag(0.01^2, 2)
> Wbeta0 <- (delta2 - 2) * diag(c(5^2, 100^2))
> ## Gibbs sampling
> MC <- 30000
> TT <- nrow(y)
> gibbsTheta <- array(0, dim = c(TT + 1, 4, MC - 1))
> gibbsV <- array(0, dim = c(2, 2, MC))
> gibbsWmu <- array(0, dim = c(2, 2, MC))
> gibbsWbeta <- array(0, dim = c(2, 2, MC))
> mod <- dlm(FF = matrix(c(1, 0), nrow = 1) %x% diag(2),
+            V = diag(2),
+            GG = matrix(c(1, 0, 1, 1), 2, 2) %x% diag(2),
+            W = bdiag(diag(2), diag(2)),
+            m0 = c(inv[1, 1], inv[1, 2], 0, 0),
+            C0 = diag(x = 1e7, nrow = 4))
> # starting values
> mod$V <- gibbsV[,, 1] <- V0 / (delta0 - 2)
```

```
24 > gibbsWmu[,, 1] <- Wmu0 / (delta1 - 2)
   > gibbsWbeta[,, 1] <- Wbeta0 / (delta2 - 2)
26 > mod$W <- bdiag(gibbsWmu[,, 1], gibbsWbeta[,, 1])
   > # MCMC loop
28 > set.seed(3420)
   > for(it in 1 : (MC - 1))
30 +     {
   +     # generate states - FFBS
32 +     modFilt <- dlmFilter(y, mod, simplify = TRUE)
   +     gibbsTheta[,, it] <- theta <- dlmBSample(modFilt)
34 +     # update V
   +     S <- crossprod(y - theta[-1, ] %*% t(mod$FF)) + V0
36 +     gibbsV[,, it+1] <- solve(rwishart(df = delta0 + 1 + TT,
   +                                       p = 2, Sigma = solve(S)))
38 +     mod$V <- gibbsV[,, it+1]
   +     # update Wmu and Wbeta
40 +     theta.center <- theta[-1, ] - (theta[-(TT + 1), ]  %*%
   +                                    t(mod$GG))
42 +     SS1 <- crossprod(theta.center)[1 : 2, 1 : 2] + Wmu0
   +     SS2 <- crossprod(theta.center)[3 : 4, 3 : 4] + Wbeta0
44 +     gibbsWmu[,, it+1] <- solve(rwishart(df =delta1 + 1 + TT,
   +                                         Sigma = solve(SS1)))
46 +     gibbsWbeta[,, it+1] <- solve(rwishart(df = delta2 + 1 + TT,
   +                                           Sigma = solve(SS2)))
48 +     mod$W <- bdiag(gibbsWmu[,, it+1], gibbsWbeta[,, it+1])
   +     }
```

We set the number of MCMC samples to 30,000 and remove the first 20,000 iterations as burn-in. Some convergence diagnostic plots are shown in Figures 4.7 and 4.8. In general, mixing in the Gibbs sampling is poorer if the parameters are strongly correlated; here, the prior restrictions on W_μ facilitate identifiability and MCMC mixing.

R code

```
  > burn<- 1 : 20000
2 > par(mar = c(2, 4, 1, 1) + 0.1, cex = 0.8)
  > par(mfrow=c(3,2))
4 > plot(ergMean(sqrt(gibbsV[1,1, -burn])), type="l",
  +      main="", cex.lab=1.5, ylab=expression(sigma[1]),
6 +      xlab="MCMC iteration")
  > acf(sqrt(gibbsV[1,1,-burn]),  main="")
```

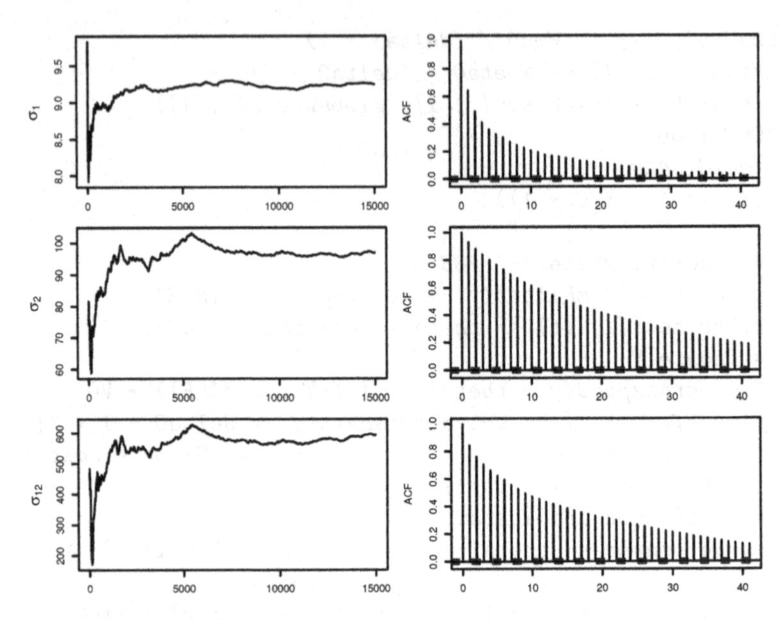

Fig. 4.7. MCMC ergodic means and autocorrelation for the observation covariance matrix V ($\sigma_1^2 = V_{11}, \sigma_2^2 = V_{22}, \sigma_{12} = V_{12}$)

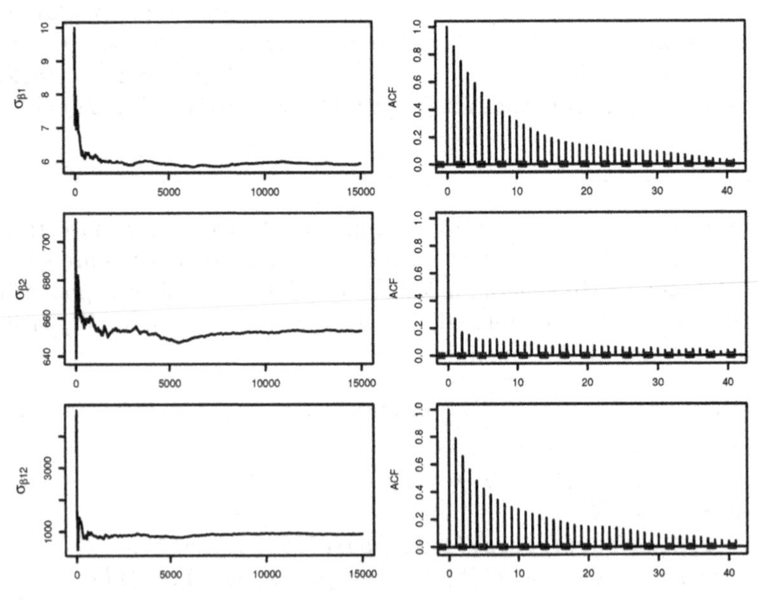

Fig. 4.8. MCMC ergodic means and autocorrelation for the covariance matrix W_β ($\sigma_{\beta,1}^2 = W_{\beta,11}, \sigma_{\beta,2}^2 = W_{\beta,22}, \sigma_{\beta,12} = W_{\beta,12}$)

The display below shows the MCMC estimates of the posterior means for the parameters in V and W_β, together with their estimated standard errors (in parenthesis), obtained from the function *mcmcMean*.

R code

```
> cbind(mcmcMean(gibbsV[1,1,-burn]),
+       mcmcMean(gibbsV[2,2,-burn]),
+       mcmcMean(gibbsV[2,1,-burn]))
```

$$\mathrm{E}(V|y_{1:T}) = \begin{bmatrix} 88.358 & 1018.469 \\ (1.1562) & (22.4865) \\ 1018.469 & 60066.43 \\ (22.4865) & (856.219) \end{bmatrix},$$

$$\mathrm{E}(W_\beta|y_{1:T}) = \begin{bmatrix} 37.399 & 396.591 \\ (0.9666) & (42.661) \\ 396.591 & 308729.71 \\ (42.661) & (2135.008) \end{bmatrix}.$$

Figure 4.9 shows the MCMC approximation of the Bayesian smoothing estimates of the level of the investments for Denmark and Spain, with marginal 5% and 95% quantiles.

The choice of an Inverse-Wishart prior on the unknown blocks of the covariance matrices has several advantages; in this exercise, computation of the full conditionals is made simple by the conjugacy properties of the Inverse-Wishart for the Gaussian model. In fact, inference on a covariance matrix is quite delicate, implying assumptions on the dependence structure of the data. We just note that the Inverse-Wishart prior may be too restrictive in modeling the prior uncertainty on the elements of the covariance matrix, as discussed earlier by Lindley (1978), and several generalizations have been proposed; some references are Dawid (1981), Brown et al. (1994), Dawid and Lauritzen (1993) in the context of graphical models, Consonni and Veronese (2003), Rajaratnam et al. (2008) and references therein.

4.5.3 A model for outliers and structural breaks

In this section we consider a generalization of the d-Inverse-Gamma model that is appropriate to account for outliers and structural breaks. As in Section 4.5.1, we assume that observations are univariate, that W_t is diagonal, and that the specification of F_t and G_t does not include any unknown parameter. To introduce the model, let us focus on observational outliers first. Structural breaks—or outliers in the state series—will be dealt with in a similar way later on. From the observation equation $Y_t = F_t\theta_t + v_t$, we see that a

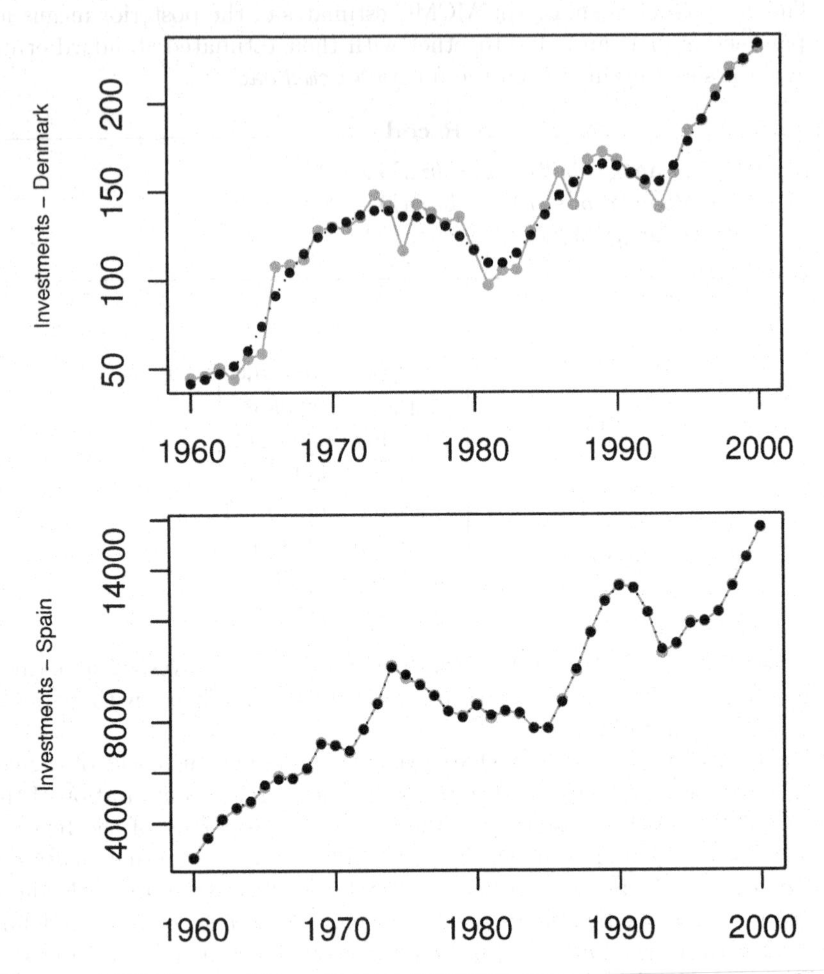

Fig. 4.9. MCMC smoothing estimates of the investments' level for Denmark and Spain, with posterior 90% probability intervals. The data are plotted in gray.

simple way to account for observations that are unusually far from their one-step-ahead predicted value is to replace the Normal distribution of v_t with a heavy-tailed distribution. The Student-t distribution family is particularly appealing in this respect for two reasons. On one hand, it can accommodate, through its degrees-of-freedom parameter, different degrees of heaviness in the tails. On the other hand, the Student-t distribution admits a simple representation as a scale mixture of Normal distributions, which allows one to treat a DLM with t-distributed observation errors as a Gaussian DLM, conditionally on the scale parameters. The obvious advantage is that all the standard algorithms for DLMs—from the Kalman filter to FFBS—can still be used, albeit conditionally. In particular, within a Gibbs sampler, one can still draw the states from their full conditional distribution using the FFBS algorithm. We assume that the v_t have Student-t distributions with $\nu_{y,t}$ degrees of freedom and common scale parameter λ_y^{-1}:

$$v_t|\lambda_y, \nu_{y,t} \overset{indep}{\sim} \mathcal{T}(0, \lambda_y^{-1}, \nu_{y,t}).$$

Introducing latent variables $\omega_{y,t}$, distributed as $\mathcal{G}(\frac{\nu_{y,t}}{2}, \frac{\nu_{y,t}}{2})$, we can equivalently write:

$$\begin{aligned} v_t|\lambda_y, \omega_{y,t} &\overset{indep}{\sim} \mathcal{N}(0, (\lambda_y \omega_{y,t})^{-1}), \\ \omega_{y,t}|\nu_{y,t} &\overset{indep}{\sim} \mathcal{G}\left(\frac{\nu_{y,t}}{2}, \frac{\nu_{y,t}}{2}\right). \end{aligned}$$

In other words, we are assuming that, given λ_y, the precisions $\phi_{y,t} = \lambda_y \omega_{y,t}$ of the observations in the DLM vary in a random fashion over time.

The latent variable $\omega_{y,t}$ in the previous representation can be informally interpreted as the degree of nonnormality of v_t. In fact, taking the $\mathcal{N}(0, \lambda_y^{-1})$ as baseline—corresponding to $\omega_{y,t} = \mathrm{E}(\omega_{y,t}) = 1$—values of $\omega_{y,t}$ lower than 1 make larger absolute values of v_t more likely. Figure 4.10 shows a plot of the

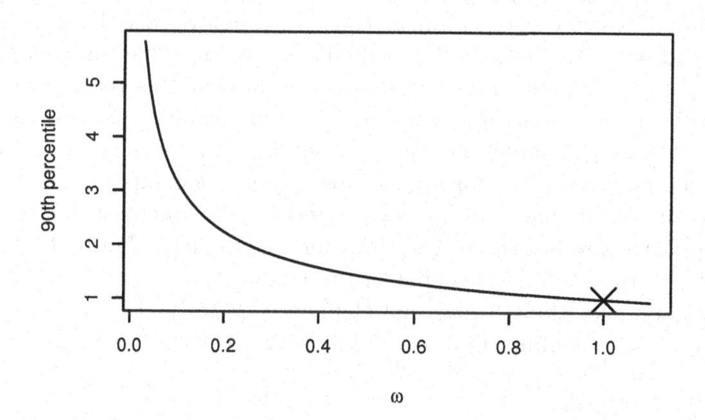

Fig. 4.10. 90th percentile of the conditional distribution of v_t as a function of $\omega_{y,t}$

90th percentile of the $\mathcal{N}(0, (\lambda_y \omega_{y,t})^{-1})$ as a function of $\omega_{y,t}$ (λ_y is selected so that the percentile is one when $\omega_{y,t}$ is one). From the previous discussion, it follows that the posterior mean of the $\omega_{y,t}$ can be used to flag possible outliers. As a prior for the precision parameter λ_y we choose a Gamma distribution with mean a_y and variance b_y,

$$\lambda_y|a_y, b_y \sim \mathcal{G}\left(\frac{a_y^2}{b_y}, \frac{a_y}{b_y}\right),$$

taking in turn a_y and b_y uniformly distributed on a large, but bounded, interval,

$$a_y \sim \mathcal{U}nif(0, A_y), \qquad b_y \sim \mathcal{U}nif(0, B_y).$$

Although the degrees-of-freedom parameter of a Student-t distribution can take any positive real value, we restrict for simplicity the set of possible values to a finite set of integers and set

$$\nu_{y,t}|p_y \overset{iid}{\sim} \mathcal{M}ult(1, p_y),$$

where $p_y = (p_{y,1}, \ldots, p_{y,K})$ is a vector of probabilities, the levels of the multinomial distribution are the integers $n_1, \ldots, n_K$, and the $\nu_{y,t}$'s are independent across t. As a convenient, yet flexible choice for $n_1, \ldots, n_K$ we use the set $\{1, 2, \ldots, 10, 20, \ldots, 100\}$. Note that for $\nu_{y,t} = 100$, v_t is approximately normally distributed, given λ_y. As a prior for p_y we adopt a Dirichlet distribution with parameter $\alpha_y = (\alpha_{y,1}, \ldots, \alpha_{y,K})$, $p_y \sim \mathcal{D}ir(\alpha_y)$. This completes the hierarchical specification of the prior distribution of the observation variances V_t. To account for possible outliers in the state components, a similar hierarchical structure is assumed for each diagonal element of W_t, i.e., for the precision parameters of the state innovations.

Note that in this model the precisions, or, equivalently, the variances, are allowed to be different at different times, although in a way that does not account for a possible correlation in time. In other words, the sequences of precisions at different times are expected to look more like independent, or exchangeable, sequences, rather than time series. For this reason the model is appropriate to account for occasional abrupt changes—corresponding to innovations having a large variance—in the state vector. For example, for polynomial and seasonal factor models, an outlier in a component of w_t corresponds to an abrupt change in the corresponding component of the state, such as a jump in the level of the series. However, the modeler does not anticipate these changes to present a clear pattern in time.

Writing $W_{t,i}$ for the ith diagonal element of W_t, $i = 1, \ldots, p$, the hierarchical prior can be summarized in the following display.

$$\begin{aligned}
V_t^{-1} &= \lambda_y \omega_{y,t}, & W_{t,i}^{-1} &= \lambda_{\theta,i}\omega_{\theta,ti}, \\
\lambda_y|a_y, b_y &\sim \mathcal{G}\left(\frac{a_y^2}{b_y}, \frac{a_y}{b_y}\right), & \lambda_{\theta,i}|a_{\theta,i}, b_{\theta,i} &\overset{indep}{\sim} \mathcal{G}\left(\frac{a_{\theta,i}^2}{b_{\theta,i}}, \frac{a_{\theta,i}}{b_{\theta,i}}\right), \\
\omega_{y,t}|\nu_{y,t} &\overset{indep}{\sim} \mathcal{G}\left(\frac{\nu_{y,t}}{2}, \frac{\nu_{y,t}}{2}\right), & \omega_{\theta,ti}|\nu_{\theta,ti} &\overset{indep}{\sim} \mathcal{G}\left(\frac{\nu_{\theta,ti}}{2}, \frac{\nu_{\theta,ti}}{2}\right), \\
a_y &\sim \mathcal{U}nif(0, A_y), & a_{\theta,i} &\overset{indep}{\sim} \mathcal{U}nif(0, A_{\theta,i}), \\
b_y &\sim \mathcal{U}nif(0, B_y), & b_{\theta,i} &\overset{indep}{\sim} \mathcal{U}nif(0, B_{\theta,i}), \\
\nu_{y,t} &\overset{indep}{\sim} \mathcal{M}ult(1; p_y) & \nu_{\theta,ti} &\overset{indep}{\sim} \mathcal{M}ult(1; p_{\theta,i}) \\
p_y &\sim \mathcal{D}ir(\alpha_y) & p_{\theta,i} &\overset{indep}{\sim} \mathcal{D}ir(\alpha_{\theta,i}),
\end{aligned}$$

with $\alpha_{\theta,i} = (\alpha_{\theta,i,1}, \ldots, \alpha_{\theta,i,K})$, $i = 1, \ldots, K$. Again, the levels of all the multinomial distributions are the integers $n_1, \ldots, n_K$.

A Gibbs sampler can be implemented to draw from the posterior distribution of parameters and states of the model specified above. Given all the unknown parameters, the states are generated at once from their joint full conditional distribution using the standard FFBS algorithm. The full conditional distributions of the parameters are easy to derive. We provide here a detailed derivation of the full conditional distribution of λ_y, as an example:

$$\begin{aligned}
\pi(\lambda_y|\ldots) &\propto \pi(y_{1:T}|\theta_{1:T}, \omega_{y,1:T}, \lambda_y) \cdot \pi(\lambda_y|a_y, b_y) \\
&\propto \prod_{t=1}^{T} \lambda_y^{\frac{1}{2}} \exp\left\{-\frac{\omega_{y,t}\lambda_y}{2}(y_t - F_t\theta_t)^2\right\} \cdot \lambda_y^{\frac{a_y^2}{b_y}-1} \exp\left\{-\lambda_y \frac{a_y}{b_y}\right\} \\
&\propto \lambda_y^{\frac{T}{2}+\frac{a_y^2}{b_y}-1} \exp\left\{-\lambda_y\left[\frac{1}{2}\sum_{t=1}^{T}\omega_{y,t}(y_t - F_t\theta_t)^2 + \frac{a_y}{b_y}\right]\right\}.
\end{aligned}$$

Therefore,

$$\lambda_y|\ldots \sim \mathcal{G}\left(\frac{a_y^2}{b_y} + \frac{T}{2}, \frac{a_y}{b_y} + \frac{1}{2}SS_y^*\right),$$

with $SS_y^* = \sum_{t=1}^{T} \omega_{y,t}(y_t - F_t\theta_t)^2$. A summary of all the full conditional distributions of the unknown parameters is shown in Table 4.1. All the full conditional distributions, except for those of $a_y, b_y, a_{\theta,i}, b_{\theta,i}$, are standard. The latter can be drawn from using ARMS. More specifically, we suggest using ARMS separately on each pair (a, b).

As an example of the use of the model discussed above, consider the time series of quarterly gas consumption in the UK from 1960 to 1986. The data are available in R as *UKgas*. A plot of the data, on the log scale, suggests a possible change in the seasonal factor around the third quarter of 1970. After taking logs, we employ a model built on a local linear trend plus seasonal component DLM to analyze the data. In this model the five-by-five variance matrix W_t has only three nonzero diagonal elements: the first refers to the level of the series, the second to the slope of the stochastic linear trend, and the third to the seasonal factor. We packaged the entire Gibbs sampler in the function *dlmGibbsDIGt*, available from the book website. The parameters $a_y, b_y, a_{\theta,1}, b_{\theta,1}, \ldots, a_{\theta,3}, b_{\theta,3}$ are taken to be uniform on $(0, 10^5)$, and the parameters of the four Dirichlet distributions of $p_y, p_{\theta,1}, p_{\theta,2}, p_{\theta,3}$ are all equal to 1/19. The posterior analysis is based on 10000 iterations, after a burn-in of 500 iterations. To reduce the autocorrelation, two extra sweeps were run between every two consecutive saved iterations, implying that the iterations of the sampler after burn-in were actually 30000.

R code

```
> y <- log(UKgas)
> set.seed(4521)
```

$$\lambda_y|\ldots \sim \mathcal{G}\left(\frac{a_y^2}{b_y}+\frac{T}{2}, \frac{a_y}{b_y}+\frac{1}{2}SS_y^*\right),$$

with $SS_y^* = \sum_{t=1}^T \omega_{y,t}(y_t - F_t\theta_t)^2$;

$$\lambda_{\theta,i}|\ldots \sim \mathcal{G}\left(\frac{a_{\theta,i}^2}{b_{\theta,i}}+\frac{T}{2}, \frac{a_{\theta,i}}{b_{\theta,i}}+\frac{1}{2}SS_{\theta,i}^*\right),$$

with $SS_{\theta,i}^* = \sum_{t=1}^T \omega_{\theta,ti}(\theta_{ti} - (G_t\theta_{t-1})_i)^2$, for $i = 1,\ldots,p$;

$$\omega_{y,t}|\ldots \sim \mathcal{G}\left(\frac{\nu_{y,t}+1}{2}, \frac{\nu_{y,t}+\lambda_y(y_t - F_t\theta_t)^2}{2}\right),$$

for $t = 1,\ldots,T$;

$$\omega_{\theta,ti}|\ldots \sim \mathcal{G}\left(\frac{\nu_{\theta,ti}+1}{2}, \frac{\nu_{\theta,ti}+\lambda_{\theta,i}(\theta_{ti} - (G_t\theta_{t-1})_i)^2}{2}\right),$$

for $i = 1,\ldots,p$ and $t = 1,\ldots,T$;

$$\pi(a_y, b_y|\ldots) \propto \mathcal{G}(\lambda_y; a_y, b_y),$$

on the set $0 < a_y < A_y, \quad 0 < b_y < B_y$;

$$\pi(a_{\theta,i}, b_{\theta,i}|\ldots) \propto \mathcal{G}(\lambda_{\theta,i}; a_{\theta,i}, b_{\theta,i}),$$

on $0 < a_{\theta,i} < A_{\theta,i}, \quad 0 < b_{\theta,i} < B_{\theta,i}$, for $i = 1,\ldots,p$;

$$\pi(\nu_{y,t} = k) \propto \mathcal{G}\left(\omega_{y,t}; \frac{k}{2}, \frac{k}{2}\right) \cdot p_{y,k},$$

on the set $\{n_1,\ldots,n_K\}$, for $t = 1,\ldots,T$;

$$\pi(\nu_{\theta,ti} = k) \propto \mathcal{G}\left(\omega_{\theta,ti}; \frac{k}{2}, \frac{k}{2}\right) \cdot p_{\theta,i,k},$$

on the set $\{n_1,\ldots,n_K\}$, for $i = 1,\ldots,p$ and $t = 1,\ldots,T$;

$$p_y|\ldots \sim \mathcal{D}ir(\alpha_y + N_y),$$

where $N_y = (N_{y,1},\ldots,N_{y,K})$ with, for each k, $N_{y,k} = \sum_{t=1}^T(\nu_{y,t} = k)$;

$$p_{\theta,i}|\ldots \sim \mathcal{D}ir(\alpha_{\theta,i} + N_{\theta,i}),$$

where $N_{\theta,i} = (N_{\theta,i,1},\ldots,N_{\theta,i,K})$ with, for each k, $N_{\theta,i,k} = \sum_{t=1}^T(\nu_{\theta,ti} = k)$, for $i = 1,\ldots,p$;

Table 4.1. Full conditional distributions for the model of Section 4.5.3

```
> MCMC <- 10500
> gibbsOut <- dlmGibbsDIGt(y, mod = dlmModPoly(2) + dlmModSeas(4),
+                          A_y = 10000, B_y = 10000, p = 3,
+                          n.sample = MCMC, thin = 2)
```

Figure 4.11, obtained with the code below, graphically summarizes the posterior means of the $\omega_{y,t}$ and $\omega_{\theta,ti}$, $t = 1,..108$, $i = 1, 2, 3$.

R code

```
> burn <- 1 : 500
> nuRange <- c(1 : 10, seq(20, 100, by = 10))
> omega_y <- ts(colMeans(gibbsOut$omega_y[-burn, ]),
+               start = start(y), freq=4)
> omega_theta <- ts(apply(gibbsOut$omega_theta[,, -burn], 1 : 2,
+                         mean), start = start(y), freq = 4)
> layout(matrix(c(1, 2, 3, 4), 4, 1, TRUE))
> par(mar = c(5.1, 4.1, 2.1, 2.1))
> plot(omega_y, type = "p", ylim = c(0, 1.2), pch = 16,
+      xlab = "", ylab = expression(omega[list(y, t)]))
> abline(h = 1, lty = "dashed")
> for (i in 1 : 3)
+ {
+     plot(omega_theta[,i], ylim=c(0,1.2), pch = 16,
+          type = "p", xlab = "",
+          ylab = bquote(omega[list(theta, t * .(i))]))
+     abline(h = 1, lty = "dashed")
+ }
```

It is clear that there are no observational outliers, except perhaps for a mild outlier in the third quarter of 1983, with $\mathrm{E}(\omega_{\theta,t3}|y_{1:T}) = 0.88$. The trend is fairly stable, in particular its slope parameter. The seasonal component, on the other hand, presents several structural breaks, particularly in the first couple of years of the seventies. The most extreme change in the seasonal component happened in the third quarter of 1971, when the corresponding ω_t had an estimated value of 0.012. It can also be seen that after that period of frequent shocks, the overall variability of the seasonal component remained higher than in the first period of observation.

From the output of the Gibbs sampler one can also estimate the unobserved components—trend and seasonal variation—of the series. Figure 4.12 provides a plot of estimated trend and seasonal component, together with 95% probability intervals. An interesting feature of a model with time-specific variances, like the one considered here, is that confidence intervals need not be of constant width—even after accounting for boundary effects. This is clearly

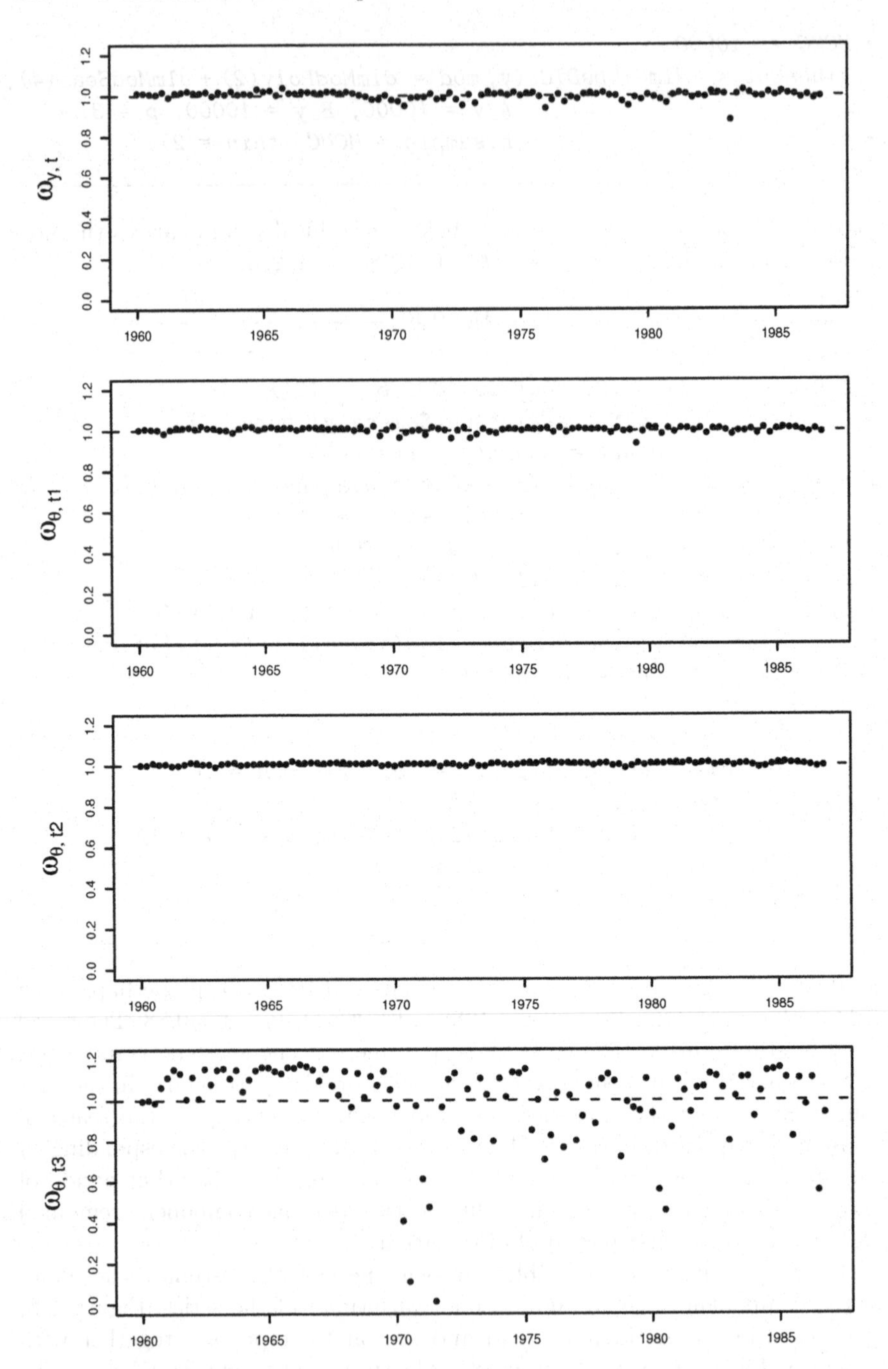

Fig. 4.11. UK gas consumption: posterior means of the ω_t's

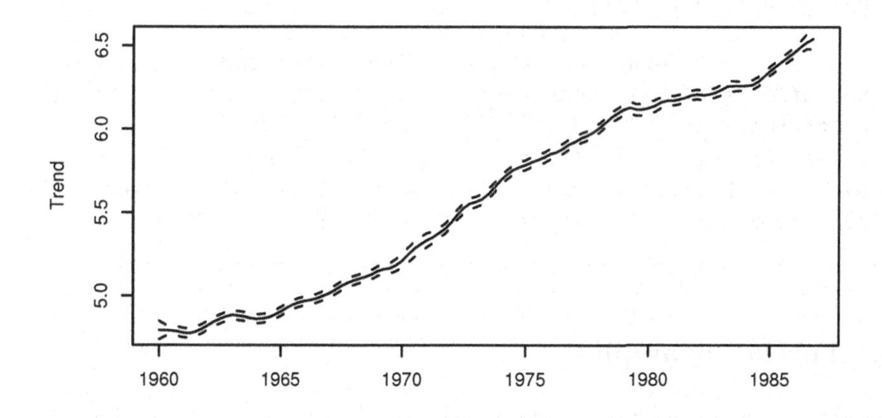

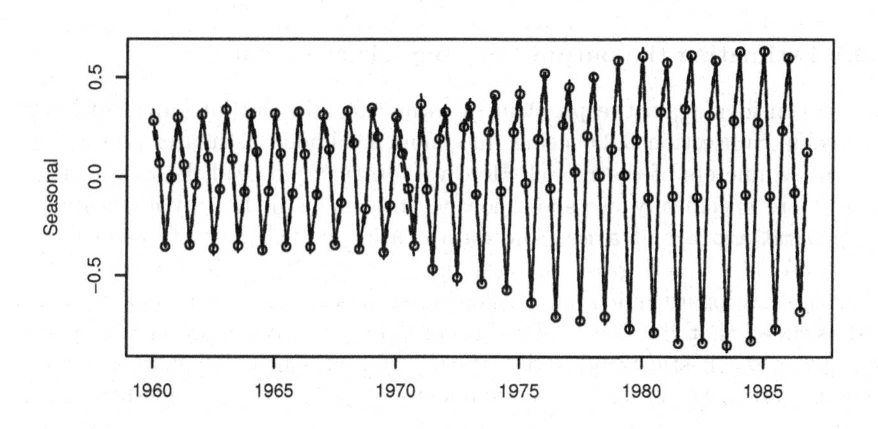

Fig. 4.12. UK gas consumption: trend and seasonal component, with 95% probability intervals

seen in the example, where the 95% probability interval for the seasonal component is wider in the period of high instability of the early seventies. The following code was used to obtain the plot.

R code

```
> thetaMean <- ts(apply(gibbsTheta, 1 : 2, mean),
+                   start = start(y),
+                   freq = frequency(y))
> LprobLim <- ts(apply(gibbsTheta, 1 : 2, quantile,
+                       probs = 0.025),
+                 start = start(y), freq = frequency(y))
```

```
> UprobLim <- ts(apply(gibbsTheta, 1 : 2, quantile,
+                      probs = 0.975),
+                start = start(y), freq = frequency(y))
> par(mfrow = c(2, 1), mar = c(5.1, 4.1, 2.1, 2.1))
> plot(thetaMean[, 1], xlab = "", ylab = "Trend")
> lines(LprobLim[, 1], lty = 2); lines(UprobLim[, 1], lty = 2)
> plot(thetaMean[, 3], xlab = "", ylab = "Seasonal", type = "o")
> lines(LprobLim[, 3], lty = 2); lines(UprobLim[, 3], lty = 2)
```

4.6 Further examples

We give further examples of Bayesian analysis via MCMC for DLMs. The first is useful to compare MLE and Bayesian estimates.

4.6.1 Estimating the output gap: Bayesian inference

Let us consider again the problem of estimating the output gap, which we treated in Subsection 3.2.6 using maximum likelihood estimates of the unknown parameters of the model. Here we illustrate, instead, Bayesian inference. More specifically, we show how to implement in R a hybrid sampler using `dlmBSample` and `arms`. The data consist of the quarterly time series of deseasonalized real gross domestic product (GDP) of the US from 1950 to 2004, on a logarithmic scale. Following standard econometric practice, we had assumed that the GDP can be decomposed into two unobservable components: a stochastic trend and a stationary component. Here we compute Bayesian estimates of the two components as well as the parameters of the model.

The stochastic trend was described by a local linear trend, while the stationary component was an AR(2) process. As a DLM, the model is the sum, in the sense discussed in Section 3.2, of a polynomial model of order two and a DLM representation of a stationary AR(2) process, observed with no error. The matrices of the resulting DLM were given by (3.36) in Section 3.2.6. The first component of the state vector represents the trend, while the third is the AR(2) stationary component. The AR parameters ϕ_1 and ϕ_2 must lie in the stationarity region $\mathcal{S}$ defined by

$$\begin{aligned} \phi_1 + \phi_2 &< 1, \\ \phi_1 - \phi_2 &> -1, \\ |\phi_2| &< 1. \end{aligned}$$

The prior we select for (ϕ_1, ϕ_2) is a product of a $\mathcal{N}(0, (2/3)^2)$ and a $\mathcal{N}(0, (1/3)^2)$, restricted to $\mathcal{S}$. In this way, the prior penalizes those values of the AR parameters close to the boundary of the stationarity region. For the three precisions, i.e., the inverses of the variances σ_μ^2, σ_δ^2, and σ_u^2, we assume independent gamma priors with mean a and variance b:

$$\mathcal{G}\left(\frac{a^2}{b}, \frac{a}{b}\right).$$

In this specific case, we set $a = 1$, $b = 1000$. A hybrid sampler can draw in turn the AR parameters from $\pi(\phi_1, \phi_2 | \sigma_\mu^2, \sigma_\delta^2, \sigma_u^2, y_{0:T})$, the states, and the three precisions from their full conditional distribution given the states and the AR parameters. In the notation used in Algorithm 4.3, $\psi_1 = (\phi_1, \phi_2)$ and $\psi_2 = ((\sigma_\mu^2)^{-1}, (\sigma_\delta^2)^{-1}, (\sigma_u^2)^{-1})$. The precisions, given the states and the AR parameters, are conditionally independent and gamma-distributed. Specifically,

$$\begin{aligned}
(\sigma_\mu^2)^{-1}|\ldots &\sim \mathcal{G}\left(\frac{a^2}{b} + \frac{T}{2}, \frac{a}{b} + \frac{1}{2}\sum_{t=1}^{T}\left(\theta_{t,1} - (G\theta_{t-1})_1\right)^2\right), \\
(\sigma_\delta^2)^{-1}|\ldots &\sim \mathcal{G}\left(\frac{a^2}{b} + \frac{T}{2}, \frac{a}{b} + \frac{1}{2}\sum_{t=1}^{T}\left(\theta_{t,2} - (G\theta_{t-1})_2\right)^2\right), \\
(\sigma_u^2)^{-1}|\ldots &\sim \mathcal{G}\left(\frac{a^2}{b} + \frac{T}{2}, \frac{a}{b} + \frac{1}{2}\sum_{t=1}^{T}\left(\theta_{t,3} - (G\theta_{t-1})_3\right)^2\right).
\end{aligned} \tag{4.24}$$

The AR parameters, given the precisions (but not the states), have a non-standard distribution and we can use ARMS to draw from their joint full conditional distribution. One can write a function to implement the sampler in R. One such function, on which the analysis that follows is based, is available from the book web site. We reproduce below the relevant part of the main loop. In the code, `theta` is a $(T + 1)$ matrix of states and `gibbsPhi` and `gibbsVars` are matrices in which the results of the simulation are stored. The states, generated in the loop, can optionally be saved, but they can also be easily generated again, given the simulated values of the AR and variance parameters.

R code

```
for (it in 1 : mcmc)
{
    ## generate AR parameters
    mod$GG[3 : 4, 3] <- arms(mod$GG[3 : 4, 3],
                             ARfullCond, AR2support, 1)
    ## generate states - FFBS
    modFilt <- dlmFilter(y, mod, simplify = TRUE)
    theta[] <- dlmBSample(modFilt)
    ## generate W
    theta.center <- theta[-1, -4, drop = FALSE] -
        (theta[-(nobs + 1), , drop = FALSE] %*% t(mod$GG))[, -4]
```

```
12      SStheta <- drop(sapply( 1 : 3, function(i)
                            crossprod(theta.center[, i])))
14      diag(mod$W)[1 : 3] <-
            1 / rgamma(3, shape = shape.theta,
16          rate = rate.theta + 0.5 * SStheta)
        ## save current iteration, if appropriate
18      if ( !(it %% every) )
        {
20          it.save <- it.save + 1
            gibbsTheta[, , it.save] <- theta
22          gibbsPhi[it.save, ] <- mod$GG[3 : 4,3]
            gibbsVars[it.save, ] <- diag(mod$W)[1 : 3]
24      }
    }
```

The 'if' statement on line 18 takes care of the thinning, saving the draw only when the iteration counter `it` is divisible by `every`. The object `SStheta` (line 12) is a vector of length 3 containing the sum of squares appearing in the full conditional distributions of the precisions (equations (4.24)). The two functions `ARfullCond` and `AR2support`, which are the main arguments of `arms` (line 5) are defined, inside the main function, as follows.

R code

```
  AR2support <- function(u)
2 {
      ## stationarity region for AR(2) parameters
4     (sum(u) < 1) && (diff(u) < 1) && (abs(u[2]) < 1)
  }
6 ARfullCond <- function(u)
  {
8     ## log full conditional density for AR(2) parameters
      mod$GG[3 : 4, 3] <- u
10    -dlmLL(y, mod) + sum(dnorm(u, sd = c(2, 1) * 0.33,
                                 log = TRUE))
12 }
```

The sampler was run using the following call, where `gdp` is a time series object containing the data.

R code

```
  outGibbs <- gdpGibbs(gdp, a.theta = 1, b.theta = 1000, n.sample =
2                     2050, thin = 1, save.states = TRUE)
```

Discarding the first 50 draws as burn in, we look at some simple diagnostic plots. The traces of the simulated variances (Figure 4.13) do not show any particular sign of a nonstationary behavior. We have also plotted the

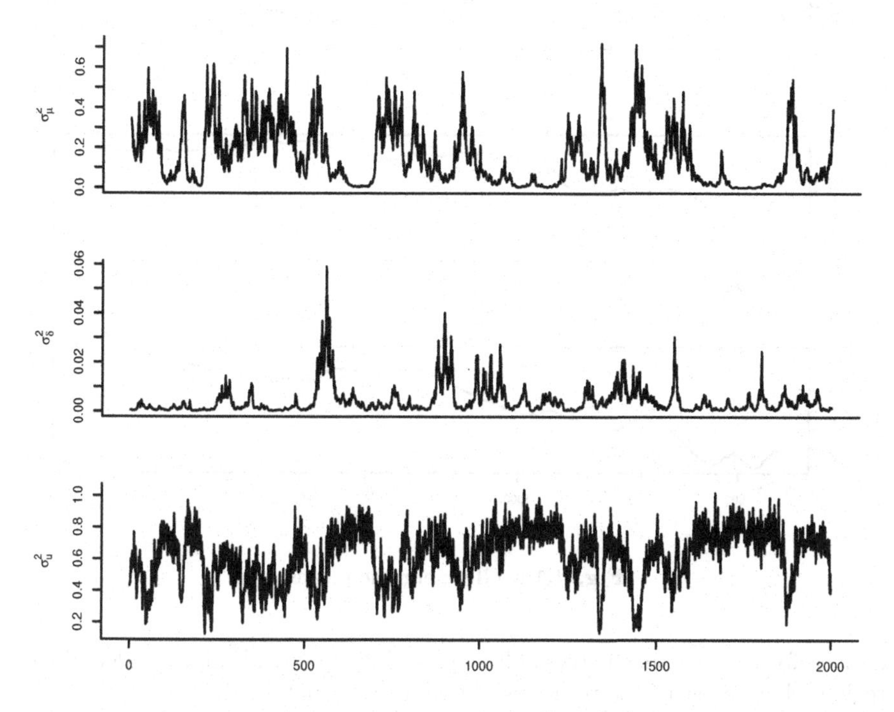

Fig. 4.13. GDP: traces of simulated variances

running ergodic means of the simulated standard deviations σ_μ σ_δ, and σ_u (Figure 4.14). The first plot shows $n^{-1}\sum_{i=1}^{n}\sigma_\mu^{(i)}$ versus i, and similarly for the second and third. In other words, this is the MC estimate of σ_μ versus the number of iterations of the sampler. The estimates look reasonably stable in the last part of the plot. (This impression was also confirmed by the results from a longer run, not shown here). The empirical autocorrelation functions of the three variances (Figure 4.15) give an idea of the degree of autocorrelation in the sampler. In the present case, the decay of the ACF is not very fast; this will reflect in a relatively large Monte Carlo standard error of the Monte Carlo esimates. Clearly, a smaller standard error can always be achieved by running the sampler longer. Similar diagnostic plots can be done for the AR parameters. The reader can find in the display below, for the three standard deviations in the model and the two AR parameters, the estimates of the posterior means and their estimated standard errors, obtained using Sokal's method (see Section 1.6). In addition, equal-tail 90% probability intervals are

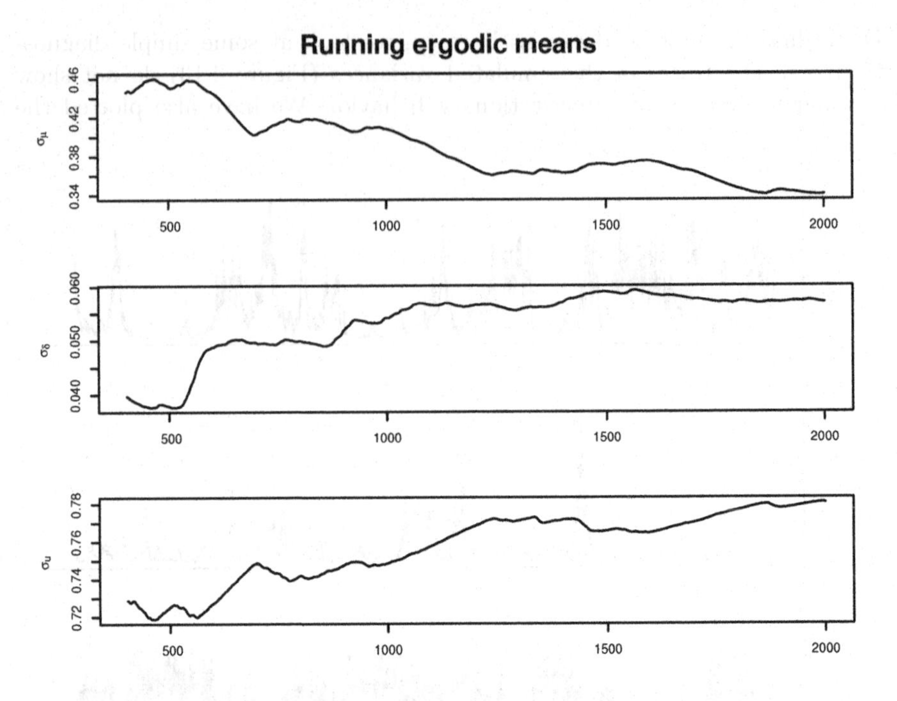

Fig. 4.14. GDP: ergodic means

derived for the five parameters. These probability intervals give an idea of the region where most of the posterior probability is contained.

R code

```
> mcmcMeans(outGibbs$phi[-burn,], names = paste("phi", 1:2))
     phi 1      phi 2
   1.3422   -0.4027
 ( 0.0112) ( 0.0120)
> apply(outGibbs$phi[-burn,], 2, quantile, probs = c(.05,.95))
        [,1]       [,2]
5%  1.174934 -0.5794382
95% 1.518323 -0.2495367
> mcmcMeans(sqrt(outGibbs$vars[-burn,]),
           names = paste("Sigma", 1:3))
  Sigma 1    Sigma 2    Sigma 3
 0.34052    0.05698    0.78059
(0.03653)  (0.00491)  (0.01766)
> apply(sqrt(outGibbs$vars[-burn,]), 2, quantile,
        probs = c(.05,.95))
          [,1]       [,2]       [,3]
```

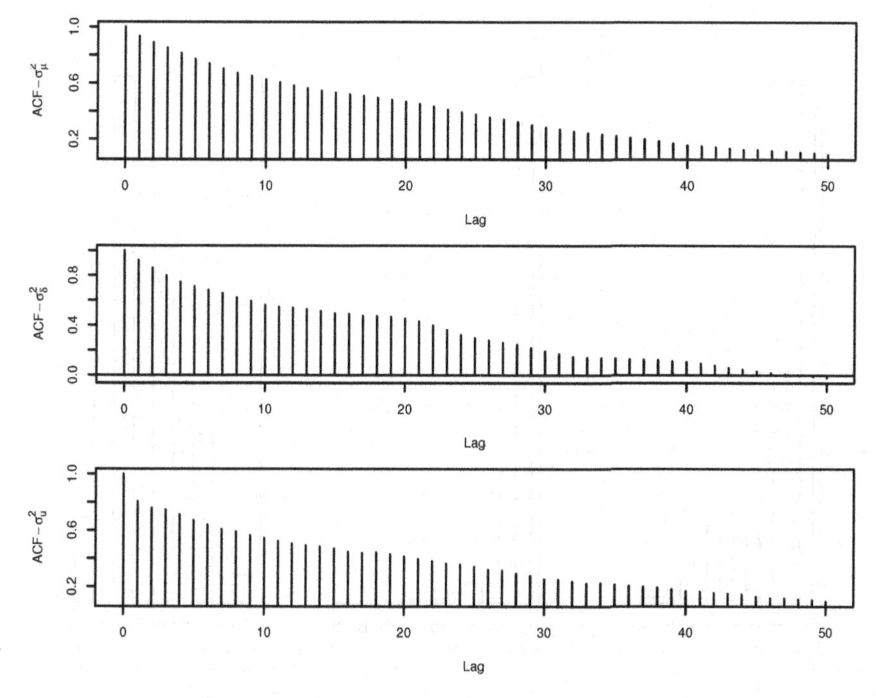

Fig. 4.15. GDP: Autocorrelation functions

```
5%  0.06792123 0.02057263 0.5596150
95% 0.65583319 0.12661949 0.9294306
```

One can also plot histograms based on the output of the sampler, to gain some insight about the shape of posterior distributions of parameters or functions thereof—at least for univariate marginal posteriors. Figure 4.16 displays the histograms of the posterior distributions of the three variances.

Scatterplots are sometimes useful to explore the shape of bivariate distributions, especially for pairs of parameters that are highly dependent on each other. Figure 4.17 displays a bivariate scatterplot of (ϕ_1, ϕ_2), together with their marginal histograms. From the picture, it is clear that there is a strong dependence between ϕ_1 and ϕ_2, which, incidentally, confirms that drawing the two at the same time was the right thing to do in order to improve the mixing of the chain.

Finally, since the sampler also included the unobservable states as latent variables, one can obtain posterior distributions and summaries of the states. In particular, in this example it is of interest to separate the trend of the GDP from the (autocorrelated) noise. The posterior mean of the trend at time t is estimated by the mean of the simulated values $\theta_{t,1}^{(i)}$. Figure 4.18 displays

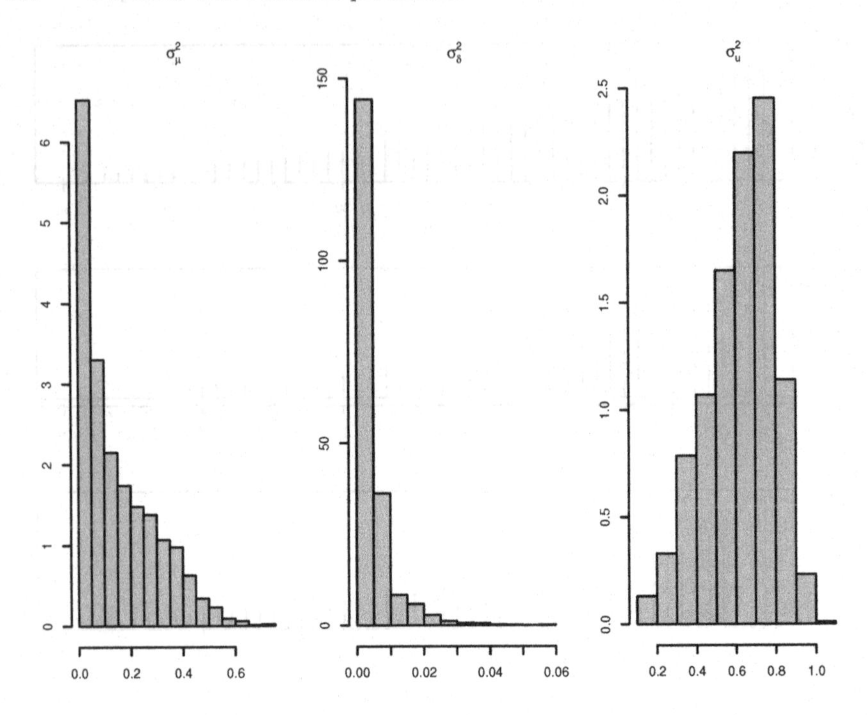

Fig. 4.16. GDP: posterior distributions of the model variances

graphically the posterior mean of the trend, together with the data, and the posterior mean of the AR(2) noise process, represented by $\theta_{t,3}$.

4.6.2 Dynamic regression

Suppose we have a time series of cross sectional data, $(Y_{i,t}, x_{i,t}), i = 1, \ldots, m$, where $Y_{i,t}$ is the value of a response variable Y corresponding to the value $x_{i,t}$ of one or more covariates X, observed over time. Typically, the interest is in estimating the regression function $m_t(x) = E(Y_t|x)$ from the cross-sectional data, at time t; moreover, one usually wants to make inference on the evolution of the regression curve over time. In Section 3.3.5 we introduced a dynamic regression model in the form of DLM for data of this nature. The model is applied here to a problem of interest in financial applications, namely, estimating the term structure of interest rates.

The problem is briefly described as follows. Let $P_t(x)$ be the price at time t of a zero-coupon bond that gives 1 euro at time to maturity x. The curve $P_t(x)$, $x \in (0, T)$, is called discount function. Other curves of interest are obtained as one-by-one transformations of the discount function; the yield curve is $\gamma_t(x) = -\log P_t(x)/x$, and the instantaneous (nominal) forward rate curve is $f_t(x) = d(-\log P_t(x)/dx) = (dP_t(x)/dx)/P_t(x)$. The yield curve, or one of its transformations, allows us to price any coupon bond as the sum

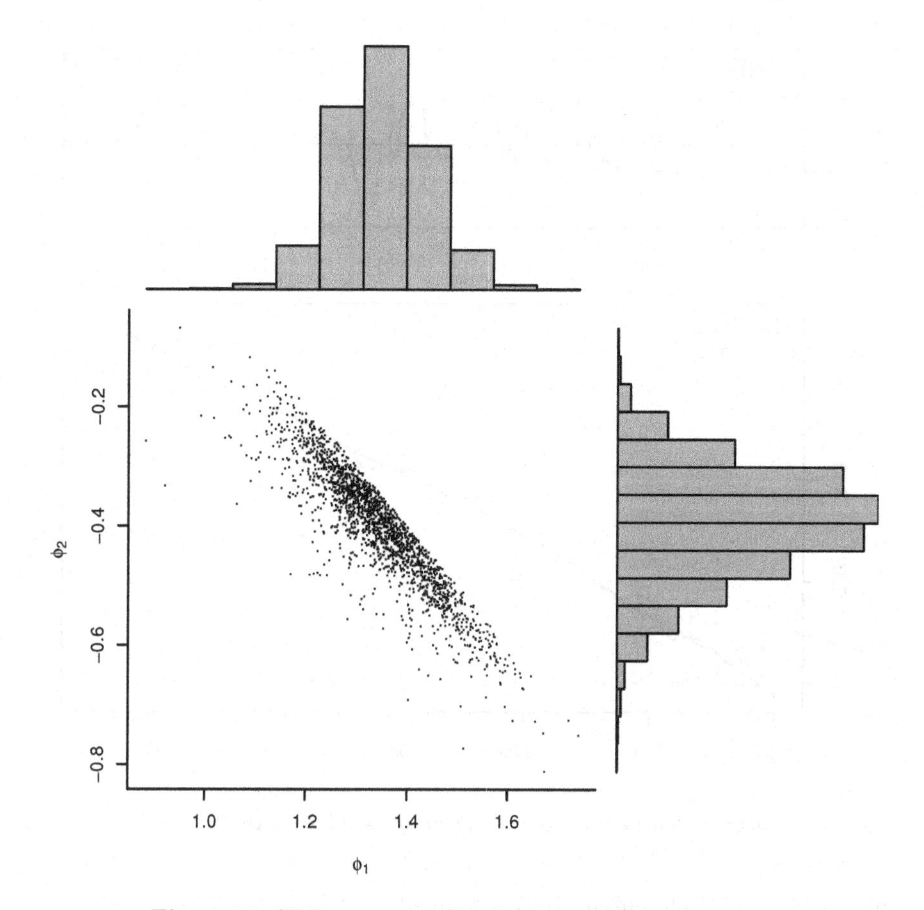

Fig. 4.17. GDP: posterior distribution of AR parameters

of the present values of future coupon and principal payments. Of course, the whole curve cannot be observed, but it has to be estimated from the bond prices observed for a finite number of times-to-maturity, $x_1, \ldots, x_m$ say. More precisely, at time t we have data $(y_{i,t}, x_i)$, $i = 1, \ldots, m$, where $y_{i,t}$ is the observed yield corresponding to time-to-maturity x_i. Due to market frictions, yields are subject to measurement error, so that the observed yields are described by

$$Y_{i,t} = \gamma_t(x_i) + v_{i,t}, \quad v_{i,t} \overset{iid}{\sim} \mathcal{N}(0, \sigma^2),\ i = 1, \ldots, m.$$

Several cross sectional models for the yield curve have been proposed in the literature; one of the most popular is the Nelson and Siegel (1987) model. In fact, Nelson and Siegel modeled the forward rate curve, as

$$f_t(x) = \beta_{1,t} + \beta_{2,t} e^{-\lambda x} + \beta_{3,t} \lambda\, x\, e^{-\lambda x},$$

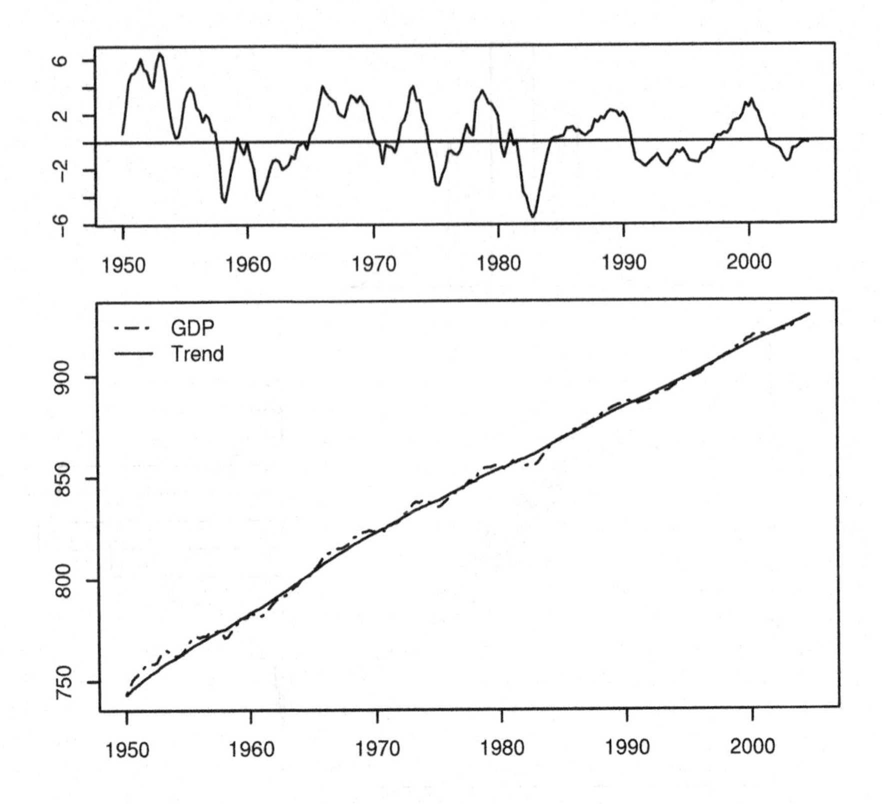

Fig. 4.18. GDP: posterior mean of trend and AR(2) noise

from which the yield curve can be obtained

$$\gamma_t(x) = \beta_{1,t} + \beta_{2,t}\frac{1-e^{-\lambda x}}{\lambda x} + \beta_{3,t}\left(\frac{1-e^{-\lambda x}}{\lambda x} - e^{-\lambda x}\right).$$

This model is not linear in the parameters $(\beta_{1,t}, \beta_{2,t}, \beta_{3,t}, \lambda)$; however, the decay parameter λ is usually approximated with a fixed value, so that the model is more simply treated as a linear model in the unknown parameters $\beta_{1,t}, \beta_{2,t}, \beta_{3,t}$. Thus, for a fixed value of λ, the model is of the form (3.43), with $k = 3$ and

$$h_1(x) = 1, \quad h_2(x) = \frac{1-e^{-\lambda x}}{\lambda x}, \quad h_3(x) = \left(\frac{1-e^{-\lambda x}}{\lambda x} - e^{-\lambda x}\right).$$

Consider the data plotted in Figure 3.19, which are monthly yields of US bonds for $m = 17$ times-to-maturity from 3 to 120 months, from January 1985 to December 2000; for a detailed description, see Diebold and Li (2006). These authors used the Nelson and Siegel cross sectional model to fit the yield curve at time t. In fact, they regard such a model as a latent factor model (see

Section 3.3.6), with $\beta_{1,t}, \beta_{2,t}, \beta_{3,t}$ playing the role of latent dynamic factors (long-term, short-term and medium-term factors), also interpreted in terms of level, slope and curvature of the yield curve. In these terms, λ determines the maturity at which the loading on the medium-term factor, or curvature, achieves its maximum; taking 30-months maturity, they fix $\lambda = 0.0609$, which is the value we take here. Given λ, cross-sectional estimates of β_t at time t

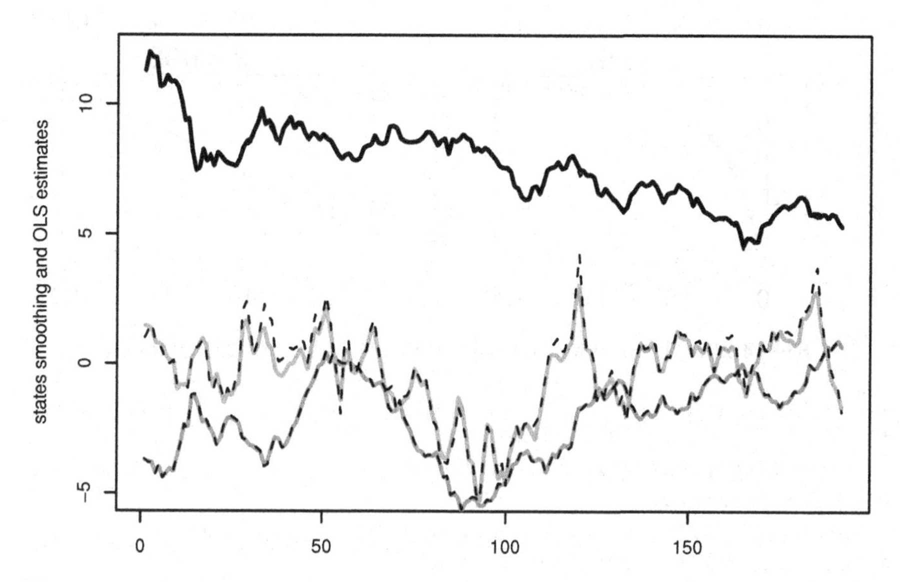

Fig. 4.19. MCMC smoothing estimates of the regression coefficients $\beta_{1,t}, \beta_{2,t}, \beta_{3,t}$ (solid lines: black, darkgray, gray respectively). The dashed lines are the month-by-month OLS estimates

are obtained by ordinary least squares (OLS), from the cross-sectional data $(y_{1,t}, \ldots, y_{m,t})$.

R code

```
  > yields <- read.table("Datasets/yields.dat")
2 > y <- yields[1 : 192, 3 : 19]; y <- as.matrix(y)
  > x <- c(3,6,9,12,15,18,21,24,30,36,48,60,72,84,96,108,120)
4 > p <- 3; m <- ncol(y); TT <- nrow(y)
  > persp(x = x, z = t(y), theta = 40, phi = 30, expand = 0.5,
6 +       col = "lightblue", ylab = "time", zlab = "yield",
  +       ltheta = 100, shade = 0.75, xlab = "maturity (months)")
8 > ### Cross-sectional model
  > lambda <- 0.0609
10 > h2 <- function(x) {(1 - exp(-lambda * x)) / (lambda * x)}
  > h3 <- function(x) {((1 - exp(-lambda * x)) / (lambda*x)) -
```

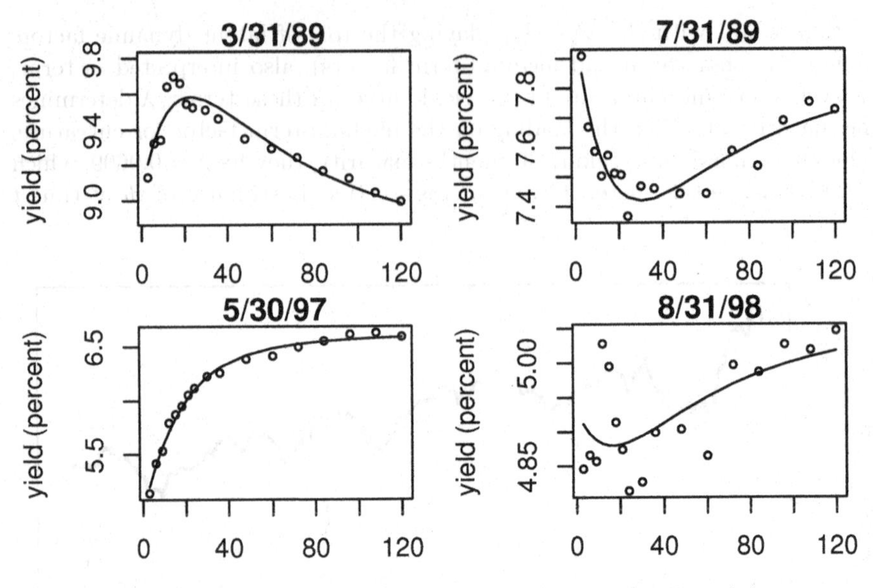

Fig. 4.20. Observed and OLS fitted yields curve, at selected dates

```
12  +                        exp(-lambda * x)}
    > X <- cbind(rep(1, m), h2(x), h3(x))
14  > ## OLS estimates
    > betahat <- solve(crossprod(X), crossprod(X, t(y)))
16  > nelsonSiegel <- function(x, beta) {
    +       beta[1] + beta[2] * h2(x)+ beta[3] * h3(x)}
18  > month <- 51
    > plot(x, y[month, ], xlab = "maturity(months)",
20  +       ylab = "yield (percent)", ylim = c(8.9,9.8),
    +       main = "yield curve on 3/31/89")
22  > lines(x, nelsonSiegel(x, betahat[, month]))
```

Month-by-month OLS estimates over time are plotted in Figure 4.19 (dashed lines). A look at the autocorrelation function of the OLS estimates $\hat{\beta}_t$ and at the estimated residual variance gives a feeling of their evolution over time (plots not shown).

R code

```
    > acf(t(betahat))
2   > yfit <- t(X %*% betahat)
    > res <- (y - yfit)^2
4   > s2 <- rowSums(res) / (m - p)
    > ts.plot(sqrt(s2))
```

The fit of the data is quite good, and the model adapts to the different shapes of the yield curve over time; see, e.g., Figure 4.20.

However, a dynamic estimate of the yield curve would give a more complete understanding of the problem. To this aim, as discussed in Section 3.3.5, a DLM can be used, introducing a state equation to describe the evolution of the regression coefficients. For example, Diebold et al. (2006) consider a VAR(1) states dynamics, also including the effects of macroeconomic variables. Petrone and Corielli (2005) propose a DLM where the state equation is derived from no-arbitrage constraints imposed on the yield curve evolution. In these papers, estimation of constant unknown parameters in the DLM matrices is obtained by MLE. Instead, here we illustrate Bayesian inference on the unknown parameters and the states of the model. For simplicity, we model $\beta_{1,t}, \beta_{2,t}, \beta_{3,t}$ as independent AR(1) processes. More specifically, the DLM we estimate is

$$\begin{aligned} Y_t &= F\theta_t + v_t, & v_t &\sim \mathcal{N}(0, V), \\ \theta_t &= G\theta_{t-1} + w_t, & w_t &\sim \mathcal{N}(0, W), \end{aligned} \tag{4.25}$$

where $Y_t = (Y_{1,t}, \cdots, Y_{m,t})'$, $\theta_t = (\beta_{1,t}, \beta_{2,t}, \beta_{3,t})'$,

$$\begin{aligned} F &= \begin{bmatrix} 1 & h_2(x_1) & h_3(x_1) \\ 1 & h2(x_2) & h_3(x_2) \\ \vdots & \vdots & \vdots \\ 1 & h_2(x_m) & h_3(x_m) \end{bmatrix}, \\ G &= \operatorname{diag}(\psi_1, \psi_2, \psi_3) \\ V &= \operatorname{diag}(\phi_{y,1}^{-1}, \cdots, \phi_{y,m}^{-1}), \\ W &= \operatorname{diag}(\phi_{\theta,1}^{-1}, \phi_{\theta,2}^{-1}, \phi_{\theta,3}^{-1}). \end{aligned}$$

Note that, while we assumed homoscedastic residuals in the cross sectional model for at each t, in the DLM above we allow different variances $\phi_{y,i}^{-1}, i = 1, \ldots, m$ for the yields at different times-to-maturity, although they are taken as time-invariant for simplicity.

As the prior, we assume that the AR parameters in the matrix G are i.i.d. Gaussian

$$\psi_j \overset{indep}{\sim} \mathcal{N}(\psi_0, \tau_0), \qquad j = 1, 2, 3,$$

with $\psi_0 = 0$ and $\tau_0 = 1$. We do not restrict the ψ_j to lie in the stationarity region. For the unknown variances, we use a d-Inverse-Gamma prior as in Section 4.5.1, so that the precisions have independent Gamma densities

$$\begin{aligned} \phi_{y,i} &\sim \mathcal{G}(\alpha_{y,i}, b_{y,i}), & i &= 1, 2, \cdots, m \\ \phi_{\theta,j} &\sim \mathcal{G}(\alpha_{\theta,j}, b_{\theta,j}), & j &= 1, \ldots, p. \end{aligned}$$

In the implementation below, we use $\alpha_{y,i} = 3, b_{y,i} = 0.01$, $i = 1, \ldots, m$ and $\alpha_{\theta,j} = 3, b_{\theta,j} = 1$, $j = 1, \ldots, p$. These choices correspond to prior guesses

$E(\phi_{y,i}^{-1}) = 0.005$, $\text{Var}(\phi_{y,i}^{-1}) = 0.005^2$, and $E(\phi_{\theta,j}^{-1}) = 0.5$, $\text{Var}(\phi_{\theta,j}^{-1}) = 0.5^2$ on the observation and state evolution variances.

The joint posterior of the states and model parameters is approximated by Gibbs sampling. Sampling the states can be done using the FFBS algorithm. The full conditionals for the parameters are as follows.

- The full conditional of $\phi_{\theta,j}$ is

$$\phi_{\theta,j}|\ldots \sim \mathcal{G}\left(\alpha_{\theta,j} + \frac{T}{2}, b_{\theta,j} + \frac{1}{2}SS_{\theta,j}\right), \qquad j = 1, 2, 3$$

  with $SS_{\theta,j} = \sum_{t=1}^{T}(\theta_{j,t} - (G\theta_{t-1})_j)^2$;
- the full conditional of the precision ϕ_{y_i} is

$$\phi_{y,i}|\ldots \sim \mathcal{G}\left(\alpha_{y,i} + \frac{T}{2}, b_{y,i} + \frac{1}{2}SS_{y,i}\right), \qquad i = 1, \cdots, m$$

  with $SS_{y,i} = \sum_{t=1}^{T}(y_{i,t} - (F\theta_t)_i)^2$;
- the full conditional of the AR parameters ψ_j is, by the theory of linear regression with a Normal prior,

$$\psi_j|\ldots \sim \mathcal{N}(\psi_{j,T}, \tau_{j,T}), \qquad j = 1, 2, 3$$

  where

$$\psi_{j,T} = \tau_{j,T}\left[\phi_{\theta,j}\sum_{t=1}^{T}\theta_{j,t-1}\theta_{j,t} + \frac{1}{\tau_{j,0}}\psi_0\right]$$

$$\tau_{j,T} = \left[\frac{1}{\tau_0} + \phi_{\theta,j}\sum_{t=1}^{T}\theta_{j,t-1}^2\right]^{-1}.$$

R code

```
> mod <- dlm(m0 = rep(0, p), C0 = 100 * diag(p),
+             FF = X, V = diag(m), GG = diag(p), W = diag(p))
> ##  Prior hyperparameters
> psi0 <- 0; tau0 <- 1
> shapeY <- 3; rateY <- .01
> shapeTheta <- 3; rateTheta <- 1
> ## MCMC
> MC <- 10000
> gibbsTheta <- array(NA, dim = c(MC, TT + 1, p))
> gibbsPsi <- matrix(NA, nrow = MC, ncol = p)
> gibbsV <- matrix(NA, nrow = MC, ncol = m)
> gibbsW <- matrix(NA, nrow = MC, ncol = p)
> ## Starting values: as specified by mod
> set.seed(3420)
> phi.init <- rnorm(3, psi0, tau0)
```

```
16 > V.init <- 1 / rgamma(1, shapeY, rateY)
   > W.init <- 1 / rgamma(1, shapeTheta, rateTheta)
18 > mod$GG <- diag(phi. init)
   > mod$V <- diag(rep(V.init, m))
20 > mod$W <- diag(rep(W.init, p))
   > ## Gibbs sampler
22 > for (i in 1 : MC)
   + {
24 +     ## generate the states by FFBS
   +     modFilt <- dlmFilter(y, mod, simplify = TRUE)
26 +     theta <- dlmBSample(modFilt)
   +     gibbsTheta[i, , ] <- theta
28 +     ## generate the W_j
   +     theta.center <- theta[-1, ] -
30 +         t(mod$GG %*% t(theta[-(TT + 1), ]))
   +     SStheta <- apply((theta.center)^2, 2, sum)
32 +     phiTheta <- rgamma(p, shape =shapeTheta + TT / 2,
   +                        rate = rateTheta + SStheta / 2)
34 +     gibbsW[i, ] <- 1 / phiTheta
   +     mod$W <- diag(gibbsW[i, ])
36 +     ## generate the V_i
   +     y.center <- y - t(mod$FF %*% t(theta[-1, ]))
38 +     SSy <- apply((y.center)^2, 2, sum)
   +     gibbsV[i, ] <- 1 / rgamma(m, shape = shapeY + TT / 2,
40 +                               rate = rateY + SSy / 2)
   +     mod$V <- diag(gibbsV[i, ])
42 +     ## generate the AR parameters psi_1, psi_2, psi_3
   +     psi.AR <- rep(NA, 3)
44 +     for (j in 1 : p)
   +     {
46 +         tau <- 1 / ((1 / tau0) + phiTheta[j] *
   +                     crossprod(theta[-(TT + 1), j]))
48 +         psi <- tau * (phiTheta[j] * t(theta[-(TT + 1), j]) %*%
   +                       theta[-1, j] + psi0 / tau0)
50 +         psi.AR[j] <- rnorm(1, psi, sd = sqrt(tau))
   +     }
52 +     gibbsPsi[i, ] <- psi.AR
   +     mod$GG <- diag(psi.AR)
54 + }
```

We generate a sample of size 10000 and discard the first 1000 draws as burn in. Diagnostic plots (not shown) indicate convergence of the MCMC chain. Below are the MCMC approximations of the posterior expectations of the AR parameters and unknown variances, with their Monte Carlo standard errors.

R code

```
> burn <- 1000
> round(mcmcMean(gibbsPsi[-(1 : burn), ]), 4)
      V.1      V.2      V.3
   0.9949   0.9883   0.9135
 (0.0000) (0.0001) (0.0003)
> round(mcmcMean(sqrt(gibbsW[-(1 : burn), ])), 4)
      V.1      V.2      V.3
   0.3137   0.3220   0.6493
 (0.0112) (0.0115) (0.0258)
> round(mcmcMean(sqrt(gibbsV[-(1 : burn), ])), 4)
      V.1      V.2      V.3      V.4      V.5      V.6
   0.1450   0.0631   0.0600   0.0851   0.0932   0.0755
 (0.0094) (0.0055) (0.0030) (0.0041) (0.0040) (0.0032)
      V.7      V.8      V.9     V.10     V.11     V.12
   0.0565   0.0520   0.0284   0.0469   0.0626   0.0807
 (0.0027) (0.0021) (0.0017) (0.0023) (0.0026) (0.0031)
     V.13     V.14     V.15     V.16     V.17
   0.0873   0.0665   0.0490   0.0503   0.0837
 (0.0032) (0.0027) (0.0026) (0.0030) (0.0034)
```

The MCMC smoothing estimates of $\theta_t = (\beta_{1,t}, \beta_{2,t}, \beta_{3,t})$ are plotted in Figure 4.19. The results are quite close to the OLS month-by-month estimates, also shown in the plot. However, this is not always the case. Roughly speaking, while the cross-sectional OLS estimates minimize the residual sum of squares at each t, in the DLM the minimization is subject to the constraints implied by the state equation. In fact, a poor specification of the state equation might result in an unsatisfactory fit of the cross-sectional data. Further developments of this example include a more thoughtful specification of the state equation, time varying observation and/or evolution variances, and the inclusion of macroeconomic variables.

4.6.3 Factor models

In this example, we use a factor model (Section 3.3.6) to extract a common stochastic trend from multiple integrated time series. The basic idea is to explain fluctuations of various markets or common latent factors that affect a set of economic or financial variables simultaneously. The data are the federal funds rate (short rate) and 30-year conventional fixed mortgage rate (long rate), obtained from Federal Reserve Bank of St. Louis[2] (see Chang et al. (2005)). The series are sampled at weekly intervals over the period April 7, 1971 through September 8, 2004. We will work with the natural logarithm of one plus the interest rate; the transformed data are plotted in Figure 4.21.

[2] Source: `http://research.stlouisfed.org/fred2/`

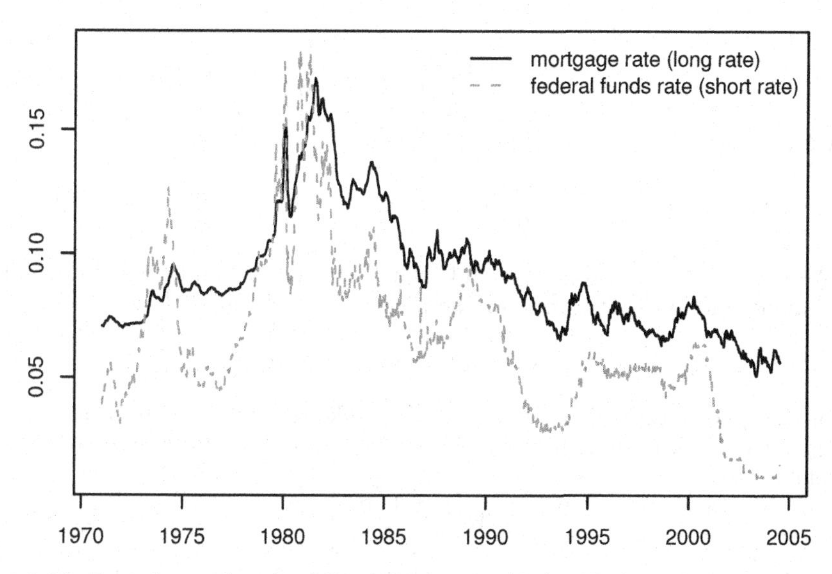

Fig. 4.21. Log of one plus the federal funds rate (FF) and the 30-year mortgage rate (WMORTG). Weekly data, 1971-2004

R code

```
> interestRate <- read.table("interestRates.dat",
+                            col.names = c("Long", "Short"))
> y <- log(1 + interestRate / 100)
> y <- ts(y, frequency = 52, start = 1971)
> ts.plot(y, lty = c(1, 2), col = c(1, "darkgray"))
> legend("topright", legend = c("mortgage rate (long rate)",
+        "federal funds rate (short rate)"),
+        col = c(1, "darkgray"), lty = c(1,2), bty = "n")
```

To extract a common stochastic trend in the bivariate time series $(Y_t = (Y_{1,t}, Y_{2,t}) : t \geq 1)$, we assume the following factor model:

$$\begin{cases} Y_t = A\mu_t + \mu_0 + v_t, & v_t \sim \mathcal{N}(0, V), \\ \mu_t = \mu_{t-1} + w_t, & w_t \sim \mathcal{N}(0, \sigma_\mu^2), \end{cases} \tag{4.26}$$

where the 2×1 matrix A is set to be $A = \begin{bmatrix}1 \ \alpha\end{bmatrix}'$ to ensure identifiability and $\mu_0 = \begin{bmatrix}0 \ \bar{\mu}\end{bmatrix}'$. The latent variable μ_t is interpreted as the common stochastic trend, here simply modeled as a random walk. In the usual DLM notation the model can be written in the form

$$
\begin{aligned}
Y_t &= F\theta_t + v_t, \\
\theta_t &= \begin{bmatrix} 1 & 0 \\ 0 & 1 \end{bmatrix} \theta_{t-1} + \begin{bmatrix} w_t \\ 0 \end{bmatrix},
\end{aligned}
\tag{4.27}
$$

with

$$
\theta_t = \left[\mu_t, \bar{\mu}\right], \quad F = \begin{bmatrix} 1 & 0 \\ \alpha & 1 \end{bmatrix}, \quad V = \begin{bmatrix} \sigma_1^2 & \sigma_{12} \\ \sigma_{12} & \sigma_2^2 \end{bmatrix}, \quad W = \text{diag}(\sigma_\mu^2, 0).
$$

We use a $\mathcal{N}(\alpha_0, \tau^2)$ prior for α and, as in Sections 4.5.1 and 4.5.2, independent Gamma and Wishart priors for the precision $1/\sigma_\mu^2$ and V^{-1},

$$
\sigma_\mu^{-2} \sim \mathcal{G}(a, b), \qquad V^{-1} \sim \mathcal{W}(\nu_0, S_0).
$$

In the specific case illustrated below, we take $\alpha_0 = 0, \tau^2 = 16$; a and b such that $\text{E}(\sigma_\mu^{-2}) = 0.01$ with $\text{Var}(\sigma_\mu^{-2}) = 1$; $\nu_0 = (\delta + 1)/2 = 2$ and $S_0 = V_0/2$, where

$$
V_0 = \begin{bmatrix} 1 & 0.5 \\ 0.5 & 4 \end{bmatrix},
$$

so that $\text{E}(V) = V_0$. The posterior distribution of the states and the parameters, given $y_{0:T}$, is proportional to

$$
\prod_{t=1}^{T} \mathcal{N}_2((y_{1,t}, y_{2,t}); (\mu_t, \alpha\mu_t + \bar{\mu})', V)\, \mathcal{N}(\bar{\mu}; m_{0,2} C_{0,22})
$$
$$
\mathcal{N}(\mu_t; \mu_{t-1}, \sigma_\mu^2)\, \mathcal{N}(\alpha; \alpha_0, \tau^2)\, \mathcal{G}(\sigma_\mu^{-2}; a, b)\, \mathcal{W}(V^{-1}; \nu_0, S_0),
$$

where $m_{0,2}$ and $C_{0,22}$ are the prior mean and variance for $\bar{\mu}$. The posterior is approximated via Gibbs sampling;

- the full conditional of α is $\mathcal{N}(\alpha_T, \tau_T^2)$, where

$$
\begin{aligned}
\tau_T^2 &= \frac{(1-\rho^2)\tau^2\sigma_2^2}{\tau^2 \sum_{t=1}^T \mu_t^2 + (1-\rho^2)\sigma_2^2} \\
\alpha_T &= \tau_T^2 \, \frac{\tau^2/\sigma_2 \sum_{t=1}^T \left(\frac{y_{2,t}-\bar{\mu}}{\sigma_2} - \rho \frac{y_{1,t}-\mu_t}{\sigma_1}\right) \mu_t + \alpha_0 (1-\rho^2)}{\tau^2\,(1-\rho^2)},
\end{aligned}
$$

  with $\rho = \sigma_{12}/(\sigma_1\sigma_2)$;
- the full conditional of the precision σ_μ^{-2} is

$$
\mathcal{G}(a + \frac{T}{2}, b + \frac{SS_\mu}{2}),
$$

  where $SS_\mu = \sum_{t=1}^T (\mu_t - \mu_{t-1})^2$;
- the full conditional of V^{-1} is

$$
\mathcal{W}\left(\frac{\delta + 1 + T}{2}, \frac{1}{2}(V_0 + \sum_{t=1}^{T} (y_t - F\theta_t)(y_t - F\theta_t)^{'})\right).
$$

R code

```
> # Prior hyperparameters
> alpha0 <- 0; tau2 <- 16
> expSigmaMu<- 0.01; varSigmaMu<- 1
> a <- (expSigmaMu^2/varSigmaMu)+2; b <- expSigmaMu*(a-1)
> delta <- 3;
> V0=matrix(c(1, 0.5, 0.5, 4), byrow=T, nrow=2)
> # Gibbs sampling
> MC <- 10000
> n <- nrow(y)
> gibbsTheta <- array(0, dim=c(n+1,2,MC-1))
> gibbsV <-array(0, dim=c(2,2,MC))
> gibbsAlpha <- rep(0,MC)
> gibbsW <- rep(0,MC)
> # model and starting values for the MCMC
> mod <- dlmModPoly(2, dW=c(1,0), C0=100*diag(2))
> gibbsAlpha[1] <- 0
> mod$FF <- rbind(c(1,0), c(gibbsAlpha[1],1))
> mod$W[1,1] <- gibbsW[1] <- 1/rgamma(1, a, rate=b)
> mod$V <- gibbsV[,,1] <- V0 /(delta-2)
> mod$GG <- diag(2)
> # MCMC loop
> for(it in 1:(MC-1))
+ {
+ # generate state- FFBS
+ modFilt <- dlmFilter(y, mod, simplify=TRUE)
+    gibbsTheta[,,it] <- theta <- dlmBSample(modFilt)
+ # update alpha
+ rho <-gibbsV[,,it][1,2]/(gibbsV[,,it][1,1]*gibbsV[,,it][2,2])^.5
+ tauT <- (gibbsV[,,it][2,2]*(1-rho^2)*tau2) /
+          (tau2 * sum(theta[-1,1]^2)+(1-rho^2)*gibbsV[,,it][2,2])
+ alphaT <- tauT * ((tau2/(gibbsV[,,it][2,2])^.5) *
+            sum(((y[,2]-theta[-1,2])/gibbsV[,,it][2,2]^.5 -
+            rho* (y[,1]-theta[-1,1])/gibbsV[,,it][1,1]^.5) *
+            theta[-1,1])+ alpha0*(1-rho^2))/(tau2*(1-rho^2))
+ mod$FF[2,1] <- gibbsAlpha[it+1] <- rnorm(1, alphaT, tauT^.5)
+ # update sigma_mu
+ SSmu <- sum( diff(theta[,1])^2)
+ mod$W[1,1] <- gibbsW[it+1] <- 1/rgamma(1, a+n/2, rate=b+SSmu/2)
+ # update V
+ S <- V0 + crossprod(y- theta[-1,] %*% t(mod$FF))
+ mod$V <- gibbsV[,,it+1] <- solve(rwishart(df=delta+1+ n,
+             Sigma=solve(S)))
+ }
```

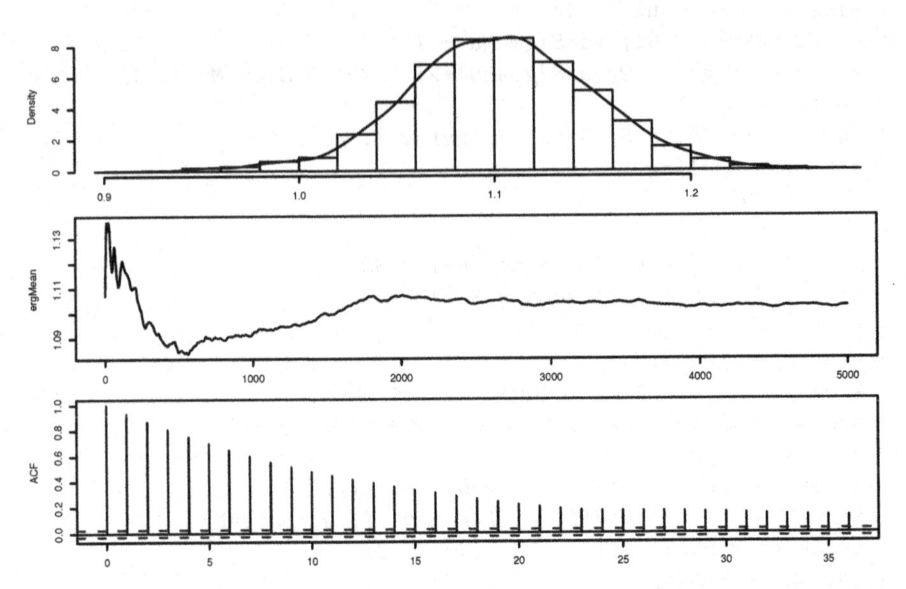

Fig. 4.22. MCMC estimate of the posterior distribution of α; running ergodic means; MCMC autocorrelation

We show the results for 10000 MCMC iterations, with a burn in of 5000 draws. Some diagnostic plots are summarized in Figures 4.22-4.23; the first panel in each figure shows the MCMC approximation of the posterior distribution of α and σ_μ.

The MCMC estimates of the parameters' posterior means are given in the display below, together with the estimated standard errors, obtained by the `dlm` function `mcmcMeans`, and the 5% and 95% quantiles of their marginal posterior distributions.

parameter	α	σ_μ	σ_1	σ_2	$\sigma_{1,2}$
posterior mean (st dev)	1.1027 (0.0036)	0.0084 (0.00001)	0.0261 (0.00001)	0.0512 (0.00001)	0.00037 (0.0000)
5% quantile	1.0262	0.0079	0.0254	0.0498	0.0003
95% quantile	1.1806	0.0089	0.0269	0.0526	0.00043

Figure 4.24 shows the data and the posterior mean of the common stochastic trend, which results very close to the long rate.

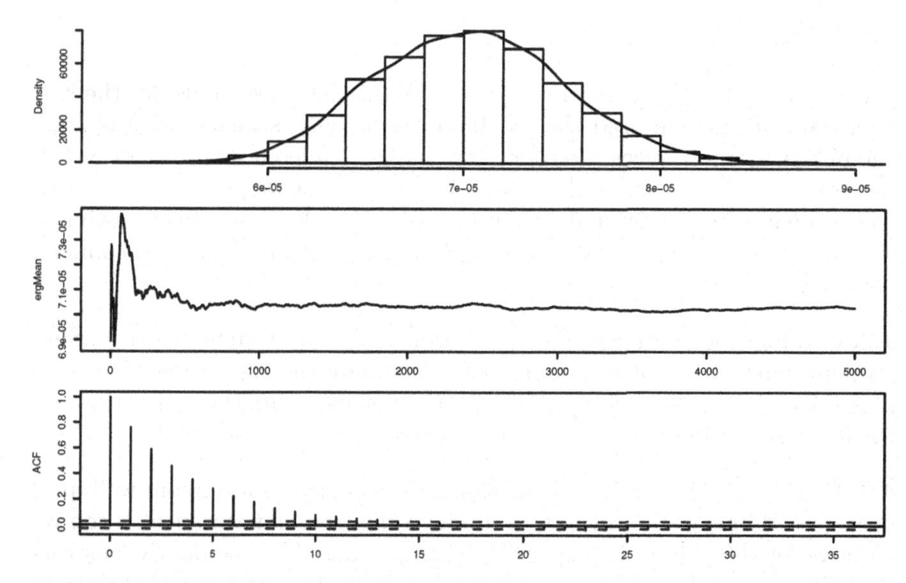

Fig. 4.23. MCMC estimate of the posterior distribution of σ_μ; running ergodic means; MCMC autocorrelation

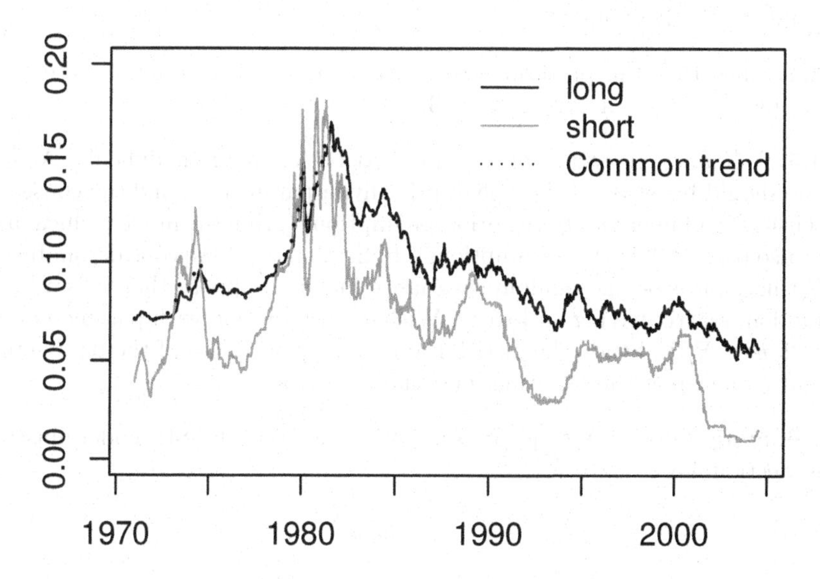

Fig. 4.24. Posterior mean of the common stochastic trend

Problems

4.1. In Chapter 2, we considered a random walk-plus-noise model for the Nile river data. There, we used the maximum likelihood estimates for the state and observation variances. Consider now Bayesian inference on the states and the unknown parameters of the model. Express conjugate priors for V and W and evaluate the posterior distribution of $(\theta_{0:T}, V, W|y_{1:T})$. Then, estimate the model using discount factors as in Sections 4.3.2 and 4.3.3, and compare the results.

4.2. Consider the DLM described in Section 4.3.1, with conjugate priors. Suppose for simplicity that θ_t is univariate. Compute the expression of $(1-\alpha)$ probability intervals for $\theta_t|y_{1:t}$. Discuss the results, comparing with the case where σ^2 is known.

4.3. Consider again the Nile data, which we modeled as a random walk plus noise (see Problem 4.1). In fact, assuming time-invariant variances is too restrictive for these data: we expect big changes caused by the dam's construction. The model described in Section 4.5.3 allows for outliers and structural breaks. Provide Bayesian estimates of this model for the Nile data (compute an MCMC approximation of the joint posterior of the states and unknown parameters, given $y_{1:T}$).

4.4. Show that the posterior distribution (4.20) is invariant for the hybrid sampler described in Algorithm 4.3.

4.5. A honest elicitation of a prior distribution is often difficult. At least, one should be aware of the role of prior hyperparameters, and the consequent sensitivity of inference to the prior assumptions (given the model, which in fact is also part of the prior assumptions). In Section 4.5.2, we consider an Inverse-Wishart prior on the random covariance matrix. Suppose that V is a (2×2) random matrix having an Inverse-Wishart distribution with parameters $(\alpha = n/2, B = \Sigma/2)$. Study the distribution for varying values of the parameters n and Σ, and plot the resulting marginal densities.

4.6. Study sensitivity to prior assumptions for the SUTSE model illustrated in Section 4.5.2.

5

Sequential Monte Carlo methods

In Chapter 2 we introduced the filtering recursion for general state space models (Proposition 2.1). The recursive nature of the algorithm that, from the filtering distribution at time $t-1$ and the observation y_t computes the filtering distribution at time t, makes it ideally suited for a large class of applications in which inference must be made online, before the data collection ends. For those types of applications one must have, at any time, an up-to-date estimate of the current state of the system. Standard examples of such online types of applications include the following: tracking the position and speed of a moving aircraft observed through a radar; monitoring the location and characteristics of a storm based on satellite data; estimating the volatility of the prices of a group of stocks from tick-to-tick data. Unfortunately, for a general state space model the integrations in (2.7) cannot be carried out analytically. DLMs are a special case for which the Kalman filter gives a closed form solution to the filtering problem. However, even in this case, as soon as a DLM contains unknown parameters in its specification, the Kalman filter alone is not enough to compute the filtering distribution and, except in a few simple cases (see Section 4.3.1) one has to resort to numerical techniques.

For off-line, or batch, inference MCMC methods can be successfully employed for DLMs with unknown parameters, as explained in Chapter 4, and can be extended to nonlinear non-Gaussian models. However, they are of limited use for online inference because any time a new observation becomes available, a totally new Markov chain has to be simulated. In other words, in an MCMC approach, the output based on $t-1$ observations cannot be used to evaluate posterior distributions based on t observations. In this sense, unlike the Kalman filter, MCMC does not lend itself easily to a sequential usage.

Early attempts to sequentially update filtering distributions for nonlinear and non-Gaussian state space models were based on some form of linearization of the state and system equations and on Gaussian approximations (for details and references, see Cappé et al.; 2007). While useful for mildly nonlinear models, an approach of this type typically performs poorly for highly nonlinear

G. Petris et al., *Dynamic Linear Models with R*, Use R, DOI: 10.1007/b135794_5,

models. Furthermore, it does not solve the problem of sequentially estimating unknown parameters, even in the simple DLM case.

In this chapter we give an account of a relatively recent simulation approach, called Sequential Monte Carlo, that has proved very successful in online filtering applications of both DLMs with unknown parameters and general nonlinear non-Gaussian state space models. Sequential Monte Carlo provides an alternative set of simulation-based algorithms to approximate complicated posterior distributions. Although not limited to time series models, it has proved extremely successful when applied to DLMs and more general state space models—especially in those applications that require frequent updates of the posterior as new data are observed. Research in sequential Monte Carlo methods is currently very active and we will not try to give here an exhaustive review of the field. Instead, we limit ourselves to a general introduction and a more specific description of a few algorithms that can be easily implemented in the context of DLMs. For more information the interested reader can consult the books by Liu (2001), Doucet et al. (2001), Del Moral (2004), and Cappé et al. (2005). The article by Cappé et al. (2007) provides a current overview of the field.

5.1 The basic particle filter

Particle filtering, which is how sequential Monte Carlo is usually referred to in applications to state space models, is easier to understand when viewed as an extension of importance sampling. For this reason we open this section with a brief recall of importance sampling.

Suppose one is interested in evaluating the expected value

$$\mathrm{E}_\pi(f(X)) = \int f(x)\pi(x)\,\mathrm{d}x. \tag{5.1}$$

If g is an *importance density* having the property that $g(x) = 0$ implies $\pi(x) = 0$, then one can write

$$\mathrm{E}_\pi(f(X)) = \int f(x)\frac{\pi(x)}{g(x)}g(x)\,\mathrm{d}x = \mathrm{E}_g(f(X)w^\star(X)),$$

where $w^\star(x) = \pi(x)/g(x)$ is the so-called *importance function.* This suggests approximating the expected value of interest by generating a random sample of size N from g and computing

$$\frac{1}{N}\sum_{i=1}^{N} f(x^{(i)})w^\star(x^{(i)}) \approx \mathrm{E}_\pi(f(X)). \tag{5.2}$$

In Bayesian applications one can typically evaluate the target density only up to a normalizing factor, i.e., only $C \cdot \pi(x)$ can be computed, for an unknown constant C. Unfortunately, this implies that also the importance function can only be evaluated up to the same factor C and (5.2) cannot be used directly. However, letting $\tilde{w}^{(i)} = Cw^\star(x^{(i)})$, if one takes $f(x) \equiv C$, then (5.2) yields

$$\frac{1}{N}\sum_{i=1}^{N} Cw^\star(x^{(i)}) = \frac{1}{N}\sum_{i=1}^{N} \tilde{w}^{(i)} \approx \mathrm{E}_\pi(C) = C. \tag{5.3}$$

Since the $\tilde{w}^{(i)}$'s are available, (5.3) provides a way of evaluating C. Moreover, for the purpose of evaluating (5.1) one does not need an explicit estimate of the constant C: in fact,

$$\begin{aligned}
\mathrm{E}_\pi(f(X)) &\approx \frac{1}{N}\sum_{i=1}^{N} f(x^{(i)})w^\star(x^{(i)}) \\
&= \frac{\frac{1}{N}\sum_{i=1}^{N} f(x^{(i)})\tilde{w}^{(i)}}{C} \approx \frac{\sum_{i=1}^{N} f(x^{(i)})\tilde{w}^{(i)}}{\sum_{i=1}^{N} \tilde{w}^{(i)}} \\
&= \sum_{i=1}^{N} f(x^{(i)})w^{(i)},
\end{aligned}$$

with $w^{(i)} = \tilde{w}^{(i)} / \sum_{j=1}^{N} \tilde{w}^{(j)}$. Note that: (1) the weights $w^{(i)}$ sum to one, and (2) the approximation $\mathrm{E}_\pi(f(X)) \approx \sum_{i=1}^{N} f(x^{(i)})w^{(i)}$ holds for every well-behaved function f. Therefore, the sample $x^{(1)}, \ldots, x^{(N)}$ with the associated weights $w^{(1)}, \ldots, w^{(N)}$ can be viewed as a discrete approximation of the target π. In other words, writing δ_x for the unit mass at x, and setting $\hat{\pi} = \sum_{i=1}^{N} w^{(i)} \delta_{x^{(i)}}$, one has $\pi \approx \hat{\pi}$.

In filtering applications, the target distribution changes every time a new observation is made, moving from $\pi(\theta_{0:t-1}|y_{1:t-1})$ to $\pi(\theta_{0:t}|y_{1:t})$. Note that the former is not a marginal distribution of the latter, even though $\theta_{0:t-1}$ are the first components of $\theta_{0:t}$. The problem then is how to efficiently update a discrete approximation of $\pi(\theta_{0:t-1}|y_{1:t-1})$ when the observation y_t becomes available, in order to obtain a discrete approximation of $\pi(\theta_{0:t}|y_{1:t})$. For every s, let us denote[1] by $\hat{\pi}_s(\theta_{0:s}|y_{1:s})$ the approximation of $\pi(\theta_{0:s}|y_{1:s})$. The updating process consists of two steps: for each point $\theta_{0:t-1}^{(i)}$ in the support of $\hat{\pi}_{t-1}$, (1) draw an additional component $\theta_t^{(i)}$ to obtain $\theta_{0:t}^{(i)}$ and, (2) update its weight $w_{t-1}^{(i)}$ to an appropriate $w_t^{(i)}$. The weighted points $(\theta_t^{(i)}, w_t^{(i)})$, $i = 1, \ldots, N$, provide the new discrete approximation $\hat{\pi}_t$. For every t, let g_t be the importance density used to generate $\theta_{0:t}$. Since at time t the observations $y_{1:t}$ are available, g_t may depend on them and we will write $g_t(\theta_{0:t}|y_{1:t})$ to make the dependence explicit. We assume that g_t can be expressed in the following form:

[1] We keep the index s in the notation $\hat{\pi}_s$ because approximations at different times can be in principle unrelated to one another, while the targets are all derived from the unique distribution of the process $\{\theta_i, y_j : i \geq 0, j \geq 1\}$.

$$g_t(\theta_{0:t}|y_{1:t}) = g_{t|t-1}(\theta_t|\theta_{0:t-1}, y_{1:t}) \cdot g_{t-1}(\theta_{0:t-1}|y_{1:t-1}).$$

This allows us to "grow" sequentially $\theta_{0:t}$ by combining $\theta_{0:t-1}$, drawn from g_{t-1} and available at time $t-1$, and θ_t, generated at time t from $g_{t|t-1}(\theta_t|\theta_{0:t-1}, y_{1:t})$. We will call the functions $g_{t|t-1}$ *importance transition densities*. Note that only the importance transition densities are needed to generate $\theta_{0:t}$. Suggestions about the selection of the importance density are provided at the end of the section. Let us consider next how to update the weights. One has, dropping the superscripts for notational simplicity:

$$\begin{aligned}
w_t &\propto \frac{\pi(\theta_{0:t}|y_{1:t})}{g_t(\theta_{0:t}|y_{1:t})} \propto \frac{\pi(\theta_{0:t}, y_t|y_{1:t-1})}{g_t(\theta_{0:t}|y_{1:t})} \\
&\propto \frac{\pi(\theta_t, y_t|\theta_{0:t-1}, y_{1:t-1}) \cdot \pi(\theta_{0:t-1}|y_{1:t-1})}{g_{t|t-1}(\theta_t|\theta_{0:t-1}, y_{1:t}) \cdot g_{t-1}(\theta_{0:t-1}|y_{1:t-1})} \\
&\propto \frac{\pi(y_t|\theta_t) \cdot \pi(\theta_t|\theta_{t-1})}{g_{t|t-1}(\theta_t|\theta_{0:t-1}, y_{1:t})} \cdot w_{t-1}.
\end{aligned}$$

Hence, for every i, after drawing $\theta_t^{(i)}$ from $g_{t|t-1}(\theta_t|\theta_{0:t-1}^{(i)}, y_{1:t})$, one can compute the unnormalized weight $\tilde{w}_t^{(i)}$ as

$$\tilde{w}_t^{(i)} = w_{t-1}^{(i)} \cdot \frac{\pi(y_t|\theta_t^{(i)}) \cdot \pi(\theta_t^{(i)}|\theta_{t-1}^{(i)})}{g_{t|t-1}(\theta_t^{(i)}|\theta_{0:t-1}^{(i)}, y_{1:t})}. \tag{5.4}$$

The fraction on the left-hand side of equation (5.4), or any quantity proportional[2] to it, is called the *incremental weight*. The final step in the updating process consists in scaling the unnormalized weights:

$$w_t^{(i)} = \frac{\tilde{w}_t^{(i)}}{\sum_{j=1}^N \tilde{w}_t^{(j)}}.$$

In practice it is often the case that, after a number of updates have been performed, a few points in the support of $\hat{\pi}_t$ have relatively large weights, while all the remaining have negligible weights. This clearly leads to a deterioration in the Monte Carlo approximation. To keep this phenomenon in control, a useful criterion to monitor over time is the *effective sample size*, defined as

$$N_{eff} = \left(\sum_{i=1}^N (w_t^{(i)})^2 \right)^{-1},$$

which ranges between N (when all the weights are equal) and one (when one weight is equal to one). When N_{eff} falls below a threshold N_0, it is advisable

[2] The proportionality constant may depend on $y_{1:t}$, but should not depend on $\theta_t^{(i)}$ or $\theta_{0:t-1}^{(i)}$ for any i.

0. Initialize: draw $\theta_0^{(1)}, \dots, \theta_0^{(N)}$ independently from $\pi(\theta_0)$ and set

$$w_0^{(i)} = N^{-1}, \qquad i = 1, \dots, N.$$

1. For $t = 1, \dots, T$:
   1.1) For $i = 1, \dots, N$:
   - Draw $\theta_t^{(i)}$ from $g_{t|t-1}(\theta_t | \theta_{0:t-1}^{(i)}, y_{1:t})$ and set

$$\theta_{0:t}^{(i)} = (\theta_{0:t-1}^{(i)}, \theta_t^{(i)})$$

   - Set

$$\tilde{w}_t^{(i)} = w_{t-1}^{(i)} \cdot \frac{\pi(\theta_t^{(i)}, y_t | \theta_{t-1}^{(i)})}{g_{t|t-1}(\theta_t^{(i)} | \theta_{0:t-1}^{(i)}, y_{1:t})}.$$

   1.2) Normalize the weights:

$$w_t^{(i)} = \frac{\tilde{w}_t^{(i)}}{\sum_{j=1}^N \tilde{w}_t^{(j)}}.$$

   1.3) Compute

$$N_{\mathit{eff}} = \left(\sum_{i=1}^N (w_t^{(i)})^2 \right)^{-1}.$$

   1.4) If $N_{\mathit{eff}} < N_0$, resample:
   - Draw a sample of size N from the discrete distribution

$$\mathrm{P}(\theta_{0:t} = \theta_{0:t}^{(i)}) = w_t^{(i)}, \qquad i = 1, \dots, N,$$

   and relabel this sample

$$\theta_{0:t}^{(1)}, \dots, \theta_{0:t}^{(N)}.$$

   - Reset the weights: $w_t^{(i)} = N^{-1}$, $i = 1, \dots, N$.

   1.5) Set $\hat{\pi}_t = \sum_{i=1}^N w_t^{(i)} \delta_{\theta_{0:t}^{(i)}}$.

Algorithm 5.1: Summary of the particle filter algorithm

to perform a *resampling* step. This can be done in several different ways. The simplest, called multinomial resampling, consists of drawing a random sample of size N from $\hat{\pi}_t$ and using the sampled points, with equal weights, as the new discrete approximation of the target. The resampling step does not change the expected value of the approximating distribution $\hat{\pi}_t$, but it increases its Monte Carlo variance. In trying to keep the variance increase as small as possible, researchers have developed other resampling algorithms, more efficient than multinomial resampling in this respect. Of these, one of the most commonly used is residual resampling. It consists of creating, for

$i = 1, \ldots, N$, $\lfloor N w_t^{(i)} \rfloor$ copies of $\theta_{0:t}^{(i)}$ deterministically first, and then adding R_i copies of $\theta_{0:t}^{(i)}$, where $(R_1, \ldots, R_N)$ is a random vector having a multinomial distribution. The size and probability parameters are given by $N - M$ and $(\bar{w}^{(1)}, \ldots, \bar{w}^{(N)})$ respectively, where

$$M = \sum_{i=1}^{N} \lfloor N w_t^{(i)} \rfloor,$$

$$\bar{w}^{(i)} = \frac{N w_t^{(i)} - \lfloor N w_t^{(i)} \rfloor}{N - M}, \qquad i = 1, \ldots, N.$$

Algorithm 5.1 contains a summary of the basic particle filter. Let us stress once again the sequential character of the algorithm. Each pass of the outermost "for" loop represents the updating from $\hat{\pi}_{t-1}$ to $\hat{\pi}_t$ following the observation of the new data point y_t. Therefore, at any time $t \leq T$ one has a working approximation $\hat{\pi}_t$ of the current filtering distribution.

At time t, a discrete approximation of the filtering distribution $\pi(\theta_t | y_{0:t})$ is immediately obtained as a marginal distribution of $\hat{\pi}_t$. More specifically, if $\hat{\pi}_t = \sum_{i=1}^{N} w^{(i)} \delta_{\theta_{0:t}^{(i)}}$, we only need to discard the first t components of each path $\theta_{0:t}^{(i)}$, leaving only $\theta_t^{(i)}$, to obtain

$$\pi(\theta_t | y_{1:t}) \approx \sum_{i=1}^{N} w^{(i)} \delta_{\theta_t^{(i)}}.$$

As a matter of fact, particle filter is most frequently viewed, as the name itself suggests, as an algorithm to update sequentially the filtering distribution. Note that, as long as the transition densities $g_{t|t-1}$ are Markovian, the incremental weights in (5.4) only depend on $\theta_t^{(i)}$ and $\theta_{t-1}^{(i)}$, so that, if the user is only interested in the filtering distribution, the previous components of the path $\theta_{0:t}^{(i)}$ can be safely discarded. This clearly translates into substantial savings in terms of storage space. Another, more fundamental, reason to focus on the filtering distribution is that the discrete approximation provided by $\hat{\pi}_t$ is likely to be more accurate for the most recent components of $\theta_{0:t}$ than for the initial ones. To see why this is the case, consider, for a fixed $s < t$, that the $\theta_s^{(i)}$'s are generated at a time when only $y_{0:s}$ is available, so that they may well be far from the center of their smoothing distribution $\pi(\theta_s | y_{0:t})$, which is conditional on $t - s$ additional observations.

We conclude this section with practical guidelines to follow in the selection of the importance transition densities. In the context of DLMs, as well as for more general state space models, two are the most used importance transition densities. The first is $g_{t|t-1}(\theta_t | \theta_{0:t-1}, y_{1:t}) = \pi(\theta_t | \theta_{t-1})$, i.e., the actual transition density of the Markov chain of the states. It is clear that in this way all the particles are drawn from the prior distribution of the states, without accounting for any information provided by the observations. The simulation of the particles and the calculation of the incremental weights are straightforward. However, most of the times the generated particles will fall in regions of low posterior density. The consequence will be an inaccurate discrete representation of the posterior density and a high Monte Carlo variance for the estimated posterior expected values. For these reasons we discourage the use of the prior as importance density. A more efficient approach, which accounts for the observations in the importance transition densities, consists of generating θ_t from its conditional distribution given θ_{t-1} and y_t. This distribution is sometimes referred to as the *optimal importance kernel*. In view of the conditional independence structure of the model, this is the same as the conditional distribution of θ_t given $\theta_{0:t-1}$ and $y_{1:t}$. Therefore, in this way one is generating θ_t from the target (conditional) distribution. However, since θ_{t-1} was not drawn from the current target, the particles $\theta_{0:t}^{(i)}$ are not draws from the target distribution[3] and the incremental importance weights need to be evaluated. Applying standard results about Normal models, it is easily seen that for a DLM the optimal importance kernel $g_{t|t-1}$ is a Normal density with mean and variance given by

$$
\begin{aligned}
\mathrm{E}(\theta_t|\theta_{t-1}, y_t) &= G_t\theta_{t-1} + W_t F_t' \Sigma_t^{-1}(y_t - F_t G_g \theta_{t-1}),\\
\mathrm{Var}(\theta_t|\theta_{t-1}, y_t) &= W_t - W_t F_t' \Sigma_t^{-1} F_t W_t,
\end{aligned}
$$

where $\Sigma_t = F_t W_t F_t' + V_t$. Note that for time-invariant DLMs the conditional variance above does not depend on t and can therefore be computed once and for all at the beginning of the process. The incremental weights, using this importance transition density, are proportional to the conditional density of y_t given $\theta_{t-1} = \theta_{t-1}^{(i)}$, i.e., to the $\mathcal{N}(F_t G_t \theta_{t-1}^{(i)}, \Sigma_t)$ density, evaluated at y_t.

5.1.1 A simple example

To illustrate the practical usage of the basic particle filter described in the previous section and to assess its accuracy, we present here a very simple example based on 100 observations simulated from a known DLM. The data are generated from a local level model with system variance $W = 1$, observation variance $V = 2$, and initial state distribution $\mathcal{N}(10, 9)$. We save the observations in *y*. Note the use of *dlmForecast* to simulate from a given model.

R code

```
> ### Generate data
> mod <- dlmModPoly(1, dV = 2, dW = 1, m0 = 10, C0 = 9)
```

[3] The reason for this apparent paradox is that the target distribution changes from time $t-1$ to time t. When one generates θ_{t-1}, the observation y_t is not used.

```
> n <- 100
> set.seed(23)
> simData <- dlmForecast(mod = mod, nAhead = n, sampleNew = 1)
> y <- simData$newObs[[1]]
```

In our implementation of the particle filter we set the number of particles to 1000, and we use the optimal importance kernel as importance transition density. As discussed in the previous section, for a DLM this density is easy to use both in terms of generating from it and for updating the particle weights. To keep things simple, instead of the more efficient residual resampling, we use plain multinomial resampling, setting the threshold for a resampling step to 500—that is, whenever the effective sample size drops below one half of the number of particles, we resample.

R code

```
> ### Basic Particle Filter - optimal importance density
> N <- 1000
> N_0 <- N / 2
> pfOut <- matrix(NA_real_, n + 1, N)
> wt <- matrix(NA_real_, n + 1, N)
> importanceSd <- sqrt(drop(W(mod) - W(mod)^2 /
+                          (W(mod) + V(mod))))
> predSd <- sqrt(drop(W(mod) + V(mod)))
> ## Initialize sampling from the prior
> pfOut[1, ] <- rnorm(N, mean = m0(mod), sd = sqrt(C0(mod)))
> wt[1, ] <- rep(1/N, N)
> for (it in 2 : (n + 1))
+ {
+     ## generate particles
+     means <- pfOut[it - 1, ] + W(mod) *
+         (y[it - 1] - pfOut[it - 1, ]) / (W(mod) + V(mod))
+     pfOut[it, ] <- rnorm(N, mean = means, sd = importanceSd)
+     ## update the weights
+     wt[it, ] <- dnorm(y[it - 1], mean = pfOut[it - 1, ],
+                       sd = predSd) * wt[it - 1, ]
+     wt[it, ] <- wt[it, ] / sum(wt[it, ])
+     ## resample, if needed
+     N.eff <- 1 / crossprod(wt[it, ])
+     if ( N.eff < N_0 )
+     {
+         ## multinomial resampling
+         index <- sample(N, N, replace = TRUE, prob = wt[it, ])
+         pfOut[it, ] <- pfOut[it, index]
+         wt[it, ] <- 1 / N
```

```
30 +      }
   + }
```

For a completely specified DLM the Kalman filter can be used to derive exact filtering means and variances. In Figure 5.1 we compare the exact fil-

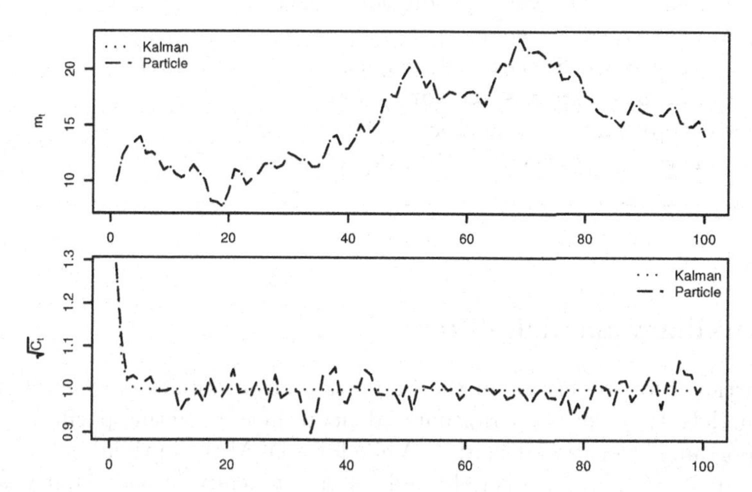

Fig. 5.1. Top: comparison of filtering state estimates computed with the Kalman filter and particle filter. Bottom: comparison of filtering standard deviations computed with the Kalman filter and particle filter

tering means and standard deviations, obtained using the Kalman filter, with the Monte Carlo approximations of the same quantities obtained using the particle filter algorithm. In terms of the filtering mean, the particle filter gives a very accurate approximation at any time (the two lines are barely distinguishable). The approximations to the filtering standard deviations are less precise, although reasonably close to the true values. The precision can be increased by increasing the number of particles in the simulation. The plots were obtained with the following code.

R code

```
  > ## Compare exact filtering distribution with PF approximation
2 > modFilt <- dlmFilter(y, mod)
  > thetaHatKF <- modFilt$m[-1]
4 > sdKF <- with(modFilt, sqrt(unlist(dlmSvd2var(U.C, D.C))))[-1]
  > pfOut <- pfOut[-1, ]
6 > wt <- wt[-1, ]
  > thetaHatPF <- sapply(1 : n, function(i)
8 +                           weighted.mean(pfOut[i, ], wt[i, ]))
  > sdPF <- sapply(1 : n, function(i)
```

```
10 +                   sqrt(weighted.mean((pfOut[i, ] -
   +                            thetaHatPF[i])^2, wt[i, ])))
12 > plot.ts(cbind(thetaHatKF, thetaHatPF),
   +          plot.type = "s", lty = c("dotted", "longdash"),
14 +          xlab = "", ylab = expression(m[t]))
   > legend("topleft", c("Kalman", "Particle"),
16 +         lty = c("dotted", "longdash"), bty = "n")
   > plot.ts(cbind(sdKF, sdPF), plot.type = "s",
18 +          lty = c("dotted", "longdash"), xlab = "",
   +          ylab = expression(sqrt(C[t])))
20 > legend("topright", c("Kalman", "Particle"),
   +         lty = c("dotted", "longdash"), bty = "n")
```

5.2 Auxiliary particle filter

The particle filter described in the previous section applies to general state space models. However, its performance depends heavily on the specification of the importance transition densities. While for a DLM the optimal importance kernel can be obtained explicitely and its use typically provides fairly good approximations to the filtering distributions, for a general state space model this is not the case, and devising effective importance transition densities is a much harder problem. The *auxiliary particle filter* algorithm was proposed by Pitt and Shephard (1999) to overcome this difficulty. While not really needed for fully specified DLMs, an extension of the algorithm, due to Liu and West (2001), turns out to be very useful even in the DLM case when the model contains unknown parameters. For this reason we present Pitt and Shephard's auxiliary particle filter here, followed in the next section by Liu and West's extension to deal with unknown model parameters.

Suppose that at time $t-1$ a discrete approximation $\hat{\pi}_{t-1} = \sum_{i=1}^{N} w_{t-1}^{(i)} \delta_{\theta_{0:t-1}^{(i)}}$ to the joint smoothing distribution $\pi(\theta_{0:t-1}|y_{1:t-1})$ is available. The goal is to update the approximate smoothing distribution when a new data point is observed or, in other words, to obtain a discrete approximation $\hat{\pi}_t$ to the joint smoothing distribution at time t, $\pi(\theta_{0:t}|y_{1:t})$. We have:

$$\begin{aligned}
\pi(\theta_{0:t}|y_{1:t}) &\propto \pi(\theta_{0:t}, y_t|y_{1:t-1}) \\
&= \pi(y_t|\theta_{0:t}, y_{1:t-1}) \cdot \pi(\theta_t|\theta_{0:t-1}, y_{1:t-1}) \cdot \pi(\theta_{0:t-1}|y_{1:t-1}) \\
&= \pi(y_t|\theta_t) \cdot \pi(\theta_t|\theta_{t-1}) \cdot \pi(\theta_{0:t-1}|y_{1:t-1}) \\
&\approx \pi(y_t|\theta_t) \cdot \pi(\theta_t|\theta_{t-1}) \cdot \hat{\pi}_{t-1}(\theta_{0:t-1}) \\
&= \sum_{i=1}^{N} w_{t-1}^{(i)} \pi(y_t|\theta_t) \pi(\theta_t|\theta_{t-1}^{(i)}) \delta_{\theta_{0:t-1}^{(i)}} .
\end{aligned}$$

Note that the last expression is an unnormalized distribution for $\theta_{0:t}$, which is discrete in the first t components and continuous in the last, θ_t. This distribution, which approximates $\pi(\theta_{0:t}|y_{1:t})$, can be taken to be our target for an importance sampling step. The target being a mixture distribution, a standard approach to get rid of the summation is to introduce a latent variable I, taking values in $\{1, \ldots, N\}$, such that:

$$\begin{aligned} P(I = i) &= w_{t-1}^{(i)}, \\ \theta_{0:t}|I = i &\sim C\pi(y_t|\theta_t)\pi(\theta_t|\theta_{t-1}^{(i)})\delta_{\theta_{0:t-1}^{(i)}}. \end{aligned}$$

Thus extended, the target becomes

$$\pi^{\text{aux}}(\theta_{0:t}, i|y_{1:t}) \propto w_{t-1}^{(i)}\pi(y_t|\theta_t)\pi(\theta_t|\theta_{t-1}^{(i)})\delta_{\theta_{0:t-1}^{(i)}}.$$

The importance density suggested by Pitt and Shephard for this target is

$$g_t(\theta_{0:t}, i|y_{1:t}) \propto w_{t-1}^{(i)}\pi(y_t|\hat{\theta}_t^{(i)})\pi(\theta_t|\theta_{t-1}^{(i)})\delta_{\theta_{0:t-1}^{(i)}},$$

where $\hat{\theta}_t^{(i)}$ is a central value, such as the mean or the mode, of $\pi(\theta_t|\theta_{t-1} = \theta_{t-1}^{(i)})$. A sample from g_t is easily obtained by iterating, for $k = 1, \ldots, N$, the following two steps.

1. Draw a classification variable I_k, with
$$P(I_k = i) \propto w_{t-1}^{(i)}\pi(y_t|\hat{\theta}_t^{(i)}), \qquad i = 1, \ldots, N.$$
2. Given $I_k = i$, draw
$$\theta_t^{(k)} \sim \pi(\theta_t|\theta_{t-1}^{(i)})$$
and set $\theta_{0:t}^{(k)} = (\theta_{0:t-1}^{(i)}, \theta_t^{(k)})$.

The importance weight of the kth draw from g_t is proportional to

$$\tilde{w}_t^{(k)} = \frac{w_{t-1}^{(I_k)}\pi(y_t|\theta_t^{(k)})\pi(\theta_t^{(k)}|\theta_{t-1}^{(k)})}{w_{t-1}^{(I_k)}\pi(y_t|\hat{\theta}_t^{(k)})\pi(\theta_t^{(k)}|\theta_{t-1}^{(k)})} = \frac{\pi(y_t|\theta_t^{(k)})}{\pi(y_t|\hat{\theta}_t^{(k)})}.$$

After normalizing the $\tilde{w}_t^{(k)}$'s and discarding the classification variables I_k's, we finally obtain the discrete approximation to the joint smoothing distribution at time t:

$$\hat{\pi}_t(\theta_{0:t}) = \sum_{i=1}^{N} w_t^{(i)}\delta_{\theta_{0:t}^{(i)}} \approx \pi(\theta_{0:t}|y_{1:t}).$$

As with the standard algorithm of Section 5.1, a resampling step is commonly applied in case the effective sample size drops below a specified threshold. A summary of the auxiliary particle filter is provided in Algorithm 5.2

```
0. Initialize: draw θ_0^(1),...,θ_0^(N) independently from π(θ_0) and set
```

$$w_0^{(i)} = N^{-1}, \qquad i = 1, \dots, N.$$

```
1. For t = 1,...,T:
  1.1) For k = 1,...,N:
```

- Draw I_k, with $\mathrm{P}(I_k = i) \propto w_{t-1}^{(i)} \pi(y_t|\hat{\theta}_t^{(i)})$.
- Draw $\theta_t^{(k)}$ from $\pi(\theta_t|\theta_{t-1} = \theta_{t-1}^{(I_k)})$ and set

$$\theta_{0:t}^{(k)} = \left(\theta_{0:t-1}^{(I_k)}, \theta_t^{(k)}\right).$$

- Set

$$\tilde{w}_t^{(k)} = \frac{\pi(y_t|\theta_t^{(k)})}{\pi(y_t|\hat{\theta}_t^{(k)})}.$$

1.2) Normalize the weights:

$$w_t^{(i)} = \frac{\tilde{w}_t^{(i)}}{\sum_{j=1}^N \tilde{w}_t^{(j)}}.$$

1.3) Compute

$$N_{\mathit{eff}} = \left(\sum_{i=1}^N \left(w_t^{(i)}\right)^2\right)^{-1}.$$

1.4) If $N_{\mathit{eff}} < N_0$, resample:

- Draw a sample of size N from the discrete distribution

$$\mathrm{P}\left(\theta_{0:t} = \theta_{0:t}^{(i)}\right) = w_t^{(i)}, \qquad i = 1, \dots, N,$$

and relabel this sample

$$\theta_{0:t}^{(1)}, \dots, \theta_{0:t}^{(N)}.$$

- Reset the weights: $w_t^{(i)} = N^{-1}$, $i = 1, \dots, N$.

1.5) Set $\hat{\pi}_t = \sum_{i=1}^N w_t^{(i)} \delta_{\theta_{0:t}^{(i)}}$.

Algorithm 5.2: Summary of the auxiliary particle filter algorithm

The main advantage of the auxiliary particle filter over the simple direct algorithm described in the previous section consists in the fact that it allows to use the one-step prior distribution $\pi(\theta_t|\theta_{t-1})$ to draw θ_t without losing much efficiency. Loosely speaking, when drawing from g_t, the role of the first step is to preselect a conditioning θ_{t-1} that is likely to evolve into a highly plausible θ_t in light of the new observation y_t. In this way possible conflicts between prior—$\pi(\theta_t|\theta_{t-1})$—and likelihood—$\pi(y_t|\theta_t)$—are minimized. It should be emphasized that for a general state space model deriving and drawing from the optimal instrumental kernel is often unfeasible, unlike in the DLM case, while the prior distribution is almost always available. Therefore, the ingenious use of the latter done by the auxiliary particle filter algorithm combines efficiency and simplicity.

5.3 Sequential Monte Carlo with unknown parameters

In real applications the model almost invariably contains unknown parameters that need to be estimated from the data. Denoting again by ψ the vector of unknown parameters, the target distribution at time t for a sequential Monte Carlo algorithm is therefore in this case $\pi(\theta_{0:t}, \psi | y_{1:t})$. As detailed in Section 4.4, a (weighted) sample from the forecast distributions can be easily obtained once a (weighted) sample from the joint posterior distribution is available. On the other hand, the filtering distribution and the posterior distribution of the parameter can be trivially obtained by marginalization. A simple-minded approach to sequential Monte Carlo for a model with an unknown parameter is to extend the state vector to include ψ as part of it, defining the trivial dynamics $\psi_t = \psi_{t-1}$ $(= \psi)$. In this way a relatively simple DLM typically becomes a nonlinear and nonnormal state space model. However, the most serious drawback is that, applying the general algorithm of Section 5.1 (or the auxiliary particle filter of Section 5.2), the values $\psi_t^{(i)}$, $i = 1, \dots, N$, are those drawn at time $t = 0$, since there is no evolution for this fictitious state. In other words, $\psi_t^{(i)} = \psi_0^{(i)}$ for every i and t, so that the $\psi_t^{(i)}$'s, drawn from the prior distribution, are typically not representative of the posterior distribution at a later time $t > 0$. It is true that, as the particle filter algorithm is sequentially applied, the weights are adjusted to reflect the changes of the target distributions. However, this can only account for the relative weights: if the $\psi_t^{(i)}$'s happen to be all in the tails of the marginal target $\pi(\psi | y_{1:t})$, the discrete approximation provided by the algorithm will always be a poor one. There is, in view of the previous considerations, a need to "refresh" the sampled values of ψ in order to follow the evolution of the posterior distribution. This can be achieved by discarding the current values of ψ each time the target changes and generating new ones. Among the different available methods, probably the most commonly used is the one proposed by Liu and West (2001) and described below, which extends the auxiliary particle filter. Fearnhead (2002), Gilks and Berzuini (2001) and Storvik (2002) propose interesting alternative algorithms.

The idea of Liu and West essentially consists of constructing an approximate target distribution at time t that is continuous not only in θ_t, but also in ψ, so that using importance sampling one draws values of ψ from a continuous importance density, effectively forgetting about the values of ψ used in the discrete approximation at time $t-1$. Consider the discrete approximation available at time $t-1$:

$$\hat{\pi}_{t-1}(\theta_{0:t-1}, \psi) = \sum_{i=1}^{N} w_{t-1}^{(i)} \delta_{(\theta_{0:t-1}^{(i)}, \psi^{(i)})} \approx \pi(\theta_{0:t-1}, \psi | y_{0:t-1}).$$

Marginally,

$$\hat{\pi}_{t-1}(\psi) = \sum_{i=1}^{N} w_{t-1}^{(i)} \delta_{\psi^{(i)}} \approx \pi(\psi | y_{0:t-1}). \tag{5.5}$$

Liu and West suggest replacing each point mass $\delta_{\psi^{(i)}}$ with a Normal distribution, so that the resulting mixture becomes a continuous distribution. A naive way of doing so would be to replace $\delta_{\psi^{(i)}}$ with a Normal centered at $\psi^{(i)}$. However, while preserving the mean, this would increase the variance of the approximating distribution. To see that this is the case, let $\bar{\psi}$ and Σ be the mean vector and variance matrix of ψ under $\hat{\pi}_{t-1}$, and let

$$\tilde{\pi}_{t-1}(\psi) = \sum_{i=1}^{N} w_{t-1}^{(i)} \mathcal{N}(\psi; \psi^{(i)}, \Lambda).$$

Introducing a latent classification variable I for the component of the mixture an observation comes from, we have

$$\begin{aligned} \mathrm{E}(\psi) &= \mathrm{E}(\mathrm{E}(\psi | I)) = \mathrm{E}(\psi^{(I)}) \\ &= \sum_{i=1}^{N} w_{t-1}^{(i)} \psi^{(i)} = \bar{\psi}; \\ \mathrm{Var}(\psi) &= \mathrm{E}(\mathrm{Var}(\psi | I)) + \mathrm{Var}(\mathrm{E}(\psi | I)) \\ &= \mathrm{E}(\Lambda) + \mathrm{Var}(\psi^{(I)}) \\ &= \Lambda + \Sigma > \Sigma, \end{aligned}$$

where expected values and variances are with respect to $\tilde{\pi}_{t-1}$. However, by changing the definition of $\tilde{\pi}_{t-1}$ to

$$\tilde{\pi}_{t-1}(\psi) = \sum_{i=1}^{N} w_{t-1}^{(i)} \mathcal{N}(\psi; m^{(i)}, h^2 \Sigma),$$

with $m^{(i)} = a\psi^{(i)} + (1-a)\bar{\psi}$ for some a in $(0, 1)$ and $a^2 + h^2 = 1$, we have

$$\begin{aligned} \mathrm{E}(\psi) &= \mathrm{E}(\mathrm{E}(\psi | I)) = \mathrm{E}(a\psi^{(I)} + (1-a)\bar{\psi}) \\ &= a\bar{\psi} + (1-a)\bar{\psi} = \bar{\psi}; \\ \mathrm{Var}(\psi) &= \mathrm{E}(\mathrm{Var}(\psi | I)) + \mathrm{Var}(\mathrm{E}(\psi | I)) \\ &= \mathrm{E}(h^2 \Sigma) + \mathrm{Var}(a\psi^{(I)} + (1-a)\bar{\psi}) \\ &= h^2 \Sigma + a^2 \mathrm{Var}(\psi^{(I)}) = h^2 \Sigma + a^2 \Sigma = \Sigma. \end{aligned}$$

Thus, ψ has the same first and second moment under $\tilde{\pi}_{t-1}$ and $\hat{\pi}_{t-1}$. Albeit this is true for any a in $(0, 1)$, in practice Liu and West recommend to set

$a = (3\delta - 1)/(2\delta)$ for a "discount factor" δ in $(0.95, 0.99)$, which corresponds to an a in $(0.974, 0.995)$. The very same idea can be applied even in the presence of $\theta_{0:t-1}$ to the discrete distribution $\hat{\pi}_{t-1}(\theta_{0:t-1}, \psi)$, leading to the extension of $\tilde{\pi}_{t-1}$ to a joint distribution for $\theta_{0:t-1}$ and ψ:

$$\tilde{\pi}_{t-1}(\theta_{0:t-1}, \psi) = \sum_{i=1}^{N} w_{t-1}^{(i)} \mathcal{N}(\psi; m^{(i)}, h^2 \Sigma) \delta_{\theta_{0:t-1}^{(i)}}.$$

Note that $\tilde{\pi}_{t-1}$ is discrete in $\theta_{0:t-1}$, but continuous in ψ. From this point onward, the method parallels the development of the auxiliary particle filter. After the new data point y_t is observed, the distribution of interest becomes

$$\begin{aligned}
\pi(\theta_{0:t}, \psi | y_{1:t}) &\propto \pi(\theta_{0:t}, \psi, y_t | y_{1:t-1}) \\
&= \pi(y_t | \theta_{0:t}, \psi, y_{1:t-1}) \cdot \pi(\theta_t | \theta_{0:t-1}, \psi, y_{1:t-1}) \cdot \pi(\theta_{0:t-1}, \psi | y_{1:t-1}) \\
&= \pi(y_t | \theta_t, \psi) \cdot \pi(\theta_t | \theta_{t-1}, \psi) \cdot \pi(\theta_{0:t-1}, \psi | y_{1:t-1}) \\
&\approx \pi(y_t | \theta_t, \psi) \cdot \pi(\theta_t | \theta_{t-1}, \psi) \cdot \tilde{\pi}_{t-1}(\theta_{0:t-1}, \psi) \\
&= \sum_{i=1}^{N} w_{t-1}^{(i)} \pi(y_t | \theta_t, \psi) \pi(\theta_t | \theta_{t-1}^{(i)}, \psi) \mathcal{N}(\psi; m^{(i)}, h^2 \Sigma) \delta_{\theta_{0:t-1}^{(i)}}.
\end{aligned}$$

Similarly to what we did in Section 5.2, we can introduce an auxiliary classification variable I such that:

$$\begin{aligned}
\mathrm{P}(I = i) &= w_{t-1}^{(i)}, \\
\theta_{0:t}, \psi | I = i &\sim C \pi(y_t | \theta_t, \psi) \pi(\theta_t | \theta_{t-1}^{(i)}, \psi) \mathcal{N}(\psi; m^{(i)}, h^2 \Sigma) \delta_{\theta_{0:t-1}^{(i)}}.
\end{aligned}$$

Note that the conditional distribution in the second line is continuous in θ_t and ψ, and discrete in $\theta_{0:t-1}$—in fact, degenerate on $\theta_{0:t-1}^{(i)}$. With the introduction of the random variable I, the auxiliary target distribution for the importance sampling update becomes

$$\pi^{\mathrm{aux}}(\theta_{0:t}, \psi, i | y_{1:t}) \propto w_{t-1}^{(i)} \pi(y_t | \theta_t, \psi) \pi(\theta_t | \theta_{t-1}^{(i)}, \psi) \mathcal{N}(\psi; m^{(i)}, h^2 \Sigma) \delta_{\theta_{0:t-1}^{(i)}}.$$

As an importance density, a convenient choice is

$$\begin{aligned}
g_t(\theta_{0:t}, \psi, i | y_{1:t}) \propto w_{t-1}^{(i)} &\pi(y_t | \theta_t = \hat{\theta}_t^{(i)}, \psi = m^{(i)}) \pi(\theta_t | \theta_{t-1}^{(i)}, \psi) \\
&\mathcal{N}(\psi; m^{(i)}, h^2 \Sigma) \delta_{\theta_{0:t-1}^{(i)}},
\end{aligned}$$

where $\hat{\theta}_t^{(i)}$ is a central value, such as the mean or the mode, of $\pi(\theta_t | \theta_{t-1} = \theta_{t-1}^{(i)}, \psi = m^{(i)})$. A sample from g_t can be obtained by iterating, for $k = 1, \ldots, N$, the following three steps.

1. Draw a classification variable I_k, with

$$\mathrm{P}(I_k = i) \propto w_{t-1}^{(i)} \pi(y_t | \theta_t = \hat{\theta}_t^{(i)}, \psi = m^{(i)}), \qquad i = 1, \ldots, N.$$

2. Given $I_k = i$, draw $\psi \sim \mathcal{N}(m^{(i)}, h^2\Sigma)$ and set $\psi^{(k)} = \psi$.
3. Given $I_k = i$ and $\psi = \psi^{(k)}$, draw

$$\theta_t^{(k)} \sim \pi(\theta_t|\theta_{t-1} = \theta_{t-1}^{(i)}, \psi = \psi^{(k)})$$

and set $\theta_{0:t}^{(k)} = (\theta_{0:t-1}^{(i)}, \theta_t^{(k)})$.

The importance weight of the kth draw from g_t is proportional to

$$\begin{aligned}\tilde{w}_t^{(k)} &= \frac{w_{t-1}^{(I_k)}\pi(y_t|\theta_t = \theta_t^{(k)}, \psi = \psi^{(k)})\pi(\theta_t^{(k)}|\theta_{t-1}^{(k)}, \psi^{(k)})\mathcal{N}(\psi^{(k)}; m^{(I_k)}, h^2\Sigma)}{w_{t-1}^{(I_k)}\pi(y_t|\theta_t = \hat{\theta}_t^{(I_k)}, \psi = m^{(I_k)})\pi(\theta_t^{(k)}|\theta_{t-1}^{(k)}, \psi^{(k)})\mathcal{N}(\psi^{(k)}; m^{(I_k)}, h^2\Sigma)} \\ &= \frac{\pi(y_t|\theta_t = \theta_t^{(k)}, \psi = \psi^{(k)})}{\pi(y_t|\theta_t = \hat{\theta}_t^{(I_k)}, \psi = m^{(I_k)})}.\end{aligned}$$

Renormalizing the weights, we obtain the approximate joint posterior distribution at time t

$$\hat{\pi}_t(\theta_{0:t}, \psi) = \sum_{i=1}^N w_t^{(i)} \delta_{(\theta_{0:t}^{(i)}, \psi^{(i)})} \approx \pi(\theta_{0:t}, \psi|y_{1:t}).$$

As was the case with the particle filter algorithms described in the previous sections, also in this case a resampling step can be applied whenever the effective sample size drops below a specified threshold. Algorithm 5.3 provides a convenient summary of the procedure.

Let us point out that, in order for the mixture of normals approximation of the posterior distribution at time $t-1$ to make sense, the parameter ψ has to be expressed in a form that is consistent with such a distribution—in particular, the support of a one-dimensional parameter must be the entire real line. For example, variances can be parametrized in terms of their log, probabilities in terms of their logit, and so on. In other words, and according to the suggestion by Liu and West, each parameter must be transformed so that the support of the distribution of the transformed parameter is the entire real line. A simpler alternative is to use a mixture of nonnormal distributions, appropriately selected so that their support is the same as that of the distribution of the parameter. For example, if a model parameter represents an unknown probability, and therefore its support is the interval $(0, 1)$, then one can consider approximating the discrete distribution obtained by the particle filter at time $t-1$ with a mixture of beta distributions instead of a mixture of normals, proceeding in all other respects as described above. Let us elaborate more on this simple example. Suppose ψ is an unknown parameter in $(0, 1)$. Denote by $\mu(\alpha, \beta)$ and $\sigma^2(\alpha, \beta)$ respectively the mean and variance of a beta distribution with parameters α and β. One can set, for each $i = 1, \ldots, N$,

$$\begin{aligned}\mu^{(i)} &= \mu(\alpha^{(i)}, \beta^{(i)}) = a\psi^{(i)} + (1-a)\bar{\psi}, \\ \sigma^{2(i)} &= \sigma^2(\alpha^{(i)}, \beta^{(i)}) = h^2\Sigma,\end{aligned} \tag{5.6}$$

0. Initialize: draw $(\theta_0^{(1)}, \psi^{(1)}), \ldots, (\theta_0^{(N)}, \psi^{(N)})$ independently from $\pi(\theta_0)\pi(\psi)$. Set $w_0^{(i)} = N^{-1}$, $i = 1, \ldots, N$, and

$$\hat{\pi}_0 = \sum_{i=1}^{N} w_0^{(i)} \delta_{(\theta_0^{(i)}, \psi^{(i)})}.$$

1. For $t = 1, \ldots, T$:
   1.1) Compute $\bar{\psi} = \mathrm{E}_{\hat{\pi}_{t-1}}(\psi)$ and $\Sigma = \mathrm{Var}_{\hat{\pi}_{t-1}}(\psi)$. For $i = 1, \ldots, N$, set

$$m^{(i)} = a\psi^{(i)} + (1-a)\bar{\psi},$$
$$\hat{\theta}_t^{(i)} = \mathrm{E}(\theta_t | \theta_{t-1} = \theta_{t-1}^{(i)}, \psi = m^{(i)}).$$

   1.2) For $k = 1, \ldots, N$:
   - Draw I_k, with $\mathrm{P}(I_k = i) \propto w_{t-1}^{(i)} \pi(y_t | \theta_t = \hat{\theta}_t^{(i)}, \psi = m^{(i)})$.
   - Draw $\psi^{(k)}$ from $\mathcal{N}(m^{(I_k)}, h^2 \Sigma)$.
   - Draw $\theta_t^{(k)}$ from $\pi(\theta_t | \theta_{t-1} = \theta_{t-1}^{(I_k)}, \psi = \psi^{(k)})$ and set

$$\theta_{0:t}^{(k)} = \left(\theta_{0:t-1}^{(I_k)}, \theta_t^{(k)}\right).$$

   - Set

$$\tilde{w}_t^{(k)} = \frac{\pi(y_t | \theta_t = \theta_t^{(k)}, \psi = \psi^{(k)})}{\pi(y_t | \theta_t = \hat{\theta}_t^{(I_k)}, \psi = m^{(I_k)})}.$$

   1.3) Normalize the weights:

$$w_t^{(i)} = \frac{\tilde{w}_t^{(i)}}{\sum_{j=1}^{N} \tilde{w}_t^{(j)}}.$$

   1.4) Compute

$$N_{\mathit{eff}} = \left(\sum_{i=1}^{N} (w_t^{(i)})^2\right)^{-1}.$$

   1.5) If $N_{\mathit{eff}} < N_0$, resample:
   - Draw a sample of size N from the discrete distribution

$$\mathrm{P}\big((\theta_{0:t}, \psi) = (\theta_{0:t}^{(i)}, \psi^{(i)})\big) = w_t^{(i)}, \qquad i = 1, \ldots, N,$$

   and relabel this sample

$$(\theta_{0:t}^{(1)}, \psi^{(1)}), \ldots, (\theta_{0:t}^{(N)}, \psi^{(N)}).$$

   - Reset the weights: $w_t^{(i)} = N^{-1}$, $i = 1, \ldots, N$.

   1.6) Set $\hat{\pi}_t = \sum_{i=1}^{N} w_t^{(i)} \delta_{(\theta_{0:t}^{(i)}, \psi^{(i)})}$.

Algorithm 5.3: Summary of Liu and West's algorithm

and solve for the pair $(\alpha^{(i)}, \beta^{(i)})$. The equations above can be explicitely solved for the parameters $\alpha^{(i)}$ and $\beta^{(i)}$, giving

$$\alpha^{(i)} = \frac{(\mu^{(i)})^2(1-\mu^{(i)})}{\sigma^{2(i)}} - \mu^{(i)},$$
$$\beta^{(i)} = \frac{\mu^{(i)}(1-\mu^{(i)})^2}{\sigma^{2(i)}} - (1-\mu^{(i)}).$$

It is straightforward to show that the mixture

$$\sum_{i=1}^{N} w_{t-1}^{i} \mathcal{B}(\psi; \alpha^{(i)}, \beta^{(i)}), \tag{5.7}$$

has the same mean and variance as (5.5), that is $\bar{\psi}$ and Σ.

On the same theme, consider the case of a positive unknown parameter ψ, such as a variance. Then we can consider a mixture of gamma distributions instead of a mixture on Normals. We can solve, for $i = 1, \ldots, N$ the system of equations (5.6) where $\alpha^{(i)}$ and $\beta^{(i)}$ are this time the parameters of a gamma distribution. The explicit solution is in this case

$$\alpha^{(i)} = \frac{(\mu^{(i)})^2}{\sigma^{2(i)}},$$
$$\beta^{(i)} = \frac{\mu^{(i)}}{\sigma^{2(i)}},$$

And the mixture

$$\sum_{i=1}^{N} w_{t-1}^{i} \mathcal{G}(\psi; \alpha^{(i)}, \beta^{(i)})$$

has mean $\bar{\psi}$ and variance Σ.

When the unknown parameter ψ is a vector it may not be easy to find a parametric family of multivariate distributions $f(\psi; \gamma)$ and proceed as we did in the examples above with the beta and gamma distributions, using a moment matching condition to come up with a continuous mixture that has the first mean and variance as the discrete particle approximation (5.5). When this is the case one can usually adopt the same moment matching approach marginally and consider a mixture of product densities. More specifically, consider a parameter $\psi = (\psi_1, \psi_2)$ and let

$$f(\psi; \gamma) = f_1(\psi_1; \gamma_1) f_2(\psi_2; \gamma_2), \qquad \gamma = (\gamma_1, \gamma_2),$$

where the parameter γ_j can be set in such a way that $f_j(\cdot|\gamma_j)$ has a specific mean and variance ($j = 1, 2$). Let, with an obvious notation,

$$\bar{\psi} = \begin{bmatrix} \bar{\psi}_1 \\ \bar{\psi}_2 \end{bmatrix}, \qquad \Sigma = \begin{bmatrix} \Sigma_1 & \Sigma_{12} \\ \Sigma_{21} & \Sigma_2 \end{bmatrix}.$$

```
1.1) For j = 1,2 and i = 1,...,N:
```
- Compute $\bar{\psi}_j = \mathrm{E}_{\hat{\pi}_{t-1}}(\psi_j)$ and $\Sigma_j = \mathrm{Var}_{\hat{\pi}_{t-1}}(\psi_j)$. Set

$$\begin{aligned} \mu_j^{(i)} &= a\psi_j^{(i)} + (1-a)\bar{\psi}_j, \\ \sigma_j^{(i)} &= h^2 \Sigma_j, \\ \mu^{(i)} &= (\mu_1^{(i)}, \mu_2^{(i)}), \\ \hat{\theta}_t^{(i)} &= \mathrm{E}(\theta_t | \theta_{t-1} = \theta_{t-1}^{(i)}, \psi = \mu^{(i)}). \end{aligned}$$

- Solve for $\gamma_j^{(i)}$ the system of equations

$$\begin{aligned} \mathrm{E}_{f_j(\cdot;\gamma_j^{(i)})}(\psi_j) &= \mu_j^{(i)}, \\ \mathrm{Var}_{f_j(\cdot;\gamma_j^{(i)})}(\psi_j) &= \sigma_j^{(i)}. \end{aligned}$$

```
1.2) For k = 1,...,N:
```
- Draw I_k, with $\mathrm{P}(I_k = i) \propto w_{t-1}^{(i)} \pi(y_t | \theta_t = \hat{\theta}_t^{(i)}, \psi = \mu^{(i)})$.
- For $j = 1, 2$, draw $\psi_j^{(k)}$ from $f_j(\cdot; \gamma_j^{(I_k)})$
- Draw $\theta_t^{(k)}$ from $\pi(\theta_t | \theta_{t-1} = \theta_{t-1}^{(I_k)}, \psi = \psi^{(k)})$ and set

$$\theta_{0:t}^{(k)} = (\theta_{0:t-1}^{(I_k)}, \theta_t^{(k)}).$$

- Set

$$\tilde{w}_t^{(k)} = \frac{\pi(y_t | \theta_t = \theta_t^{(k)}, \psi = \psi^{(k)})}{\pi(y_t | \theta_t = \hat{\theta}_t^{(I_k)}, \psi = m^{(I_k)})}.$$

Algorithm 5.4: Changes to Liu and West's algorithm when using product kernels

Then we can set, for $i = 1, \dots, N$,

$$\begin{aligned} \mu_j^{(i)} &= \int \psi_j f_j(\psi_j; \gamma_j^{(i)})\, \mathrm{d}\psi_j = a\psi_j^{(i)} + (1-a)\bar{\psi}_j, \\ \sigma_j^{2(i)} &= \int (\psi_j - \mu_j^{(i)})^2 f_j(\psi_j; \gamma_j^{(i)})\, \mathrm{d}\psi_j = h^2 \Sigma, \end{aligned} \tag{5.8}$$

and solve for $\gamma_j^{(i)}$ $(j = 1, 2)$. These are the same equations as (5.6) for the marginal distributions of ψ_1 and ψ_2. Finally, we can consider the mixture

$$\sum_{i=1}^{N} w_{t-1}^{(i)} f_1(\psi_1; \gamma_1^{(i)}) f_2(\psi_2; \gamma_2^{(i)}). \tag{5.9}$$

This will have the same mean as (5.5), $\bar{\psi}$, and the same marginal variances, Σ_1 and Σ_2. As far as the covariance is concerned, a simple calculation shows that under the mixture distribution (5.9) ψ_1 and ψ_2 have covariance $a^2\Sigma_{12}$. Since in the practical applications of Liu and West's method a is close to one, it follows that $a^2\Sigma_{12} \approx \Sigma_{12}$. In summary, for a bivariate parameter ψ, the mixture (5.9) provides a countinuous approximation to (5.5) that matches first moments, marginal second moments, and covariances up to a factor $a^2 \approx 1$. Using this product kernel approximation instead of the mixture of normals originally proposed by Liu and West, parts 1.1 and 1.2 of Algorithm 5.3 have to be changed accordingly, as shown in Algorithm 5.4.

To conclude the discussion of mixtures of product kernels within Liu and West's approach, let us note that the two components ψ_1 and ψ_2 may also be multivariate. Furthermore, the technique described can be generalized in an obvious way to product kernels containing more than two factors.

5.3.1 A simple example with unknown parameters

As a simple application of particle filtering for models containing unknown parameters, we go back to the example discussed in Section 5.1.1, this time assuming that both the system and observation variances are unknown. Since we have two unknown positive parameters, we are going to use at any time t products of Gamma kernels in the mixture approximation to the posterior distribution of the parameters. In the notation of the previous section, we have $\psi_1 = V$, $\psi_2 = W$, and $f_j(\psi_j;\gamma_j)$ is a gamma density for $j = 1, 2$, where $\gamma_j = (\alpha_j, \beta_j)$ is the standard vector parameter of the gamma distribution (see Appendix A). We use the same data that we simulated in Section 5.1.1. We choose independent uniform priors on $(0, 10)$ for both V and W. By looking at a plot of the data, the upper limit 10 for the variances seems more than enough for the interval to contain the true value of the parameter. Within these boundaries, a uniform prior does not carry any particularly strong information about the unknown variances. The reason for not choosing a more spread out prior distribution is that the particle filter algorithm initially generates the particles from the prior and, if the prior puts little probability in regions having high likelihood, most of the particles will be discarded after just one or two steps. Note that we are not arguing for selecting a prior based on the particular numerical method that one uses to evaluate the posterior distribution. On the contrary, we think that in this case a uniform prior on a finite interval better represents our belief about the variances than, say, a prior with infinite variance. After all, after plotting the data, who will seriously consider the possibility of V being larger than 100? Or 1000?

R code

```
  > ### PF with unknown parameters: Liu and West
2 > N <- 10000
  > a <- 0.975
4 > set.seed(4521)
  > pfOutTheta <- matrix(NA_real_, n + 1, N)
6 > pfOutV <- matrix(NA_real_, n + 1, N)
```

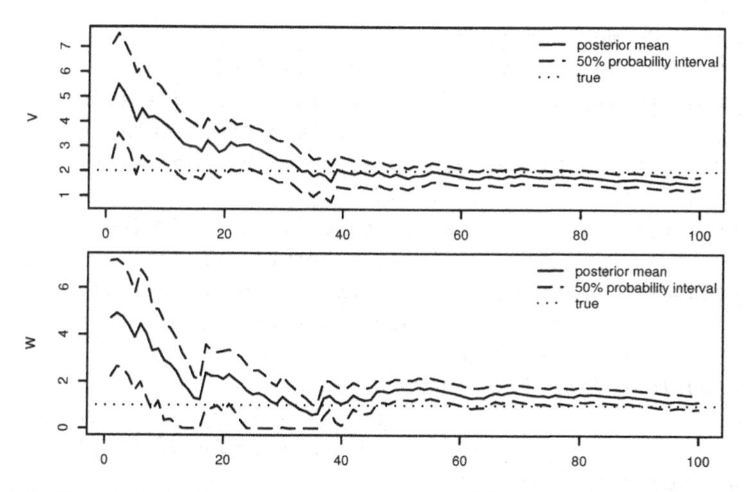

Fig. 5.2. Sequential estimation obtained via particle filter of V (top) and W (bottom)

```
> pfOutW <- matrix(NA_real_, n + 1, N)
> wt <- matrix(NA_real_, n + 1, N)
> ## Initialize sampling from the prior
> pfOutTheta[1, ] <- rnorm(N, mean = m0(mod),
+                          sd = sqrt(C0(mod)))
> pfOutV[1, ] <-  runif(N, 0, 10)
> pfOutW[1, ] <-  runif(N, 0, 10)
> wt[1, ] <- rep(1/N, N)
> for (it in 2 : (n + 1))
+ {
+     ## compute means and variances of the particle
+     ## cloud for V and W
+     meanV <- weighted.mean(pfOutV[it - 1, ], wt[it - 1, ])
+     meanW <- weighted.mean(pfOutW[it - 1, ], wt[it - 1, ])
+     varV <- weighted.mean((pfOutV[it - 1, ] - meanV)^2,
+                           wt[it - 1, ])
+     varW <- weighted.mean((pfOutW[it - 1, ] - meanW)^2,
+                           wt[it - 1, ])
+     ## compute the parameters of Gamma kernels
+     muV <- a * pfOutV[it - 1, ] + (1 - a) * meanV
+     sigma2V <- (1 - a^2) * varV
+     alphaV <- muV^2 / sigma2V
+     betaV <- muV / sigma2V
+     muW <- a * pfOutW[it - 1, ] + (1 - a) * meanW
+     sigma2W <- (1 - a^2) * varW
+     alphaW <- muW^2 / sigma2W
```

```
+       betaW <- muW / sigma2W
+       ## draw the auxiliary indicator variables
+       probs <- wt[it - 1,] * dnorm(y[it - 1], sd = sqrt(muV),
+                                   mean = pfOutTheta[it - 1, ])
+       auxInd <- sample(N, N, replace = TRUE, prob = probs)
+       ## draw the variances V and W
+       pfOutV[it, ] <- rgamma(N, shape = alphaV[auxInd],
+                              rate = betaV[auxInd])
+       pfOutW[it, ] <- rgamma(N, shape = alphaW[auxInd],
+                              rate = betaW[auxInd])
+       ## draw the state theta
+       pfOutTheta[it, ] <- rnorm(N, mean =
+                                 pfOutTheta[it - 1, auxInd],
+                                 sd = sqrt(pfOutW[it, ]))
+       ## compute the weights
+       wt[it, ] <- exp(dnorm(y[it - 1],
+                             mean = pfOutTheta[it, ],
+                             sd = sqrt(pfOutV[it, ]),
+                             log = TRUE) -
+                       dnorm(y[it - 1],
+                             mean = pfOutTheta[it - 1, auxInd],
+                             sd = sqrt(muV[auxInd]),
+                             log = TRUE))
+       wt[it, ] <- wt[it, ] / sum(wt[it, ])
+ }
```

5.4 Concluding remarks

We stressed in this chapter that the particle filter is very useful in online inference, where it can be used to recursively update a posterior distribution when the Kalman filter is not available because the model contains unknown parameters or otherwise—for example, the model is nonlinear. While this is certainly true, we would like to add a word of caution about the practical usage of particle filtering techniques in genuine sequential applications.

With very few exceptions that do not cover the samplers presented above, all the available asymptotic results hold when the time horizon T is fixed and the number of particles N is let to go to infinity. Furthermore, in order to obtain Monte Carlo approximations of similar quality for different time horizons T_1 and T_2, we need to use a number of particles proportional to T_i. The implication of these results is that if we start a particle filter with N particles, and we keep running it as new observations become available, the quality of the approximation will eventually deteriorate, making the particle approximation useless in the long run. The reason is that, even if we only use the particle approximation of the marginal posterior at time t, $\hat{\pi}_t(\theta_t)$, we are effectively targeting the joint posterior distribution $\pi(\theta_{0:t}|y_{1:t})$, so we are trying to track a distribution in an increasingly large number of dimensions. Another intuitive way to explain the deterioration of the particle filter approximation over time is to consider that the approximation at time t is based on the approximation at time $t-1$, so that the errors accumulate.

A possible practical solution, for applications that require a sequential updating of a posterior destribution over an unbounded time horizon, is to run an MCMC sampler every T sampling intervals to draw a sample from the posterior distribution at that time—possibly based on the most recent kT data points only, with $k \gg 1$—and use this sample to start the particle filter updating scheme for the next T sampling intervals. In this way one can run the MCMC off-line, while at the same time keeping updating the posterior using a particle filter. For example, in tracking the domestic stock market, one can run an MCMC over the weekend, when the data flow stops, and use a particle filter with hourly data, say, to update the posterior during the working week.

A similar idea can be used to initialize the particle filter when the prior distribution is diffuse. In this case, as we previously noticed, starting the particle filter with a sample from the prior will produce particles that are going to be off-target after just one or two updating steps. Instead, one can run an MCMC based on a small initial stretch of data in order to start the particle filter from a fairly stable particle cloud.

A

Useful distributions

Bernoulli distribution

A Bernoulli random variable is the indicator function of an event or, in other words, a discrete random variable whose only possible values are zero and one. If $X \sim \mathcal{B}e(p)$,

$$P(X = 1) = 1 - P(X = 0) = p.$$

The probability mass function is

$$\mathcal{B}e(x;p) = \begin{cases} 1-p & \text{if } x = 0, \\ p & \text{if } x = 1. \end{cases}$$

Normal distribution

Arguably the most used (and abused) probability distribution. Its density is

$$\mathcal{N}(x;\mu,\sigma^2) = \frac{1}{\sqrt{2\pi\sigma^2}} \exp\left\{-\frac{(x-\mu)^2}{2\sigma^2}\right\},$$

with expected value μ and variance σ^2.

Beta distribution

The support of a Beta distribution is the interval (0,1). For this reason it is often used as prior distribution for an unknown probability. The distribution is parametrized in terms of two positive parameters, a and b, and is denoted by $\mathcal{B}(a,b)$. Its density is

$$\mathcal{B}(x;a,b) = \frac{\Gamma(a+b)}{\Gamma(a)\Gamma(b)} x^{a-1}(1-x)^{b-1}, \qquad 0 < x < 1,$$

For a random variable $X \sim \mathcal{B}(a,b)$ we have

$$\mathrm{E}(X) = \frac{a}{a+b}, \qquad \mathrm{Var}(X) = \frac{ab}{(a+b)^2(a+b+1)}.$$

A multivariate generalization is provided by the Dirichlet distribution.

Gamma distribution

A random variable X has a Gamma distribution, with parameters (a, b), if it has density

$$\mathcal{G}(x; a, b) = \frac{b^a}{\Gamma(a)} x^{a-1} \exp(-bx), \qquad x > 0$$

where a and b are positive parameters. We find that

$$\mathrm{E}(X) = \frac{a}{b}, \qquad \mathrm{Var}(X) = \frac{a}{b^2}.$$

If $a > 1$, there is a unique mode at $(a-1)/b$. For $a = 1$, the density reduces to the (negative) exponential distribution with parameter b. For $(a = k/2, b = 1/2)$ it is a Chi-square distribution with k degrees of freedom, $\chi^2(k)$.

If $X \sim \mathcal{G}(a, b)$, the density of $Y = 1/X$ is called Inverse-Gamma, with parameters (a, b), and we have $\mathrm{E}(Y) = b/(a-1)$ if $a > 1$ and $\mathrm{Var}(Y) = b^2/((a-1)^2(a-2))$ if $a > 2$.

Student-t distribution

If $Z \sim \mathcal{N}(0,1), U \sim \chi^2(k), k > 0$ and Z and U are independent, then the random variable $T = Z/\sqrt{U/k}$ has a (central) *Student-t distribution* with k degrees of freedom, with density

$$f(t; k) = c \left(1 + \frac{t^2}{k}\right)^{-\frac{k+1}{2}},$$

where $c = \Gamma((k+1)/2)/(\Gamma(k/2)\sqrt{k\pi})$. We write $T \sim \mathcal{T}(0, 1, k)$ or simply $T \sim \mathcal{T}_k$.

It is clear from the definition that the density is positive on the whole real line and symmetric around the origin. It can be shown that, as k increases to infinity, the density converges to a standard Normal density at any point. We have

$$\mathrm{E}(X) = 0 \qquad \text{if } k > 1,$$

$$\mathrm{Var}(X) = \frac{k}{k-2} \qquad \text{if } k > 2.$$

If $T \sim \mathcal{T}(0, 1, k)$, then $X = \mu + \sigma T$ has a Student-t distribution, with parameters (μ, σ^2) and k degrees of freedom; we write $X \sim \mathcal{T}(\mu, \sigma^2, k)$. Clearly $\mathrm{E}(X) = \mu$ if $k > 1$ and $\mathrm{Var}(X) = \sigma^2 \frac{k}{k-2}$ if $k > 2$.

Normal-Gamma distribution

Let (X, Y) be a bivariate random vector. If $X|Y = y \sim \mathcal{N}(\mu, (n_0 y)^{-1})$, and $Y \sim \mathcal{G}(a, b)$, then we say that (X, Y) has a Normal-Gamma density with parameters (μ, n_0^{-1}, a, b) (where of course $\mu \in \mathbb{R}$, $n_0, a, b \in \mathbb{R}^+$). We write $(X, Y) \sim \mathcal{NG}(\mu, n_0^{-1}, a, b)$ The marginal density of X is a Student-t, $X \sim \mathcal{T}(\mu, (n_0 \frac{a}{b})^{-1}, 2a)$.

Multivariate Normal distribution

A continuous random vector $Y = (Y_1, \ldots, Y_k)'$ has a k-variate Normal distribution with parameters $\mu = (\mu_1, \ldots, \mu_k)'$ and Σ, where $\mu \in \mathbb{R}^k$ and Σ is a symmetric positive-definite matrix, if it has density

$$\mathcal{N}_k(y; \mu, \Sigma) = |\Sigma|^{-1/2}(2\pi)^{-k/2} \exp\left\{-\frac{1}{2}(y-\mu)'\Sigma^{-1}(y-\mu)\right\}, \qquad y \in \mathbb{R}^k$$

where $|\Sigma|$ denotes the determinant of the matrix Σ. We write

$$Y \sim \mathcal{N}_k(\mu, \Sigma).$$

Clearly, if $k = 1$, so that Σ is a scalar, the $\mathcal{N}_k(\mu, \Sigma)$ reduces to the univariate Normal density.

We have $\mathrm{E}(Y_i) = \mu_i$ and, denoting by $\sigma_{i,j}$ the elements of Σ, $\mathrm{Var}(Y_i) = \sigma_{i,i}$ and $\mathrm{Cov}(Y_i, Y_j) = \sigma_{i,j}$. The inverse of the covariance matrix Σ, $\Phi = \Sigma^{-1}$ is the precision matrix of Y.

Several results are of interest; their proof can be found in any multivariate analysis textbook (see, e.g. Barra and Herbach; 1981, pp.92,96).

1. If $Y \sim \mathcal{N}_k(\mu, \Sigma)$ and X is a linear transformation of Y, that is $X = AY$ where A is a $n \times k$ matrix, then $X \sim \mathcal{N}_k(A\mu, A\Sigma A')$.
2. Let X and Y be two random vectors, with covariance matrices Σ_X and Σ_Y, respectively. Let Σ_{YX} be the covariance between Y and X, i.e. $\Sigma_{YX} = \mathrm{E}((Y - \mathrm{E}(Y))(X - \mathrm{E}(X))')$. The covariance between X and Y is then $\Sigma_{XY} = \Sigma'_{YX}$. Suppose that Σ_X is nonsingular. Then it can be proved that the joint distribution of (X, Y) is Gaussian if and only if the following conditions are satisfied:
   (i) X has a Gaussian distribution;
   (ii) the conditional distribution of Y given $X = x$ is a Gaussian distribution whose mean is
   $$\mathrm{E}(Y|X = x) = \mathrm{E}(Y) + \Sigma_{YX}\Sigma_X^{-1}(x - \mathrm{E}(X))$$
   and whose covariance matrix is
   $$\Sigma_{Y|X} = \Sigma_Y - \Sigma_{YX}\Sigma_X^{-1}\Sigma_{XY}.$$

Multinomial distribution

Consider a set of n independent and identically distributed observations taking values in a finite label set $\{L_1, L_2, \ldots, L_k\}$. Denote by p_i the probability of an observation being equal to L_i, $i = 1, \ldots, k$. The vector of label counts $X = (X_1, \ldots, X_k)$, where X_i is the number of observations equal to L_i ($i = 1, \ldots, k$) has a Multinomial distribution, whose probability mass function is

$$\mathcal{M}ult(x_1, \ldots, x_k; n, p) = \frac{n!}{x_1! \ldots x_k!} p_1^{x_1} \cdots p_k^{x_k},$$

where $p = (p_1, \ldots, p_k)$ and the counts $x_1, \ldots, x_k$ satisfy the constraint $\sum x_i = n$.

Dirichlet distribution

The Dirichlet distribution is a multivariate generalization of the Beta distribution. Consider a parameter vector $a = (a_1, \ldots, a_k)$. The Dirichlet distribution $\mathcal{D}ir(a)$ has $k-1$-dimensional density

$$\mathcal{D}ir(x_1, \ldots, x_{k-1}; a) = \frac{\Gamma(a_1 + \cdots + a_k)}{\Gamma(a_1) \ldots \Gamma(a_k)} x_1^{a_1 - 1} \ldots x_{k-1}^{a_{k-1} - 1} \left(1 - \sum_{i=1}^{k-1} x_i\right)^{a_k - 1},$$

$$\text{for} \quad \sum_{i=1}^{k-1} x_i < 1, \quad x_i > 0, \quad i = 1 \ldots, k-1.$$

Wishart distribution

Let W be a symmetric positive-definite matrix of random variables $w_{i,j}$, $i, j = 1, \ldots, k$. The distribution of W is the joint distribution of its entries (in fact, the distribution of the $k(k+1)/2$-dimensional vector of the distinct entries). We say that W has a *Wishart* distribution with parameters α and B ($\alpha > (k-1)/2$ and B a symmetric, positive-definite matrix), if it has density

$$\mathcal{W}_k(W; \alpha, B) = c|W|^{\alpha - (k+1)/2} \exp\big(-\operatorname{tr}(BW)\big),$$

where $c = |B|^\alpha / \Gamma_k(\alpha)$, $\Gamma_k(\alpha) = \pi^{k(k-1)/4} \prod_{i=1}^k \Gamma((2\alpha + 1 - i)/2)$ is the *generalized gamma function* and $\operatorname{tr}(\cdot)$ denotes the trace of a matrix argument. We write $W \sim \mathcal{W}_k(\alpha, B)$ or just $W \sim \mathcal{W}(\alpha, B)$. We have

$$\mathrm{E}(W) = \alpha B^{-1}.$$

The Wishart distribution arises in sampling from a multivariate Gaussian distribution. If $(Y_1, \ldots, Y_n)$, $n > 1$, is a random sample from a multivariate normal distribution $\mathcal{N}_k(\mu, \Sigma)$ and $\bar{Y} = \sum_{i=1}^n Y_i/n$, then $\bar{Y} \sim \mathcal{N}_k(\mu, \Sigma/n)$ and

$$S = \sum_{i=1}^n (Y_i - \bar{Y})(Y_i - \bar{Y})'$$

is independent of $\bar{Y}$ and has a Wishart distribution $\mathcal{W}_k((n-1)/2, \Sigma^{-1}/2)$. In particular, if $\mu = 0$, then

$$W = \sum_{i=1}^n Y_i Y_i' \sim \mathcal{W}_k\left(\frac{n}{2}, \frac{1}{2}\Sigma^{-1}\right),$$

whose density (for $n > k-1$) is

$$f(w; n, \Sigma) \propto |W|^{\frac{n-k-1}{2}} \exp\left\{-\frac{1}{2}\operatorname{tr}(\Sigma^{-1} W)\right\}.$$

In fact, the Wishart distribution is usually parametrized in n and Σ, as in the expression above; then the parameter n is called *degrees of freedom*. Note that $\mathrm{E}(W) = n\Sigma$. We used the parametrization in α and B for analogy with the Gamma distribution; indeed, if $k = 1$, so that B is a scalar, then $\mathcal{W}_1(\alpha, B)$ reduces to the Gamma density $\mathcal{G}(\cdot; \alpha, B)$.

The following properties of the Wishart distribution can be proved. Let $W \sim \mathcal{W}_k(\alpha = n/2, B = \Sigma^{-1}/2)$ and $Y = AWA'$, where A is an $(m \times k)$ matrix of real numbers $(m \leq k)$. Then Y has a Wishart distribution of dimension m with parameters α and $\frac{1}{2}(A\Sigma A)^{-1}$, if the latter exists. In particular, if W and Σ conformably partition into

$$W = \begin{bmatrix} W_{1,1} & W_{1,2} \\ W_{2,1} & W_{2,2} \end{bmatrix}, \qquad \Sigma = \begin{bmatrix} \Sigma_{1,1} & \Sigma_{1,2} \\ \Sigma_{2,1} & \Sigma_{2,2} \end{bmatrix},$$

where $W_{1,1}$ and $\Sigma_{1,1}$ are $h \times h$ matrices $(1 \leq h < k)$, then

$$W_{1,1} \sim \mathcal{W}_h \left(\alpha = \frac{n}{2}, \frac{1}{2} \Sigma_{1,1}^{-1} \right).$$

This property allows to compute the marginal distribution of the elements on the diagonal of W; for example, if $k = 2$ and $A = (1, 0)$, then $Y = w_{1,1} \sim \mathcal{G}(\alpha = n/2, \sigma_{1,1}^{-1}/2)$, where $\sigma_{1,1}$ is the first element of the diagonal of Σ. It follows that $w_{1,1}/\sigma_{1,1} \sim \chi^2(n)$. Then,

$$\mathrm{E}(w_{1,1}) = n\sigma_{1,1}, \quad \mathrm{Var}(w_{1,1}) = 2n\sigma_{1,1}^2.$$

More generally, it can be proved that

$$\mathrm{Var}(w_{i,j}) = n(\sigma_{i,j}^2 + \sigma_{i,i}\sigma_{j,j}), \quad \mathrm{Cov}(w_{i,j}, w_{l,m}) = n(\sigma_{i,l}\sigma_{j,m} + \sigma_{i,m}\sigma_{j,l}).$$

If $W \sim \mathcal{W}_k(\alpha = n/2, B = \Sigma^{-1}/2)$, then $V = W^{-1}$ has an *Inverse-Wishart* distribution and

$$\mathrm{E}(V) = \mathrm{E}(W^{-1}) = \left(\alpha - \frac{k+1}{2} \right)^{-1} B = \frac{1}{n-k-1} \Sigma^{-1}.$$

Multivariate Student-t distribution

If Y is a p-variate random vector with $Y \sim \mathcal{N}_p(0, \Sigma)$ and $U \sim \chi^2(k)$, with Y and U independent, then $X = \frac{Y}{\sqrt{U/k}} + \mu$ has a p-variate Student-t distribution, with parameters (μ, Σ) and $k > 0$ degrees of freedom, with density

$$f(x) = c \, [1 - \frac{1}{k} \, (x - \mu)' \Sigma^{-1} (x - \mu)]^{-(k+p)/2} \quad, x \in \mathbb{R}^p \, ,$$

where $c = \Gamma((k+p)/2)/(\Gamma(k/2)\pi^{p/2}k^{p/2}|\Sigma|^{1/2}$. We write $X \sim \mathcal{T}(\mu, \Sigma, k)$. For $p = 1$ it reduces to the univariate Student-t distribution. We have

$$\mathrm{E}(X) = \mu \text{ if } k > 1$$

$$\mathrm{Var}(X) = \Sigma \frac{k}{k-2} \text{ if } k > 2.$$

Multivariate Normal-Gamma distribution

Let (X, Y) be a random vector, with $X|Y = y \sim \mathcal{N}_m(\mu, (N_0 y)^{-1})$, and $Y \sim Ga(a, b)$. Then we say that (X, Y) has a Normal-Gamma density with parameters (μ, N_0^{-1}, a, b), denoted as $(X, Y) \sim \mathcal{NG}(\mu, N_0^{-1}, a, b)$.

The marginal density of X is a multivariate Student-t, $X \sim \mathcal{T}(\mu, (N_0 \frac{a}{b})^{-1}, 2a)$, so that $\mathrm{E}(X) = \mu$ and $\mathrm{Var}(X) = N_0^{-1} b/(a-1)$.

B

Matrix algebra: Singular Value Decomposition

Let M be a $p \times q$ matrix and let $r = \min\{p, q\}$. The singular value decomposition (SVD) of M consists in a triple of matrices (U, D, V) with the following properties:

(i) U is a $p \times p$ orthogonal matrix;
(ii) V is a $q \times q$ orthogonal matrix;
(iii) D is a $p \times q$ matrix with entries $D_{ij} = 0$ for $i \neq j$;
(iv) $UDV' = M$.

If M is a square matrix, D is a diagonal matrix. If, in addition, M is nonnegative definite, then $D_{ii} \geq 0$ for every i. In this case one can define a diagonal matrix S by setting $S_{ii} = \sqrt{D_{ii}}$, so that $M = US^2V'$. It can be shown that if M is also symmetric, such as, for example, a variance matrix, then $M = US^2U'$. M is invertible if and only if $S_{ii} > 0$ for every i. The SVD has many applications in numerical linear algebra. For example, it can be used to compute a square root[1] of a variance matrix M, i.e., a square matrix N such that $M = N'N$. In fact, if $M = US^2U'$, it is enough to set $N = SU'$. The inverse of M can also be easily computed from its SVD, provided M is invertible. In fact, it is immediate to verify that $M^{-1} = US^{-2}U'$. Note also that $S^{-1}U'$ is a square root of M^{-1}. More generally, for a noninvertible M, a generalized inverse M^- is a matrix with the property that $MM^-M = M$. The generalized inverse of a variance matrix can be found by defining the diagonal matrix S^-,

$$S_{ii}^- = \begin{cases} S_{ii}^{-1} & \text{if } S_{ii} > 0, \\ 0 & \text{if } S_{ii} = 0, \end{cases}$$

and setting $M^- = U(S^-)^2V'$.

In package dlm the SVD is used extensively to compute filtering and smoothing variances in a numerically stable way, see Wang et al. (1992) and

[1] Note that our definition of matrix square root differs slightly from the most common one based on the relation $M = M^{\frac{1}{2}}M^{\frac{1}{2}}$.

Zhang and Li (1996) for a complete discussion of the algorithms used. For example, consider the filtering recursion used to compute C_t. The calculation can be broken down in three steps:

(i) compute $R_t = G_t C_{t-1} G_t' + W_t$;
(ii) compute[2] $C_t^{-1} = F_t' V_t^{-1} F_t + R_t^{-1}$;
(iii) invert C_t^{-1}.

Suppose $(U_{C,t-1}, S_{C,t-1})$ are the components of the SVD of C_{t-1}, so that $C_{t-1} = U_{C,t-1} S_{C,t-1}^2 U_{C,t-1}'$ and $N_{W,t}$ is a square root of W_t. Define the $2p \times p$ partitioned matrix

$$M = \begin{bmatrix} S_{C,t-1} U_{C,t-1}' G_t' \\ N_{W,t} \end{bmatrix}$$

and let (U, D, V) be its SVD. The crossproduct of M is

$$\begin{aligned} M'M &= VDU'UDV' = VD^2V' \\ &= G_t U_{C,t-1} S_{C,t-1}^2 U_{C,t-1}' G_t' + N_{W,t}' N_{W,t} \\ &= G_t C_{t-1} G_t' + W_t = R_t. \end{aligned}$$

Since V is orthogonal and D^2 is diagonal, (V, D^2, V) is the SVD of R_t and we can set $U_{R,t} = V$ and $S_{R,t} = D$. Now consider a square root $N_{V,t}$ of V_t^{-1}, define the $(m+p) \times p$ partitioned matrix

$$M = \begin{bmatrix} N_{V,t} F_t U_{R,t} \\ S_{R,t}^{-1} \end{bmatrix}$$

and let (U, D, V) be its SVD. The crossproduct of M is

$$\begin{aligned} M'M &= VDU'UDV' = VD^2V' \\ &= U_{R,t}' F_t' N_{V,t}' N_{V,t} F_t U_{R,t} + S_{R,t}^{-2}. \end{aligned}$$

Premultiplying by $U_{R,t}$ and postmultiplying by $U_{R,t}'$ we obtain

$$\begin{aligned} U_{R,t} M'M U_{R,t}' &= U_{R,t} V D^2 V' U_{R,t}' \\ &= U_{R,t} U_{R,t}' F_t' N_{V,t}' N_{V,t} F_t U_{R,t} U_{R,t}' + U_{R,t} S_{R,t}^{-2} U_{R,t}' \\ &= F_t' V_t^{-1} F_t + R_t^{-1} = C_t^{-1}. \end{aligned}$$

---

[2] The expression for C_t^{-1} follows from the formulas in Proposition 2.2 by applying the general matrix equality

$$(A + BEB')^{-1} = A^{-1} - A^{-1}B(B'A^{-1}B + E^{-1})^{-1}B'A^{-1},$$

in which A and E are nonsingular matrices of orders m and n and B is a $m \times n$ matrix.

Since $U_{R,t}V$ is the product of two orthogonal matrices of order p, it is itself an orthogonal matrix. Therefore $U_{R,t}VD^2V'U'_{R,t}$ gives the SVD of C_t^{-1}: the "U" matrix of the SVD of C_t^{-1} is $U_{R,t}V$ and the "S" matrix is D. It follows that for the SVD of C_t we have $U_{C,t} = U_{R,t}V$ and $S_{C,t} = D^{-1}$.

Index

References

Akaike, H. (1974a). Markovian representation of stochastic processes and its application to the analysis of autoregressive moving average processes, *Annals of the Institute of Statistical Mathematics* **26**: 363–387.

Akaike, H. (1974b). Stochastic theory of minimal realization, *IEEE Trans. on Automatic Control* **19**: 667–674.

Amisano, G. and Giannini, C. (1997). *Topics in Structural VAR Econometrics*, 2nd edn, Springer, Berlin.

Anderson, B. and Moore, J. (1979). *Optimal Filtering*, Prentice-Hall, Englewood Cliffs.

Aoki, M. (1987). *State Space Modeling of Time Series*, Springer Verlag, New York.

Barndorff-Nielsen, O., Cox, D. and Klüppelberg, C. (eds) (2001). *Complex Stochastic Systems*, Chapman & Hall, London.

Barra, J. and Herbach, L. H. (1981). *Mathematical Basis of Statistics*, Academic Press, New York.

Bawens, L., Lubrano, M. and Richard, J.-F. (1999). *Bayesian inference in Dynamic Econometric Models*, Oxford University Press, New York.

Bayes, T. (1763). *An essay towards solving a problem in the doctrine of chances.* Published posthumously in *Phil. Trans. Roy. Stat. Soc. London*, **53**, 370–418 and **54**, 296–325. Reprinted in *Biometrika* **45** (1958), 293–315, with a biographical note by G.A. Barnard. Reproduced in Press (1989), 185–217.

Berger, J. (1985). *Statistical Decision Theory and Bayesian Analysis*, Springer, Berlin.

Bernardo, J. (1979a). Expected information as expected utility, *Annals of Statistics* pp. 686–690.

Bernardo, J. (1979b). Reference posterior distributions for Bayesian inference (with discussion), *Journal of the Royal Statistical Society, Series B* pp. 113–147.

Bernardo, J. and Smith, A. (1994). *Bayesian Theory*, Wiley, Chichester.

Berndt, R. (1991). *The Practice of Econometrics*, Addison-Wesley.

Bollerslev, T. (1986). Generalized autoregressive conditional heteroskedasticity, *Journal of Econometrics* **31**: 307–327.

Box, G., Jenkins, G. and Reinsel, G. (2008). *Time Series Analysis: Forecasting and Control*, 4th edn, Wiley, New York.

Brandt, P. (2008). *MSBVAR: Markov-Switching Bayesian Vector Autoregression Models*. R package version 0.3.2.
URL: *http://www.utdallas.edu/~pbrandt/*

Brown, P., Le, N. and Zidek, J. (1994). Inference for a covariance matrix, *in* P. Freeman and e. A.F.M. Smith (eds), *Aspects of Uncertainty: A Tribute to D. V. Lindley*, Wiley, Chichester, pp. 77–92.

Caines, P. (1988). *Linear Stochastic Systems*, Wiley, New York.

Campbell, J., Lo, A. and MacKinley, A. (1996). *The Econometrics of Financial Markets*, Princeton University Press, Princeton.

Canova, F. (2007). *Methods for Applied Macroeconomic Research*, Princeton University Press, Princeton.

Cappé, O., Godsill, S. and Moulines, E. (2007). An overview of existing methods and recent advances in sequential Monte Carlo, *Proceedings of the IEEE* **95**: 899–924.

Cappé, O., Moulines, E. and Rydén, T. (2005). *Inference in Hidden Markov Models*, Springer, New York.

Carmona, R. A. (2004). *Statistical analysis of financial data in S-plus*, Springer-Verlag, New York.

Caron, F., Davy, M., A., D., Duflos, E. and Vanheeghe, P. (2008). Bayesian inference for linear dynamic models with dirichlet process mixtures, *IEEE Transactions on Signal Processing* **56**: 71–84.

Carter, C. and Kohn, R. (1994). On Gibbs sampling for state space models, *Biometrika* **81**: 541–553.

Chang, Y., Miller, J. and Park, J. (2005). Extracting a common stochastic trend: Theories with some applications, *Technical report*, Rice University.
URL: *http://ideas.repec.org/p/ecl/riceco/2005-06.html.*

Chatfield, C. (2004). *The Analysis of Time Series*, 6th edn, CRC-Chapman & Hall, London.

Cifarelli, D. and Muliere, P. (1989). *Statistica Bayesiana*, Iuculano Editore, Pavia. (In Italian).

Consonni, G. and Veronese, P. (2001). Conditionally reducible natural exponential families and enriched conjugate priors, *Scandinavian Journal of Statistics* **28**: 377–406.

Consonni, G. and Veronese, P. (2003). Enriched conjugate and reference priors for the wishart family on symmetric cones, *Annals of Statistics* **31**: 1491–1516.

Cowell, R., Dawid, P., Lauritzen, S. and Spiegelhalter, D. (1999). *Probabilistic networks and expert systems*, Springer-Verlag, New York.

D'Agostino, R. and Stephens, M. (eds) (1986). *Goodness-of-fit Techniques*, Dekker, New York.

Dalal, S. and Hall, W. (1983). Approximating priors by mixtures of conjugate priors, *J. Roy. Statist. Soc. Ser. B* **45**: 278–286.

Dawid, A. (1981). Some matrix-variate distribution theory: Notational considerations and a bayesian application, *Biometrika* **68**: 265–274.

Dawid, A. and Lauritzen, S. (1993). Hyper-markov laws in the statistical analysis of decomposable graphical models, *Ann. Statist.* **21**: 1272–1317.

de Finetti, B. (1970a). *Teoria della probabilità I*, Einaudi, Torino. English translation as *Theory of Probability I* in 1974, Wiley, Chichester.

de Finetti, B. (1970b). *Teoria della probabilità II*, Einaudi, Torino. English translation as *Theory of Probability II* in 1975, Wiley, Chichester.

De Finetti, B. (1972). *Probability, Induction and Statistics*, Wiley, Chichester.

DeGroot, M. (1970). *Optimal Statistical Decisions*, McGraw Hill, New York.

Del Moral, P. (2004). *Feyman-Kac Formulae: Genealogical and Interacting Particle Systems with Applications*, Springer-Verlag, New York.

Diaconis, P. and Ylvisaker, D. (1985). Quantifying prior opinion, *in* J. Bernardo, M. deGroot, D. Lindley and A. Smith (eds), *Bayesian Statistics 2*, Elsevier Science Publishers B.V. (North Holland), pp. 133–156.

Diebold, F. and Li, C. (2006). Forecasting the term structure of government bond yields, *Journal of Econometrics* **130**: 337–364.

Diebold, F., Rudebuschb, G. and Aruoba, S. (2006). The macroeconomy and the yield curve: A dynamic latent factor approach, *Journal of Econometrics* **131**: 309–338.

Doan, T., Litterman, R. and Sims, C. (1984). Forecasting and conditional projection using realistic prior distributions, *Econometric Reviews* **3**: 1–144.

Doucet, A., De Freitas, N. and Gordon, N. (eds) (2001). *Sequential Monte Carlo Methods in Practice*, Springer, New York.

Durbin, J. and Koopman, S. (2001). *Time Series Analysis by State Space Methods*, Oxford University Press, Oxford.

Engle, R. (1982). Autoregressive conditional heteroskedasticity with estimates of the variance of UK inflation, *Econometrica* **50**: 987–1008.

Engle, R. and Granger, C. (1987). Co-integration and error correction: Representation, estimation, and testing, *Econometrica* **55**: 251–276.

Fearnhead, P. (2002). Markov chain Monte Carlo, sufficient statistics, and particle filter, *Journal of Computational and Graphical Statistics* **11**: 848–862.

Forni, M., Hallin, M., Lippi, M. and Reichlin, L. (2000). The generalized dynamic factor model: Identification and estimation, *Review of Economics and Statistics* **82**: 540–552.

Frühwirth-Schnatter, S. and Kaufmann, S. (2008). Model-based clustering of multiple time series, *Journal of Business and Economic Statistics* **26**: 78–89.

Früwirth-Schnatter, S. (1994). Data augmentation and dynamic linear models, *Journal of Time Series Analysis* **15**: 183–202.

Früwirth-Schnatter, S. and Wagner, H. (2008). Stochastic model specification search for Gaussian and non-Gaussian state space models, *IFAS Research Papers, 2008-36* .

Gamerman, D. and Migon, H. (1993). Dynamic hierarchical models, *Journal of the Royal Statistical Society, Series B* **55**: 629–642.

Gelman, A., Carlin, J., Stern, H. and Rubin, D. (2004). *Bayesian Data Analysis*, 2nd edn, Chapman & Hall/CRC, Boca Raton.

Geweke, J. (2005). *Contemporary Bayesian Econometrics and Statistics*, Wiley, Hoboken.

Gilbert, P. (2008). *Brief User's Guide: Dynamic Systems Estimation.* **URL:** *http: www. bank-banque-canada. ca/pgilbert/.*

Gilks, W. and Berzuini, C. (2001). Following a moving target – Monte Carlo inference for dynamic Bayesian models, *Journal of the Royal Statistical Society, Series B* **63**: 127–146.

Gilks, W., Best, N. and Tan, K. (1995). Adaptive rejection Metropolis sampling within Gibbs sampling (Corr: 97V46 p541-542 with R.M. Neal), *Applied Statistics* **44**: 455–472.

Gilks, W. and Wild, P. (1992). Adaptive rejection sampling for Gibbs sampling, *Applied Statistics* **41**: 337–348.

Gourieroux, C. and Monfort, A. (1997). *Time Series and Dynamic Models*, Cambridge University Press, Cambridge.

Granger, C. (1981). Some properties of time series data and their use in econometric model specification, *Journal of Econometrics* **16**: 150–161.

Gross, J. (n.d.). *Nortest: Tests for Normality.* R package version 1.0.

Hannan, E. and Deistler, M. (1988). *The Statistical Theory of Linear Systems*, Wiley, New York.

Harrison, P. and Stevens, C. (1976). Bayesian forecasting (with discussion), *Journal of the Royal Statistical Society, Series B* **38**: 205–247.

Harvey, A. (1989). *Forecasting, Structural Time Series Models and the Kalman filter*, Cambridge University Press, Cambridge.

Hastings, W. (1970). Monte Carlo sampling methods using Markov chains and their applications, *Biometrika* **57**: 97–109.

Hutchinson, C. (1984). The Kalman filter applied to aerospace and electronic systems, *Aerospace and Electronic Systems, IEEE Transactions on Aerospace and Electronic Systems* **AES-20**: 500–504.

Hyndman, R. (2008). *Forecast: Forecasting Functions for Time Series.* R package version 1.14.
URL: *http://www.robhyndman.com/Rlibrary/forecast/.*

Hyndman, R. (n.d.). Time Series Data Library.
URL: *http://www.robjhyndman.com/TSDL.*

Hyndman, R., Koehler, A., Ord, J. and Snyder, R. (2008). *Forecasting with Exponential smoothing*, Springer, Berlin.

Jacquier, E., Polson, N. and Rossi, P. (1994). Bayesian analysis of stochastic volatility models (with discussion), *Journal of Business and Economic Statistics* **12**: 371–417.

Jazwinski, A. (1970). *Stochastic Processes and Filtering Theory*, Academic Press, New York.

Jeffreys, H. (1998). *Theory of Probability*, 3rd edn, Oxford University Press, New York.

Johannes, M. and Polson, N. (2009). MCMC methods for continuous-time financial econometrics, *in* Y. Ait-Sahalia and L. Hansen (eds), *Handbook of Financial Econometrics*, Elsevier. (To appear).

Kalman, R. (1960). A new approach to linear filtering and prediction problems, *Trans. of the AMSE - Journal of Basic Engineering (Series D)* **82**: 35–45.

Kalman, R. (1961). On the general theory of control systems, *Proc. IFAC Congr Ist.* **1**: 481–491.

Kalman, R. (1968). Contributions to the theory of optimal control, *Bol. Soc. Mat. Mexicana* **5**: 558–563.

Kalman, R. and Bucy, R. (1963). New results in linear filtering and prediction theory, *Trans. of the AMSE – Journal of Basic Engineering (Series D)* **83**: 95–108.

Kalman, R., Ho, Y. and Narenda, K. (1963). Controllability of linear dynamical systems, *in* J. Lasalle and J. Diaz (eds), *Contributions to Differential Equations*, Vol. 1, Wiley Interscience.

Kim, C.-J. and Nelson, C. (1999). *State Space Models with Regime Switching*, MIT Press, Cambridge.

Kolmogorov, A. (1941). Interpolation and extrapolation of stationary random sequences, *Bull. Moscow University, Ser. Math.* **5**.

Künsch, H. (2001). State space and hidden Markov models, *in* O. Barndorff-Nielsen, D. Cox and C. Klüppelberg (eds), *Complex stochastic systems*, Chapman & Hall/CRC, Boca Raton, pp. 109–173.

Kuttner, K. (1994). Estimating potential output as a latent variable, *Journal of Business and Economic Statistics* **12**: 361–68.

Landim, F. and Gamerman, D. (2000). Dynamic hierarchical models; an extension to matrix-variate observations, *Computational Statistics and Data Analysis* **35**: 11–42.

Laplace, P. (1814). *Essai Philosophique sur les Probabilitiès*, Courcier, Paris. The 5th edn (1825) was the last revised by Laplace. English translation in 1952 as *Philosophical Essay on Probabilities*, Dover, New York.

Lau, J. and So, M. (2008). Bayesian mixture of autoregressive models, *Computational Statistics and Data Analysis* **53**: 38–60.

Lauritzen, S. (1981). Time series analysis in 1880: A discussion of contributions made by T.N. Thiele, *International Statist. Review* **49**: 319–331.

Lauritzen, S. (1996). *Graphical Models*, Oxford University Press, Oxford.

Lindley, D. (1978). The Bayesian approach (with discussion), *Scandinavian Journal of Statistics* **5**: 1–26.

Lindley, D. and Smith, A. (1972). Bayes estimates for the linear model, *Journal of the Royal Statistical Society, Series B* **34**: 1–41.

Lipster, R. and Shiryayev, A. (1972). Statistics of conditionally Gaussian random sequences, *Proceedings of the Sixth Berkeley Symposium on Mathematical Statistics and Probability*, Univ. California Press, Berkeley.

Litterman, R. (1986). Forecasting with Bayesian vector autoregressions – five years of experience, *Journal of Business and Economic Statistics* **4**: 25–38.

Liu, J. (2001). *Monte Carlo Strategies in Scientific Computing*, Springer, New York.

Liu, J. and West, M. (2001). Combined parameter and state estimation in simulation-based filtering, *in* A. Doucet, N. De Freitas and N. Gordon (eds), *Sequential Monte Carlo Methods in Practice*, Springer, New York.

Ljung, G. and Box, G. (1978). On a measure of lack of fit in time series models, *Biometrika* **65**: 297–303.

Lütkepohl, H. (2005). *New Introduction to Multiple Time Series Analysis*, Springer-Verlag, Berlin.

Maybeck, P. (1979). *Stochastic Models, Estimation and Control*, Vol. 1 and 2, Academic Press, New York.

Metropolis, N., Rosenbluth, A., Rosenbluth, M., Teller, A. and Teller, E. (1953). Equations of state calculations by fast computing machines, *Journal of Chemical Physics* **21**: 1087–1091.

Migon, H., Gamerman, D., Lopez, H. and Ferreira, M. (2005). Bayesian dynamic models, *in* D. Day and C. Rao (eds), *Handbook of Statistics*, Vol. 25, Elsevier B.V., chapter 19, pp. 553–588.

Morf, M. and Kailath, T. (1975). Square-root algorithms for least-squares estimation, *IEEE Trans. Automatic Control* **AC-20**: 487–497.

Muliere, P. (1984). Modelli lineari dinamici, *Studi Statistici (8)*, Istituto di Metodi Quantitativi, Bocconi University, Milan. (In Italian).

O'Hagan, A. (1994). *Bayesian Inference*, Kendall's Advanced Theory of Statistics, 2B, Edward Arnold, London.

Oshman, Y. and Bar-Itzhack, I. (1986). Square root filtering via covariance and information eigenfactors, *Automatica* **22**: 599–604.

Petris, G. and Tardella, L. (2003). A geometric approach to transdimensional Markov chain Monte Carlo, *Canadian Journal of Statistics* **31**: 469–482.

Pfaff, B. (2008a). *Analysis of Integrated and Cointegrated Time Series with R*, 2nd edn, Springer, New York.

Pfaff, B. (2008b). Var, svar and svec models: Implementation within R package vars, *Journal of Statistical Software* **27**.
URL: *http://www.jstatsoft.org/v27/i04/*.

Pitt, M. and Shephard, N. (1999). Filtering via simulation: Auxiliary particle filters, *Journal of the American Statistical Association* **94**: 590–599.

Plackett, R. (1950). Some theorems in least squares, *Biometrika* **37**: 149–157.

Poirier, D. (1995). *Intermediate Statistics and Econometrics: a Comparative Approach*, MIT Press, Cambridge.

Pole, A., West, M. and Harrison, J. (n.d.). *Applied Bayesian forecasting and time series analysis*, Chapman & Hall, New York.

Prakasa Rao, B. (1999). *Statistical Inference for Diffusion Type Processes*, Oxford University Press, New York.

Rabiner, L. and Juang, B. (1993). *Fundamentals of Speech Recognition*, Prentice-Hall, Englewood Cliffs.

Rajaratnam, B., Massam, H. and Carvalho, C. (2008). Flexible covariance estimation in graphical Gaussian models, *Annals of Statistics* **36**: 2818–2849.

Reinsel, G. (1997). *Elements of Multivariate Time Series Analysis*, 2nd edn, Springer-Verlag, New York.

Robert, C. (2001). *The Bayesian Choice*, 2nd edn, Springer-Verlag, New York.

Robert, C. and Casella, G. (2004). *Monte Carlo Statistical Methods*, 2nd edn, Springer, New York.

Rydén, T. and Titterington, D. (1998). Computational Bayesian analysis of hidden Markov models, *J. Comput. Graph. Statist.* **7**: 194–211.

Savage, L. (1954). *The Foundations of Statistics*, Wiley, New York.

Schervish, M. (1995). *Theory of Statistics*, Springer-Verlag, New York.

Shephard, N. (1994). Partial non-Gaussian state space models, *Biometrika* **81**: 115–131.

Shephard, N. (1996). Statistical aspects of ARCH and stochastic volatility, *in* D. Cox, D. Hinkley and O. Barndorff-Nielsen (eds), *Time Series Models with Econometric, Finance and other Applications*, Chapman and Hall, London, pp. 1–67.

Shumway, R. and Stoffer, D. (2000). *Time Series Analysis and its Applications*, Springer-Verlag, New York.

Sokal, A. (1989). *Monte Carlo Methods in Statistical Mechanics: Foundations and New Algorithms*, Cours de Troisiéme Cycle de la Physique en Suisse Romande, Lausanne.

Stein, C. (1956). Inadmissibility of the usual estimator for the mean of a multivariate normal distribution, *in* J. Neyman (ed.), *Proceedings of the Third Berkeley Symposium on Mathematical Statistics and Probability, Volume 1*, University of California Press, Berkeley, pp. 197–206.

Storvik, G. (2002). Particle filters for State-Space models with the presence of unknown static parameters, *IEEE Transactions on Signal Processing* **50**: 281–289.

Theil, H. (1966). *Applied Economic Forecasting*, North Holland, Amsterdam.

Thiele, T. (1880). Om anvendelse af mindste kvadraters methode i nogle tilflde, hvor en komplikation af visse slags uensartede tilfldige fejlkilder giver fejlene en "systematisk" karakter, *Det Kongelige Danske Videnskabernes Selskabs Skrifter – Naturvidenskabelig og Mathematisk Afdeling* pp. 381–408. English Transalation in: Thiele: Pioneer in Statistics, S. L. Lauritzen, Oxford University Press (2002).

Tierney, L. (1994). Markov chain for exploring posterior distributions (with discussion), *Annals of Statistics* **22**: 1701–1786.

Uhlig, H. (1994). On singular Wishart and singular multivariate beta distributions, *Annals of Statistics* **22**: 395–405.

Venables, W. and Ripley, B. (2002). *Modern Applied Statistics with S*, 4th edn, Springer-Verlag, New York.

Wang, L., Liber, G. and Manneback, P. (1992). Kalman filter algorithm based on singular value decomposition, *Proc. of the 31st Conf. on Decision and Control*, pp. 1224–1229.

West, M. and Harrison, J. (1997). *Bayesian Forecasting and Dynamic Models*, 2nd edn, Springer, New York.

West, M., Harrison, J. and Migon, H. (1985). Dynamic generalized linear models and Bayesian forecasting, *Journal of the American Statistical Association* **80**: 73–83.

Wiener, N. (1949). *The Extrapolation, Intepolation and Smoothing of Stationary Time Series*, Wiley, New York.

Wold, H. (1938). *A Study in the Analysis of Stationary Time Series*, Almquist and Wiksell, Uppsala.

Wuertz, D. (2008). *fBasics: Rmetrics – Markets and Basic Statistics.* R package version 280.74.
URL: *http://www.rmetrics.org.*

Zellner, A. (1971). *An Introduction to Bayesian Inference in Econometrics*, Wiley, New York.

Zhang, Y. and Li, R. (1996). Fixed-interval smoothing algorithm based on singular value decomposition, *Proceedings of the 1996 IEEE International Conference on Control Applications*, pp. 916–921.